Communications in Computer and Information Science 1953

Rationale
The CCIS series is devoted to the publication of proceedings of computer science conferences. Its aim is to efficiently disseminate original research results in informatics in printed and electronic form. While the focus is on publication of peer-reviewed full papers presenting mature work, inclusion of reviewed short papers reporting on work in progress is welcome, too. Besides globally relevant meetings with internationally representative program committees guaranteeing a strict peer-reviewing and paper selection process, conferences run by societies or of high regional or national relevance are also considered for publication.

Topics
The topical scope of CCIS spans the entire spectrum of informatics ranging from foundational topics in the theory of computing to information and communications science and technology and a broad variety of interdisciplinary application fields.

Information for Volume Editors and Authors
Publication in CCIS is free of charge. No royalties are paid, however, we offer registered conference participants temporary free access to the online version of the conference proceedings on SpringerLink (http://link.springer.com) by means of an http referrer from the conference website and/or a number of complimentary printed copies, as specified in the official acceptance email of the event.

CCIS proceedings can be published in time for distribution at conferences or as post-proceedings, and delivered in the form of printed books and/or electronically as USBs and/or e-content licenses for accessing proceedings at SpringerLink. Furthermore, CCIS proceedings are included in the CCIS electronic book series hosted in the SpringerLink digital library at http://link.springer.com/bookseries/7899. Conferences publishing in CCIS are allowed to use Online Conference Service (OCS) for managing the whole proceedings lifecycle (from submission and reviewing to preparing for publication) free of charge.

Publication process
The language of publication is exclusively English. Authors publishing in CCIS have to sign the Springer CCIS copyright transfer form, however, they are free to use their material published in CCIS for substantially changed, more elaborate subsequent publications elsewhere. For the preparation of the camera-ready papers/files, authors have to strictly adhere to the Springer CCIS Authors' Instructions and are strongly encouraged to use the CCIS LaTeX style files or templates.

Abstracting/Indexing
CCIS is abstracted/indexed in DBLP, Google Scholar, EI-Compendex, Mathematical Reviews, SCImago, Scopus. CCIS volumes are also submitted for the inclusion in ISI Proceedings.

How to start
To start the evaluation of your proposal for inclusion in the CCIS series, please send an e-mail to ccis@springer.com.

Prodipto Das · Shahin Ara Begum ·
Rajkumar Buyya
Editors

Advanced Computing, Machine Learning, Robotics and Internet Technologies

First International Conference, AMRIT 2023
Silchar, India, March 10–11, 2023
Revised Selected Papers, Part I

Editors
Prodipto Das
Assam University
Silchar, India

Shahin Ara Begum
Assam University
Silchar, India

Rajkumar Buyya
The University of Melbourne
Melbourne, VIC, Australia

ISSN 1865-0929 ISSN 1865-0937 (electronic)
Communications in Computer and Information Science
ISBN 978-3-031-47223-7 ISBN 978-3-031-47224-4 (eBook)
https://doi.org/10.1007/978-3-031-47224-4

This Springer imprint is published by the registered company Springer Nature Switzerland AG
The registered company address is: Gewerbestrasse 11, 6330 Cham, Switzerland

Preface

This book contains contributions of computer scientists, engineers, academicians and researchers from three different countries and various states of India, who have participated in the First International Conference on Advanced Computing, Machine Learning, Robotics and Internet Technologies (AMRIT 2023), held at Assam University, Silchar, Assam, India during 10–11 March 2023. The conference focused on four major thrust areas, *viz.* Advanced Computing, Machine Learning, Robotics and Internet Technologies.

Assam University, Silchar is a Central University established by an Act of the Indian Parliament (Act XXIII of 1989) and came into existence in 1994. Situated in the Barak Valley of southern Assam, the University is nestled in a sprawling 600-acre campus at Dargakona, about 23 km from Silchar town. In 2007, Assam University opened a 90-acre satellite campus at Diphu in the Karbi Anglong Hills District of Assam, thereby making quality higher education accessible to a wider section of society. With student strength about 5000 on both campuses, the University is a veritable melting pot of diverse communities, ideas and creativity. There are 43 postgraduate departments under 16 schools of study on the Silchar campus and 10 departments on the Diphu campus, offering a wide range of programmes geared towards equipping students and research scholars with knowledge, skills, experience and confidence. Apart from the two campuses of the University, there are 73 affiliated and permitted colleges in the five districts of south Assam, which together constitute the jurisdiction of Assam University. Besides, there are several Centres of Study in the University dedicated to research of high quality.

The Department of Computer Science was established in July 1997. The Department offers a 5-year integrated course leading to the Degree of M. Sc. in Computer Science. Provision exists for lateral exit and entry. The Department also offers a 2-year M. Sc. in Computer Science and a Ph. D. Programme. The thrust areas of the Department are Artificial Intelligence, Machine Learning, Natural Language Processing, Image Processing, Data Science, Soft Computing Techniques, Computer Networks and Security, Computer Architecture and Algorithms. The placement pattern of the passed-out students has been encouraging. Alumni of the Department of Computer Science of Assam University are now to be found in organisations such as IBM, TCS, Infosys, Oracle, HCL, WiproTech, Cognizant and many others including educational and research institutions in India and abroad.

AMRIT 2023 was the First International Conference in the Department of Computer Science. The conference aimed to gain significant interest among researchers, academicians, professionals and students in the region, covering topics of Computer Science and its current trends of research. AMRIT 2023 brought together researchers, educators, students, practitioners, technocrats and policymakers from academia, government, industry and non-governmental organizations and provided a platform for research collaboration, networking and presentation of recent research findings in the fields of Computer Science and allied subjects.

This event was a part of the Silver Jubilee Celebration of the Department. The Department has been providing a remarkable service to the region for the last 25 years and has reached a prestigious milestone with many achievements and remarkable success. The Department has successfully organised many national and regional programmes in the recent past on various research topics. The Department witnessed a grand National Conference, CTCS 2010, with funding from prestigious organisations like such as, DIT (now MeiTY), DoNER etc. In the recent past the research work in the Department in various thrust areas has touched a new level. The first International Conference AMRIT 2023 has initiated the journey to further heights. Five keynote addresses by reputed academicians, five high-quality invited talks by academicians and scientists from ISRO and BARC, one tutorial, technical sessions, poster presentations and industry-academia interactions were part of the conference. The participants and attendees of the conference came from Australia, the Philippines, Malaysia, Bangladesh and from different nationally reputed universities and organisations. Paper presenters, research scholars and student participants attend the conference both in physical and in virtual mode.

In the conference, a total of one hundred and ten papers were received, out of which forty-seven papers were accepted provisionally for this book volume. The review process was a single-blind peer review model. A total of 42 reviewers reviewed the papers with an average of 2–3 papers per reviewer. The reviewers were from various institutes and organisations with very high academic repute. The papers selected for the book volume were presented during the conference by the respective authors. The contents of the forty-seven papers are of high quality and cover new areas of Computer Science and allied subjects. The primary distinguishing features of this book are the latest novel research in the theme areas of the conference. This book contains systematic material for understanding the latest trends in Computer Science research and various challenging issues of Computer Science research and their solutions. We hope that this book volume will provide a valuable resource to the growing research community in the field of Computer Science.

The First International Conference on Advanced Computing, Machine Learning, Robotics and Internet Technologies (AMRIT 2023) at Assam University, Silchar was supported by SERB, Department of Science and Technology (DST), Govt. of India, Oil and Natural Gas Corporation, Silchar Asset, India and Assam University, Silchar, India.

March 2023

Prodipto Das
Shahin Ara Begum
Rajkumar Buyya

Organization

Conference Committee Members

Chief Patron

Rajive Mohan Pant (Vice-chancellor)	Assam University, India

Patron

Karabi Dutta Choudhury (Dean)	AESoPS, Assam University, India

General Chair

Shahin Ara Begum	Assam University, India

Technical Chair

Bipul Syam Purkayastha	Assam University, India

Publication Chair

Pankaj Kumar Deva Sarma	Assam University, India

Organising Chair

Prodipto Das	Assam University, India

Joint Organising Chairs

Saptarshi Paul	Assam University, India
Purnendu Das	Assam University, India
Biswa Ranjan Roy	Assam University, India
Debasish Roy	Assam University, India

Organising Members

Arindam Roy	Assam University, India
Rakesh Kumar	Assam University, India
Indrani Das	Assam University, India
Bhagwan Sahay Meena	Assam University, India
Rahul Kumar Chawda	Assam University, India

International Advisory Committee

Rajkumar Bhuyya	University of Melbourne, Australia
Neil P. Balba	Lyceum of the Philippines University – Laguna, The Philippines
Md. Fokray Hossain	Daffodil International University, Bangladesh
Mark Shen	National Central University, Taiwan
Is-Haka Mkwawa	University of Plymouth, UK

National Advisory Committee

Tanmoy Som	IIT-BHU, India
J. K. Mandal	Kalyani University, India
Utpal Roy	Visva-Bharati, India
Jamal Hussain	Mizoram University, India
N. P. Maity	Mizoram University, India
Rashmi Bhardwaj	Indraprastha University, India
Susmita Sur-Kolay	ISI Kolkata, India
M. K. Ghosh	ISRO, India
Shirshendu Das	IIT, Hyderabad, India
Samarjit Borah	SMIT, India
Ferdous Ahmed Barbhuiya	IIIT Guwahati, India
Nabanita Das	ISI Kolkata, India
Anajana Kakati Mahanta	Guwahati University, India
Smriti Kumar Sinha	Tezpur University, India

Technical Program Committee

Rajkumar Banoth	University of Texas at San Antonio, USA
Shayak Chakraborty	Florida State University, USA
Anjay Kumar Mishra	Madan Bhandari Memorial College, Nepal
Ricardo Saavedra	Azteca University, Mexico
S. Sharma	Texas A&M University Texarkana, USA
Goi Bok Min	Universiti Tunku Abdul Rahman, Malaysia

Keynote Speakers

Rajkumar Buyya	University of Melbourne, Australia
Neil Perez Balba	Lyceum of the Philippines University – Laguna, The Philippines
Md. Forhad Rabbi	Shahjalal University of Science and Technology, Bangladesh
Susmita Sur-Kolay	Indian Statistical Institute, India
Prithwijit Guha	Indian Institute of Technology Guwahati, India

Invited Speakers

Mrinal Kanti Ghose	GLA University & ISRO, India
Debabrata Datta	Heritage Institute of Technology & Bhabha Atomic Research Centre, India
P. Shivakumar	University of Malaya, Malaysia
Utpal Roy	Visva-Bharati University, India
Prantosh Kumar Paul	Raiganj University, India

Neoteric Frontiers in Cloud, Edge, and Quantum Computing (Keynote)

Rajkumar Buyya[1,2]

[1] Cloud Computing and Distributed Systems (CLOUDS) Lab, University of Melbourne, Australia
[2] Manjrasoft Pvt Ltd., Melbourne, Australia

Abstract. Computing is being transformed into a model consisting of services that are delivered in a manner similar to utilities such as water, electricity, gas, and telephony. In such a model, users access services based on their requirements without regard to where the services are hosted or how they are delivered. The cloud computing paradigm has turned this vision of "computing utilities" into a reality. It offers infrastructure, platform, and software as services, which are made available to consumers as subscription-oriented services. Cloud application platforms need to offer.

(1) APIs and tools for rapid creation of elastic applications and
(2) A runtime system for deployment of applications on geographically distributed Data Centre infrastructures (with Quantum computing nodes) in a seamless manner.

The Internet of Things (IoT) paradigm enables seamless integration of the cyber-and-physical worlds and opens opportunities for creating new classes of applications for domains such as smart cities, smart robotics, and smart healthcare. The emerging Fog/Edge computing paradigms support latency-sensitive/real-time IoT applications with a seamless integration of network-wide resources all the way from Edge to the Cloud.

This keynote presentation covers:

(a) 21st-century vision of computing and identifies various IT paradigms promising to deliver the vision of computing utilities;
(b) Innovative architecture for creating elastic Clouds integrating edge resources and managed Clouds;
(c) Aneka 5G, a Cloud Application Platform, for rapid development of Cloud/Big Data applications and their deployment on private/public Clouds with resource provisioning driven by SLAs;
(d) A novel FogBus software framework with Blockchain-based data-integrity management for facilitating end-to-end IoT-Fog/Edge-Cloud integration for execution of sensitive IoT applications;

(e) Experimental results on deploying Cloud and Big Data/IoT applications in engineering, health care (e.g., COVID-19), deep learning/Artificial intelligence (AI), satellite image processing, and natural language processing (mining COVID-19 research literature for new insights) on elastic Clouds;
(f) QFaaS: A Serverless Function-as-a-Service Framework for Quantum Computing; and
(g) Directions for delivering our 21st-century vision along with pathways for future research in Cloud and Edge/Fog computing.

Contents – Part I

A Hybrid Framework for Implementing Modified K-Means Clustering Algorithm for Hindi Word Sense Disambiguation ... 1
Prajna Jha, Shreya Agarwal, Ali Abbas, and Tanveer J. Siddiqui

Detection of Leaf Disease Using Mask Region Based Convolutional Neural Network ... 11
D. S. Bharathi, H. Harish, M. G. Shruthi, M. Mamatha, U. Ashwitha, and A. Manasa

An Improved Machine Learning Approach for Throughput Prediction in the Next Generation Wireless Networks ... 23
Aritri Debnath and Barnali Dey

Protein Secondary Structure Prediction Without Alignment Using Graph Neural Network ... 31
Tanvir Kayser and Pintu Chandra Shill

A Lexicon-Based Approach for Sentiment Analysis of Bodo Language ... 46
Jaya Rani Mushahary, Bipul Roy, and Mandwip Baruah

An Osprey Optimization Based Efficient Controlling of Nuclear Energy-Based Power System ... 57
Prince Kumar, Kunal Kumar, Aashish Kumar Bohre, and Nabanita Adhikary

Fuzzy Association Rule Mining Techniques and Applications ... 73
Rajkamal Sarma and Pankaj Kumar Deva Sarma

Brain Tumor Detection Using VGG-16 ... 88
Taniya Nandy, Laishram Munglemkhombi Devi, and Ishita Chakraborty

Deep Learning Model for Fish Copiousness Detection to Maintain the Ecological Balance Between Marine Food Resources and Fishermen ... 96
O. M. Divya, M. Ranjitha, and K. Aruna Devi

A Study on Smart Contract Security Vulnerabilities ... 105
Vaibhav Bhajanka and Nitisha Pradhan

A GUI-Based Study of Weather Prediction Using Machine Learning Algorithms 116
Debdeep Mukherjee, Rupak Parui, and Lopamudra Dey

A Systematic Review on Latest Approaches of Automated Sleep Staging System Using Machine Intelligence Techniques 127
Suren Kumar Sahu, Santosh Kumar Satapathy, and Sudhir Kumar Mohapatra

Network Security Threats Detection Methods Based on Machine Learning Techniques 137
Bikash Kalita and Pankaj Kumar Deva Sarma

Optimized Traffic Management in Software Defined Networking 157
M. P. Ramkumar, J. Lece Elizabeth Rani, R. Jeyarohini, G. S. R. Emil Selvan, and S. Arun Karthick

Information Extraction for Design of a Multi-feature Hybrid Approach for Pronominal Anaphora Resolution in a Low Resource Language 169
Shreya Agarwal, Prajna Jha, Ali Abbas, and Tanveer J. Siddiqui

Signature-Based Batch Auditing Verification in Cloud Resource Pool 181
Paromita Goswami, Munmi Gogoi, Somen Debnath, and Ajoy Kumar Khan

Genetic Algorithm Based Anomaly Detection for Intrusion Detection 194
Sushilata D. Mayanglambam, Nainesh Hulke, and Rajendra Pamula

Machine Learning Based Malware Identification and Classification in PDF: A Review Paper 204
Vaishali and Nahita Pathania

A Survey on Lung Cancer Detection and Location from CT Scan Using Image Segmentation and CNN 213
K. Hari Priya, Suryatheja Alladi, Saidesh Goje, M. Nithin Reddy, and Himanshu Nama

Bi-directional Long Short-Term Memory with Gated Recurrent Unit Approach for Next Word Prediction in Bodo Language 221
Ajit Das, Abhijit Baruah, and Sudipta Roy

Authorship Attribution for Assamese Language Documents: Initial Results 232
Smriti Priya Medhi and Shikhar Kumar Sarma

Load Balancing and Energy Efficient Routing in Software-Defined Networking 243
M. P. Ramkumar, J. Lece Elizabeth Rani, R. Jeyarohini, G. S. R. Emil Selvan, and Rahul Shiva Konar

Sentiment Analysis: Indian Languages Perspective 262
Abhishek A. Vichare and Satishkumar L. Varma

Author Index 277

Contents – Part II

Comparative Analysis of Different Machine Learning Based Techniques for Crop Recommendation 1
Rohit Kumar Kasera, Deepak Yadav, Vineet Kumar, Aman Chaudhary, and Tapodhir Acharjee

Arduino Based Multipurpose Solar Powered Agricultural Robot Capable of Ploughing, Seeding, and Monitoring Plant Health 14
Hrituparna Paul, Anubhav Pandey, and Saurabh Pandey

Galactic Simulation: Visual Perception of Anisotropic Dark Matter 25
Anand Kushwah, Tushar Rajora, Divyansh Singh, Satwik Pandey, and Eva Kaushik

Protocol Anomaly Detection in IIoT 37
S. S. Prasanna, G. S. R. Emil Selvan, and M. P. Ramkumar

Agricultural Informatics & ICT: The Foundation, Issues, Challenges and Possible Solutions—*A Policy Work* 47
P. K. Paul, Mustafa Kayyali, and Ricardo Saavedra

A Guava Leaf Disease Identification Application 61
Nikhil James, Kunal Kumar Shriwastav, Shilpita Medhi, and Smriti Priya Medhi

Text to Image Generation Using Attentional Generative Adversarial Network 70
Supriya B. Rao, Shailesh S. Shetty, Chandra Singh, Srinivas P. M, and Anush Bekal

Attention-CoviNet: A Deep-Learning Approach to Classify Covid-19 Using Chest X-Rays 83
Thejas Karkera, Chandra Singh, Anush Bekal, Shailesh Shetty, and P. Prajwal

A Deep Learning Framework for Violence Detection in Videos Using Transfer Learning 100
Gurmeet Kaur and Sarbjeet Singh

Multi-focus Image Fusion Methods: A Review 112
Ravpreet Kaur and Sarbjeet Singh

Cache Memory and On-Chip Cache Architecture: A Survey 126
Nurulla Mansur Barbhuiya, Purnendu Das, and Bishwa Ranjan Roy

Authenticating Smartphone Users Continuously Even if the Smartphone is in the User's Pocket 139
Sandip Dutta, Soumen Roy, and Utpal Roy

Comparative Analysis of Machine Learning Algorithms for COVID-19 Detection and Prediction 147
Shiva Sai Pavan Inja, Koppala Somendra Sahil, Shanmuk Srinivas Amiripalli, Viswa Ajay Reddy, and Surya Rongala

Machine Learning Classifiers Explanations with Prototype Counterfactual 157
Ankur Kumar, Shivam Dwivedi, Aditya Mehta, and Varun Malhotra

A Systematic Study of Super-Resolution Generative Adversarial Networks: Review 170
Ravindra Singh Kushwaha and Rajan Kakkar

Stance Detection in Manipuri Editorial Article Using CRF 187
Pebam Binodini, Kishorjit Nongmeikapam, and Sunita Sarkar

Deep Learning Based Software Vulnerability Detection in Code Snippets and Tag Questions Using Convolutional Neural Networks 198
Anurag Khanra, Arvind Krishna, L. H. Jeevan Samrudh, Rahul D. Makhija, and V. R. Badri Prasad

A Comprehensive Study of the Performances of Imbalanced Data Learning Methods with Different Optimization Techniques 209
Debashis Roy, Utathya Aich, Anandarup Roy, and Utpal Roy

Smart Parking System Using Arduino and IR Sensor 229
R. Chawngsangpuii and Angelina Lalruatfeli

QuMaDe: Quick Foreground Mask and Monocular Depth Data Generation 236
Sridevi Bonthu, Abhinav Dayal, Arla Lakshmana Rao, and Sumit Gupta

Fine-Grained Air Quality with Deep Air Learning 247
Jyoti Srivastava, Neha Singh, Rahul Chakravorty, and Anu Raj

Enhancing Melanoma Skin Cancer Detection with Machine Learning and Image Processing Techniques 256
S. Mahaboob Hussain, B. V. Prasanthi, Narasimharao Kandula, Padma Jyothi Uppalapati, and Surayanarayana Dasika

Image Processing Technique and SVM for Epizootic Ulcerative Syndrome Fish Image Classification 273
Hitesh Chakravorty and Prodipto Das

Author Index 283

A Hybrid Framework for Implementing Modified K-Means Clustering Algorithm for Hindi Word Sense Disambiguation

Prajna Jha(✉), Shreya Agarwal, Ali Abbas, and Tanveer J. Siddiqui

Department of Electronics and Communication, University of Allahabad, Prayagraj, India
pragya.jha.jk@gmail.com

Abstract. In this paper, we have proposed a new method for the implementation of an unsupervised K-Means clustering algorithm for Word Sense Disambiguation in Hindi language. We created a feature vector for each instance of an ambiguous target word, as well as for the lexical information obtained by utilizing Hindi WordNet for the same word. The lexical information was obtained from synset definitions, and its glosses for a given ambiguous target word from Hindi WordNet. The seed value was selected randomly from the lexical information. The features used for the experiment were term-frequency, term-frequency inverse instance-frequency, and n-grams. We applied similarity measures such as Cosine, Jaccard, and Dice similarity measures. We performed an experiment on the Hindi dataset containing three polysemous nouns. We obtained an average accuracy of 69.6%. We also observed that the accuracy of Dice similarity measure performs better with the modified K-means clustering algorithm.

Keywords: Hindi word sense disambiguation · unsupervised classification · k-means clustering · distance-metric · purity

1 Introduction

In the modern era of globalization, it is important to overcome language barriers across different regions. Natural Language Processing (NLP) is the branch of Artificial Intelligence which essentially focuses towards development of new tools and techniques to develop multilingual computer applications. This field incorporates extraction of useful information from text, speech, etc., which is used for developing different evolving models for various applications such as Machine Translation, Text Analysis, Speech Recognition, etc. However, there are multiple challenges involved with NLP. These challenges form an open research problem for researchers' exploration. One of the key challenge is lexical ambiguity. In a given sentence, each word carries a distinct meaning, even though some words may be polysemous. A human can easily comprehend the meaning of the sentence and its individual words, but the same cannot be said for a machine. The task of resolving the sense ambiguity of a word for a given context automatically is known as Word Sense Disambiguation (WSD).

P. Das et al. (Eds.): AMRIT 2023, CCIS 1953, pp. 1–10, 2024.
https://doi.org/10.1007/978-3-031-47224-4_1

WSD is an essential task which plays an important role in various NLP applications. A word can be inherently polysemous. The senses of an ambiguous words form different classes to which they could belong.The main purpose of WSD task is the classification of an appropriate sense of a word into a distinct class, by utilizing the contextual information and lexical knowledge-base such as machine-readable dictionary, thesaurus, wordnet, etc. Many research methods and algorithms have been proposed by scientists for WSD tasks. Navigli, 2009 [1] primarily classified WSD task into supervised, and unsupervised classification techniques. The supervised approach requires a large amount of sense-annotated dataset for training, and precise tools which is a major disadvantage for low resource languages such as Hindi. Thus, to overcome the bottleneck problem of knowledge acquisition, unsupervised and semi-supervised approaches for WSD are being constantly explored. Unsupervised approaches (as discussed in Sect. 2) do not require sense-tagged corpus, and can overcome resource problems. Semi-supervised approach, on the other hand, combines the two approaches. It makes use of the limited available sense-tagged dataset. They form the seeds. The algorithm uses these seeds to learn various features and assign the correct sense for untagged data iteratively.

Our main objective is to incorporate the features of unsupervised and semi-supervised approaches and develop a modified k-means clustering algorithm for classification of sense into appropriate class of sense definition. The main contribution of this paper is that it is an extension of corpus-based framework as proposed by Singh et al., 2015 [2] by proposing an unsupervised targeted word sense disambiguation for disambiguating polysemous nouns. Our proposed algorithm relies on the fact that in a clustering approach, a cluster represents the cohesiveness among words in a given context, and the implicit relationships among the textual units. The second contribution of this paper is to evaluate in-depth the efficiency of the clustering algorithm, and similarity measures such as cosine, dice, and jaccard for searching the semantically related senses for a given word, and selecting the most appropriate sense.

In the following section (Sect. 2), we discuss the various unsupervised algorithms developed for Hindi Word Sense Disambiguation, in Sect. 3 we bring to light the framework developed by combining information from lexical knowledge-base and corpus statistics developed for WSD, and in Sect. 4, we discuss our method and outline of our experimental approach in detail. We present the result and analysis of our experiment in Sect. 5, and conclusion Sect. 6.

2 Related Works

The WSD is essential for various NLP tasks in Hindi. The sense ambiguity is present at multiple levels in Hindi language such as syntactic, semantic, and discourse level. One of the major problem for WSD in Hindi is the absence of sense annotated dataset, or if available, they are mostly domain specific. Recently, efforts are continuously being made to develop unsupervised algorithms for WSD in Hindi. The unsupervised approach consists of two main methods: clustering and graph-based approach. These methods developed and proposed by researchers in Hindi language are being primarily discussed in this section.

Mishra et al., [3] proposed unsupervised WSD using decision lists and untagged dataset alongwith manually developed seed instances used for training. The authors found that stemming, and removal of stop words increases the accuracy of the algorithm. Rastogi and Dwivedi, 2011 [4] utilized the concept of highest sense count and phrase frequency by counting the occurrence of terms and query phrases in context of hypernym, gloss, and test corpus respectively. The authors developed a test dataset of 100 Hindi queries and compared three Hindi search engines, viz. Google, Raftaar and Guruji.

Jain et al., 2013 [5] applied graph based centrality algorithms, and graph clustering algorithms to measure the effectiveness of approach on dataset consisting of 1200 open class words for disambiguation and reported 60.25%, 41.25% and 54.81% accuracies for three different sets respectively. Jain et al., 2014 [6] developed a graph representing semantic relations existing between each word pair in a sentence using Hindi WordNet. The authors interpreted the spanning tree with minimum cost corresponding to each graph. Jain and Lobiyal, 2015 [7] developed an interpretation graph for each sentence graph, and calculated network agglomeration for each interpretation graph for health and tourism Hindi dataset. The authors reported an F-score 50.20% for the health domain, and 53.01% for the tourism domain respectively.

Sudhabhingardive et al., 2015 [8] proposed most frequent sense detection from untagged corpus consisting of 80,000 polysemous nouns by developing word embeddings, and compared them with their respective sense embeddings. The sense bearing the highest similarity score was selected. The authors obtained 62.43% precision. İn [9] the authors explored the level of sense granularity for verbs according to different properties such as subject, object, compulsion and time period. The authors obtained precision of 0.97 and f-score of 0.69 by applying unsupervised word embedding for WSD. Kumari and Lobiyal, 2020 [10] proposed word2vec models based on skip-gram and continuous bag of word architectures to create word embeddings. The authors obtained an accuracy of 52% on Hindi corpus.

Patel et al., 2015 [11] applied unsupervised hierarchical clustering using Hindi WordNet with three distinct distance-metrics. The Hindi dataset consisted of 246 words from history, 1279 words from social study, and 2672 words from short stories. The authors reported a precision of 81.64% using cosine, 80.38% using Jaccard and 79.74% using Dice similarity measures respectively. Tayal et al., 2015 [12] proposed unsupervised Hyperanalogue to create language vectors for training and applied Fuzzy C-means clustering method on dataset obtained from Hindi Wikipedia and news articles. The authors obtained an accuracy of 79.16%.

3 A Hybrid Framework for Unsupervised WSD

In this paper we propose a hybrid framework to perform targeted wsd for polysemous nouns. the main objective of this framework is to analyze and develop a semantically related clustering method for grouping different classes of senses, and ıdentify the most appropriate sense of target word. for a given polysemous word w, each ınstance of w containing contextual ınformation I is being analyzed. each polysemous word w is classified into different classes according to ıts senses. we use hindi wordnet to determine distinct synonym sets/synsets for w. hence, our framework utilizes corpus statistics to learn textual features, and a lexical knowledge-base for ıdentifying the most appropriate sense of a given polysemous word w. the framework consists of the following parts:

i) A lexical knowledge-base to identify the distinct class of senses for an ambiguous word
ii) A pre-processed corpus containing instances of ambiguous words. The context-based feature vectors is created by discovering contextual information from each instance
iii) To create a vector space model using i) and ii), and to apply similarity measures to identify the most appropriate sense among its different sense classes.
iv) A clustering algorithm to group the sense classes of ***w*** according to their semantically related cluster
v) A heuristic function to determine the most appropriate cluster for sense that matches best with the given contextual information ***I***
vi) A convergence criteria for clustering algorithm. It is noted that the clustering algorithm is iterated till its convergence.

The task of determining the most appropriate sense for an ambiguous word using the above framework begins with identifying all possible sense representations of the polysemous word ***w*** to determine its cluster distribution. Each class of sense representation consists of synsets, sense definition, and its gloss derived from lexical knowledge-base. The clustering algorithm tries to identify the appropriate cluster which matches best according to the contextual information ***I*** provided by sentence instances of ***w*** via the heuristic function. The entire process of clustering is iteratively repeated until convergence criteria is achieved. At this stage, it is interpreted that each instance of ***w*** has been grouped into its most appropriate cluster. Figure 1 depicts the algorithmic steps of our proposed framework.

Input: The polysemous noun $\boldsymbol{w} \epsilon W$, where W is the finite set of ambiguous words, and the contextual information ***I***.
Output: The most fitting cluster for the disambiguation of polysemous noun ***w***.

a) Let us consider the number of clusters for ***w*** is corresponding to its sense classes S_i where i is the number of senses of ***w*** in lexical knowledge base
b) Let us consider the similarity score of the feature vector between the contextual information ***I*** and its sense representation S_i be indicated as sim_score_i
c) Repeat
    i) K=Clustering(sim_score_i)
    ii) Selecting_cluster_criteria = H(K, ***w***, ***I***), where H is the heuristic function
    iii) Update centroid by calculating mean of datapoints in each cluster

    Until convergence
d) Return most appropriate cluster with the best sense definition for ***w*** in a given context ***I***.

Fig. 1. Algorithm for the proposed framework

4 Methodology

In this section we discuss in-detail about development and implementation of our algorithm. Our framework is incremental, it processes the algorithm on a sentential basis. The experiment for our algorithm includes dataset pre-processing, sense representations obtained from Hindi WordNet, development of vector model for feature representation, application of distance-metrics, and design and implementation of modified K-means clustering for unsupervised word sense disambiguation.

A. Dataset pre-processing
The dataset is pre-processed by converting each instance into respective tokens, removing stop-words from each instances, and stemming each tokens to remove morphological inflections, and assigning part-of-speech tags to each tokens. We utilize the POS tags to use only content words (noun, verb, adjective, and adverb). Hence, the contextual information ***I*** comprises content words for a given polysemous noun ***w***.

B. Sense representation using Hindi WordNet
Hindi WordNet[1] is a lexical knowledge-base tool developed by IIT Bombay. It contains lexical information of each polysemous word in the form of synonym sets. The class of synonyms bearing similar meaning are represented in each synonym set or synset. The synsets are semantically related using hypernymy, hyponymy, meronymy, gradation, etc.

The lexical information for polysemous noun ***w*** are obtained from Hindi WordNet is used to create feature vectors for sense classes. It consists of synsets, sense definition, and their example sentences (glosses) respectively. The glosses are further preprocessed as discussed in A. For example, consider the polysemous noun for ***w*** as 'उत्तर'. In Fig. 2 the three prominent synsets, their sense definitions, and preprocessed glosses:

1. उत्तर जवाब; उतर कोई प्रश्न बात सुनकर समाधान कही हुई बात; आपने मेरे प्रश्न उत्तर नहीं दिया
2. उत्तर उत्तर दिशा उदीची शिमाल तिर्यग्दिश्; दक्षिण दिशा सामने दिशा; भारत उत्तर हिमालय पर्वत विराजमान
उत्तर; दिक्सूचक यंत्र प्रधान बिन्दु शून्य तीन सौ साठ डिग्री; उत्तर हमेशा उत्तर दिशा ओर
उत्तर; उत्तर दिशा पड़नेवाला प्रदेश; महेश उत्तर रहनेवाला
नॉर्थ नार्थ उत्तर; मेसन-डिक्सन लाइन उत्तर स्थित अमरीकी क्षेत्र; नॉर्थ मौसम खराब चल
3. उत्तर उत्तर कुमार; राजा विराट पुत्र; उत्तर बहन उत्तरा विवाह अभिमन्यु

Fig. 2. Sense representation for 'उत्तर'from Hindi WordNet

C. Development of vector model for feature representation
In our algorithm we have used the features such as term-frequency (tf), inverse-instance frequency (iif), term-frequency inverse-instance frequency (tf-iif), and n-gram such as unigram, bigram and trigram. Each of these features are derived using

[1] Hindi WordNet: https://www.cfilt.iitb.ac.in/wordnet/webhwn/.

sentence instances and sense representation for polysemous nouns. Each feature is described in brief as:

i. Term-frequency (tf): It is the most popular metric used in NLP tasks. It emphasizes the relative importance of each word in a document by measuring frequency of occurrence. It highlights the appearance of the particular word in the text relative to the total number of words in the document.

ii. Inverse instance frequency (iif): We define inverse instance frequency as – the number of instances occurring in a particular context of a polysemous word ***w*** over the total number of sentences for the same word in which different contexts are combined together. The intuition behind it is assigning importance to context words appearing in a particular context in a given instance. The formula for IIF is:

$$IIF = \log\left(\frac{N(TS)}{n(C_i)}\right) \tag{1}$$

where $N(TS)$ is the total number of sentences for a given word ***w*** in different contexts combined, and $n(C_i)$ is the number of sentences appearing with a particular context C_i.

iii. Term-frequency-inverse-instance-frequency (tf-iif): It is used to establish the importance of a word in a collection of sentence instances. The formula for Tf-iif is:

$$Tf - iif = tf * iif \tag{2}$$

where tf is the term-frequency, and iif is the inverse instance frequency across the document

iv. N-gram: In natural language processing n-gram is a specific sequence of n-items, extracted from textual contexts, speech, etc. In WSD, for each sentence instance n-gram is a combination of contiguous n-words occurring together. It is described as:

- Unigram- A unigram is a sequence of one word, and it represents a text as a sequence of bag of word. Here each word is considered as an independent entity, and frequency of each word is computed irrespective of the context and relationships between words.
- Bigram- A bigram is a contiguous sequence of two words occurring in a given instance. It is developed by sliding a window over adjacent pairs of words, and keeping a record for each word pair.
- Trigram- A trigram is a contiguous sequence of three consecutive words occurring in a given instance. It is developed by sliding a window over three adjacent words, and keeping a record of them.

All these features were used to create vector space model for mathematical calculations.

D. Similarity metrics: In NLP the similarity metrics is used to compute similarity between two vectors obtained from a vector space model. In our approach, we have used the following similarity metrics:

- Cosine similarity- It is used in WSD to measure the similarity between two text instances, represented as vectors of Tf-iif scores. It is computed by considering the dot product between two vectors divided by their magnitudes. The formula for cosine similarity is:

$$cosine_{sim}(I, S_i) = \frac{I \cdot S_i}{\|I\| \|S_i\|} \quad (3)$$

  where I is the instance vector obtained from tf-iif scores of each instance present in document, and S_i is the sense vector obtained from tf-iif scores of each sense representation.
- Jaccard similarity- In WSD, it is used to measure the similarity between specific contiguous sequences of words obtained from the n-gram model. The formula for Jaccard similarity is:

$$Jaccard_{sim}(I, S_i) = \left| \frac{I \cap S_i}{I \cup S_i} \right| \quad (4)$$

  where I and S_i are same as described for $cosine_{sim}(I, S_i)$
- Dice similarity- It is similar to Jaccard similarity, however, it assigns more importance to the intersection of two vectors. The formula for Dice similarity is:

$$Dice_{sim}(I, S_i) = \frac{2 * |I \cap S_i|}{|I| + |S_i|} \quad (5)$$

All these similarity metrics were used to determine the most appropriate cluster to which the sentence instance might belong to in the modified K-means clustering algorithm.

E. Modified K-means clustering algorithm: K-means clustering is one of the most popular unsupervised machine learning classifier. It is used to group data points into a predefined number of clusters, where the number of clusters is denoted by K.

   In our approach, we have introduced certain modifications to the existing K-means clustering algorithm. The zstep-by–step detail of modified K-means clustering algorithm are described as:

   i) Randomly initialization of K cluster centroids: We developed initial seeds using sense representations from Hindi WordNet, as described in subsection B. The number of senses S_i of a particular polysemous noun $\boldsymbol{w}$ is used to determine the number of clusters i. We initialized i cluster centroids by randomly choosing one of the seed instances.
   ii) Assignment of each data point to its nearest centroid: We applied similarity metrics, as discussed in D, to compute the data points and assigned them to their nearest centroid. It means that a sentence instance I that are most similar to sense representation S_i is assigned to i^{th} cluster. It forms the heuristic function for selecting a particular cluster.
   iii) Update each centroid by computing the mean of the data points assigned to it.

iv) The steps ii) and iii) are repeated until convergence: The modified K-means clustering algorithm is implemented iteratively until the inertia of the cluster allocation in one iteration does not change in the next iteration. It is determined by comparing the current inertia value with the previous inertia value. The algorithm is said to converge when inertia stops changing, and the final cluster allocation for each instance is returned.

Thus, we have developed a novel framework to implement the modified K-means clustering algorithm for polysemous nouns. We will implement our algorithm for polysemous nouns selected from sense-annotated Hindi dataset [13].

5 Result and Discussion

We have implemented our algorithm on 3 polysemous nouns selected from sense-annotated Hindi dataset [13]. The sense tags have been used as a reference for developing seeds for initializing centroids in modified K-means clustering algorithm. They are also used as a reference for evaluation of our algorithm. The dataset was pre-processed by tokenization, stop word removal, and POS-tagging. All the open class words were considered for context representation of the target word. The dataset description is given Table 1.

Table 1. Dataset description

# of Senses	Target word	No. Of instances
2	माँग	46
2	विधि	83
3	उत्तर	66

The vectors were generated for each instance in a document for ambiguous words, alongwith their sense representations in Hindi WordNet. We computed cosine similarity using tf-iif vectors of sentence instances, and tf-iif vectors of sense representations. We computed Jaccard and Dice similarity measures using n-gram models (unigram, bigram, and trigram). We determined the purity of each cluster for the evaluation purpose. Purity of each cluster is defined as:

$$\text{purity}_{\text{score}} = \left(\frac{\text{matching\_sentences}}{\text{len(cluster\_sentences)}} \right) * 100 \quad (6)$$

where matching_sentences is obtained by counting the number of similar sentences in a text file, and the sentences occurring in a particular sense cluster, and len(cluster_sentences) is the total number of sentences present in a cluster.

From the above Table 2, we can clearly observe that Dice similarity measure with trigrams gives the best result, when compared to cosine and Jaccard. It is due to the fact that Dice similarity checks for existence of seed words and word pairs in a set of sentences, whereas cosine similarity relies on the orientation of vectorial representation between seed and sentence instance.

Table 2. Purity of Modified K-means clustering algorithm with different similarity metrics

Words	Cosine	J_uni	J_bi	J_tri	D_uni	D_bi	D_tri
माँग	0.6522	o.6521	0.8043	0.7391	0.5652	0.5217	0.6304
विधि	0.5663	0.7349	0.5903	0.5783	0.6265	0.7229	0.8072
उत्तर	0.5151	0.6818	0.5909	0.5010	0.5303	0.6060	0.6515
Avg:	**0.5778**	**0.6896**	0.6618	0.6061	0.574	0.6168	**0.6963**

During the experiment with Modified K-means clustering on three polysemous nouns, we observe that in cases of bigrams, and trigrams the algorithm converges to a stable state of clusters more rapidly, as compared to the cosine similarity measure and unigrams.

6 Conclusion

In this paper we tried to capture the characteristic aspects of unsupervised and semi-supervised algorithms, and proposed a modified version of k-means clustering algorithm. The algorithm successfully disambiguates polysemous nouns. Our algorithm is language-independent, and it is not restricted to a particular domain. Moreover, our algorithm successfully captured the semantic information associated with ambiguous words, along with cohesiveness of context words in a sentence. The future scope of our algorithm is that it can be implemented for other Indian languages, and it can incorporate other similarity measures. The future scope also includes comparative analysis of our approach with simple or standard clustering methods.

Acknowledgements. We take this opportunity to acknowledge that the experiments and results are performed on a part of Sense Annotated Hindi Corpus available at Indian Language Technology Proliferation and Deployment Centre (TDIL). We also acknowledge the use of Hindi WordNet developed by IIT Bombay. Hindi WordNet [14, 15] is an online API that is available for students for research purposes under the condition that it is used for research only and its references are duly cited.

References

1. Navigli, R.: Word sense disambiguation: a survey. ACM Comput. Surv. **41**(2), 1–69 (2009)
2. Singh, S., Siddiqui, T.J.: Utilizing corpus statistics for Hindi word sense disambiguation. Int. Arab J. Inf. Technol. **12**(6A), 755–763 (2015)
3. Mishra, N., Yadav, S., Siddiqui, T.J.: An unsupervised approach to Hindi word sense disambiguation. In: Tiwary, U.S., Siddiqui, T.J., Radhakrishna, M., Tiwari, M.D. (eds.) Proceedings of the First International Conference on Intelligent Human Computer Interaction, pp. 327–335. Springer, New Delhi (2009). https://doi.org/10.1007/978-81-8489-203-1_32
4. Rastogi, P., and Dwivedi, S.K.: Performance comparison of word sense disambiguation (WSD) algorithm on Hindi language supporting search engines. Int. J. Comput. Sci. Iss. **8**(2), 375–379 (2011)

5. Jain, A., Yadav S., Tayal, D.: Measuring context-meaning for open class words in Hindi language. In: 6th Conference on Contemporary Computing (IEEE) , pp. 118–123, Noida, India (2013)
6. Jain, A., Lobiyal, D.K.: A new approach for unsupervised word sense disambiguation in Hindi language using graph connectivity measures. Int. J. Artif. Intell. Soft Comput. **4**(4), 318–334 (2014)
7. Jain, A., Lobiyal, D.K.: Unsupervised Hindi word sense disambiguation based on network agglomeration. In: 2nd International Conference on Computing for Sustainable Global Development (INDIACom), pp. 195–200, India (2015)
8. Bhingardive, S., Singh, D., Rudramurthy, V., Redkar, H., Bhattacharyya, P.: Unsupervised most frequent sense detection using word embeddings. In: Proceedings of the Conference of the North American Chapter of the Association for Computational Linguistics: Human Language Technologies: Association for Computational Linguistics, Denver, Colorado, pp. 1238–1243 (2015)
9. Bhingardive, S., Puduppully, R., Singh, D., Bhattacharyya, P.: Merging verb senses of Hindi wordnet using word embeddings. In: Proceedings of the 11th International Conference on Natural Language Processing: NLP Association of India, Goa, India, pp. 344–352 (2014)
10. Kumari, A., Lobiyal, D.K.: Word2vec's distributed word representation for Hindi word sense disambiguation. In: Hung, D.V., D'Souza, M. (eds.) ICDCIT 2020. LNCS, vol. 11969, pp. 325–335. Springer, Cham (2020). https://doi.org/10.1007/978-3-030-36987-3_21
11. Patel, N., Patel, B., Parikh, R., Bhatt, B.: Hierarchical clustering technique for word sense disambiguation using Hindi WordNet. In: 5th Nirma University International Conference on Engineering (IEEE), Ahmedabad, India, pp. 1–5 (2015)
12. Tayal, D.K., Ahuja, L., Chhabra, S.: Word sense disambiguation in Hindi language using hyperspace analogue to language and fuzzy c-means clustering. In: Proceedings of 12th International Conference on Natural Language Processing, Trivandrum, India, pp. 49–58. NLP Association of India (2015)
13. Sense Annotated Hindi Corpus: Indian Language Technology Proliferation and Deployment Centre. https://tdil-dc.in/index.php
14. Narayan, D., Chakrabarti, D., Pande, P., Bhattacharya, P.: An experience in building the indo WordNet- a WordNet for Hindi. In: First International Conference on Global WordNet, Mysore, India (2002)
15. Narayan, D., Bhattacharya, P.: Using verb noun association for word sense disambiguation. In: International Conference on Natural Language Processing (ICON 2002), Mumbai, India (2002)

Detection of Leaf Disease Using Mask Region Based Convolutional Neural Network

D. S. Bharathi, H. Harish(✉), M. G. Shruthi, M. Mamatha, U. Ashwitha, and A. Manasa

Department of Computer Science, Maharani Lakshmi Ammanni College for Women, Bangalore, Karnataka, India
hh.harish@gmail.com

Abstract. Agriculture plays an essential role for individuals as it is the fundamental requirement for everyone's life. Most of the time crops are identified with disease due to which there is a loss in agricultural productivity. Also because of lack of knowledge and skills cultivators are finding it difficult in detecting and rectifying the disease of crops. In India every year 17.5% of crops are lost as it gets diseased because of the usage of pests. In recent days the technology has advanced that new device has come up that are much faster and smart enough in recognizing and detecting disease in leaf that is helping the farmers in the process of monitoring the farms which reduces the job of a farmer. Hence, this approach mainly concentrates on detecting disease in the leaf using "Mask R-CNN" a deep learning technique.

Keywords: Deep Learning · Mask R-CNN · VGG Annotator · Rectified Linear Unit(ReLU)

1 Introduction

Despite generating adequate food, its not reaching people due to various aspects such as decrease in pollinators, plant diseases, climatic variations to name a few. Ailments in crop are the major reason for the deficiency in food at universal measure, which is a hazardous thing for many small- scale agriculturalists, where they mainly depended on healthful harvest of crop. Prior to this, plant clinics, agricultural enterprises, or associations was used to identify crop disease. The key attempt that experts used for detection of plant disease is by examination through naked eye. Visual discovery of ailments by researchers involves continuous examination that makes uneconomical for some agriculturalists and leads to time consuming and expensive. Many a time farmer, are not conscious of non-neighborhood diseases which compels them in consulting professionals. Therefore, an application is made possible to detect plant disease which solves problems of farmers.

Artificial Intelligence (AI) is taking a major role in the future of world and also for the economy of future. AI is enabling the machine to think like human, they provide better results and impact in all kinds of domains. All types of industries want use and utilize AI to automate a particular work using intelligence of machines. There are plenty

P. Das et al. (Eds.): AMRIT 2023, CCIS 1953, pp. 11–22, 2024.
https://doi.org/10.1007/978-3-031-47224-4_2

of algorithms which will provide good searching in large datasets, easy to track the status of datasets. The agricultural industry uses AI technology for monitoring plants, crops health to improve better productivity, Agricultural robots are developed to perform different works in farmer land, it is trained to increase the quality of crops and control weeds.

Machine Learning (ML) is a technology derived from artificial intelligence. There are more than fourteen different types of machine learning. Various categories of ML techniques includes, unsupervised machine learning, reinforcement learning, supervised learning, semi supervised learning, self- supervised and many more learning types. Machine learning in agriculture allows the planter to supervise & cure the plants individually. Machine Learning also impacts in predictive statistical data, identifying the disease, classifying the leaf and its prevention.

In the recent days deep learning (DL) algorithms has become very powerful while associated to machine learning techniques, for the reason that various deep learning techniques are being utilized in extracting the required information from the image dataset directly. Deep learning is implemented by using neurons that think like a human similar to that of humans. DL is established using various architectures of neural network which may contains numerous hidden layers to predict estimated results. The DL has many architecture, like convolutional neural networks (CNN) is applied in the field of detecting an image.

By implementing Mask R-CNN we can be able to achieve an application for detecting plant diseases. This approach uses dataset of 2500 images with 10 various types of crops, that contain both diseased and healthy images, where CNN is used for extracting features and Mask R-CNN to mask actual disease on leaf. The intention of the whole experiment is to develop a crop diseases detection using Mask R-CNN with VGG annotator tool and predefined CNN. In present days, Deep Learning is developing rapidly and will have greater impact in further days hence learning this would help in the future to implement new research topics. Along with this python, programming language is also in demand for developing new applications. Therefore, we have selected these two domains in our research paper titled "Detection of Leaf Diseases Using Region Based Convolutional Neural Network".

2 Literature Survey

There have been a number of prominent attempts made from time to time by different researchers to detect crop disease using Mask R-CNN. The literature review of the task completed is presented in this section.

Pravin et al. [1] have proposed an algorithm for object detection called Mask RCNN, together with transfer learning to discover if this algorithm stand best to detect disease in cotton leaf. The procedure they used has 4 steps namely, Dataset Collection, Data Annotation, Augmentation and Detection of Object. The dataset that they used is Plant Village dataset. For annotating images Labelling is used, which allows in creating bounding boxes around an infected areas and these annotated results were stored in XML annotation format. Various types of changes such as shifting, rotating, and mirroring implemented in the training dataset. Both Single Shot Detector (SSD) (single stage) and Mask RCNN (two-stage), was implemented on the dataset. Mask RCNN predicted most of the diseases correctly and gave an accuracy for the model as 94%. By this they concluded that Mask RCNN works good for detecting Cotton Leaf Disease.

Usman Afzaal ORCID et al. [2] proposed a model that is predominantly created on the Mask R-CNN structure, which achieves better with instance segmentation. They used ResNet as their backbone along with a technique of data expansion which allows segmentation of diseases that have been targeted under complicated ecological circumstances, they were able to get an accuracy of 82.43%.

Wen-Hao Su et al. [3] proposed a model which uses Mask-RCNN that allowed a consistent discovery of the disease in the exact location. Diseased regions were labelled and divided into sub-images. To train wheat spikes, Images with marked spikes and sub-images of every single spike with labelled unhealthy regions Mask-RCNN models for automatic image segmentation were used as ground truth data. This etiquette permitted the fast recognition of diseased portions which achieves the detection rates of 77.76%.

Anandhan K et al. [4] implemented a model for detecting the disease in leaf of rice crop. This is carried out by creating their own dataset of rice leaf and applied faster R-CNN and Mask R-CNN to have an highest accuracy of 96%.

Zia ur Rehman et al. [5] proposed a model which initially uses a mixture of contrast stretching method to improve graphical effect of an image and for detecting the infected regions, MASK RCNN was implemented. These improved images are further used for training, 2 features are used for an initially trained CNN model. While selecting important characteristics for the final classification, Kapur's entropy together with MSVM (multiple classification with cross entropy- support vector machine) approach-focused choice technique was implemented. Few fruits were chosen and initially trained using ResNet- 50 model all through transfer learning (TL), feature choice and classification. Manual annotation of the dataset uses VGG image annotator (VIA). They used Plant Village dataset and achieved the best accuracy of 96.6%.

Shijie Wang et al. [6] proposed an automatic extracting procedure which uses Mask RCNN on fruit dataset. Initially, fruits 360 was pre-processed and splits into training and test dataset. An enhanced Mask RCNN system model formation is designed using PyTorch, a deep learning structure. The bilinear interpolation method called as ROI Align is used to save the data of the feature map. Lastly, boundary precision of the segmentation mask is still enhanced by way of introducing a fully connected sheet to the mask area of the ROI production, Edge detection method Sobel was applied for predicting the target boundary and adding up the edge loss function. By this they came up with the statement where an enhanced Mask RCNN procedure performed well for yield image extraction by scoring good results.

Tejaswi Pallapothu et al. [7] recommended methodology which uses the advanced concept of computer vision and machine learning. They mainly concentrate and applies Mask-RCNN entity finding algorithm which is built on an example segmentation for identifying the pests and ailments on leaf of cotton.

Laiali Almazaydeh et al. [8] They used Mask R-CNN for building a classification system which identifies a medicinal plant. They demonstrated the progress on the classification system which uses Mask RCNN along with its support: region proposal network, RoI Pooling, RoI Align, classification & detection, segmentation. This model attained an average accuracy of 95.7%. Dataset used is Mendely Dataset. The model output is obtained by a creating a box for object detected in crop, mask, and class indicating a plant species.

Gary Storey et al. [9] presents a study that employs instance segmentation of leaf as well uses Mask R-CNN for detecting erosion disease in apple orchards. 3 distinct Mask R-CNN network supports are used. For the purpose of detecting objects, segmentation and disease detection They are then trained and evaluated. Datasets are annotated and the same is used for the training as well as estimation based on R-CNN models. This research also emphasizes that Mask R-CNN standard with ResNet-50 support offers excellent accuracy, especially in the process of detecting minute rust diseases on the leaves.

Sriram Baireddy et al. [10] In this experiment they firstly described an approach that make use of automated image analysis tools to create ground truth images that are then implemented for training a Mask R-CNN. The paper proved that Mask R-CNN model can be used prominently to detect tar spots in images of leaf surfaces. They additionally convey that the Mask R-CNN can be used for in-field images of whole leaves to capture the amount of tar spots and area of the leaf infected by the disease.

Na Yao et al. [11] implemented deep network models for segmenting and recognizing the peach diseases. By implementing this instance segmentation technique, they obtained names of diseases, location and segmentation. By using ResNetx101 the backbone network, the accuracy of segmentation increased from 0.452 to 0.463. Harish et al., 12, 13] segmented image using watershed and particle swarm optimization techniques to obtain good results. Sunanda Das et al. [14] In this paper, Mask R-CNN used for segmenting the image. Digital image processing technique for feature extraction was used for implementation. Their experimental findings proved that Mask R-CNN is best for object detection.

Vivek Tiwari et al. [15] they presented a model for identifying insects in soyabean which uses mask RCNN structure, a deep learning result for the identification and localization insects in soybean. The presented model created bounding boxes and segmentation frame for each of the insect identified in the image. It uses Resnet101 as backbone and is developed on feature pyramid network (FPN).

Deepak Kumar et al. [15] used Mask-RCNN model by taking 15,536 wheat images. Visual object tagging tool (VOTT) was used for annotating the images. Ground truth data uses labelled each leaf of wheat and mosaic virus. The backbone of Mask RCNN was resnet-50 model. The Mask-RCNN model fragments every leaf of wheat for detecting the mosaic virus on every single leaf. The model achieves an accuracy of 88.19%.

Banafshe Felfeliyan et al. [16] uses limited labeled medical image data for solving the task of learning. A deep learning training strategy built on self- supervised pretraining on unlabeled MRI scans is projected in this work. An advanced version of Mask-RCNN architecture has been fitted which proved it as the best model for object detection.

Nandhini et al. [17] have proposed a deep learning model which aims in early detection and prediction of the leaf of banian plantation which helps the farmer in brining better decisions in cultivating the crop. Implementation uses CNN along with recurrent CNN to gain a better accuracy. Harish et al. [18] has obtained better result using improved PSO segmentation. Nirmal Raj et al. [19] have done a research by studying the pre historic cultivation for disease affected data of the crop. Live images from the field have been used for the experiment. Pre-processing uses convoluted Guassian filtering and the image had been segmented using a deep active convolutional neural network during the process of training.

Yashwanth Kurmi et al. [20] have used a plant image dataset in identifying the disease of the plant leaf. They have used deep learning model to localize the region of interest which gains mean classifying accuracy and area under the descriptions curve of 95.35 and 94.7 individually. Predicts the breast cancer using PSO and SVM technique and results were compared [21].

3 Methodology

A. Data Collection

Dataset has been collected from Kaggle. Dataset consists of 6000 images of several leaves in the crops. In total 10 crops namely Apple, Bell Pepper, Cherry, Citrus, Corn, Grape, Peach, Potato, Strawberry, Tomato have been used having both healthy and diseased leaf images. The input image given by the user undergoes several steps in detecting the disease by comparing with trained dataset image. In this experiment, we have used 70% data from the dataset for training, 10% as valid and 10% for testing.

B. Convolutional Neural Network (CNN)

Convolutional Neural Network (CNN) is one of the types of artificial neural network mainly which finds its application in image recognition and processing, which is explicitly designed in processing pixel data. Here CNN is used mainly for the purpose of training the input images. In this, CNN takes leaf image as inputs, extracts, and learns the features of the image, and classifies the images based on the learned features. CNN has filters, and each filter extracts some information from the image such as edges, different kinds of shapes (vertical, horizontal, and round). All these are combined in identifying the image. It then uses ReLU as activation function.

C. Mask R - Convolutional Neural Network (CNN)

Mask R-CNN is a model that is explicitly intended for the detection of an object. This is done for image segmentation and instance segmentation. It uses selective search to generate regions from each image. For each of the leaf image, affected region is identified and marked using Mask R – CNN.

D. Visual Geometry Group (VGG) Image Annotator (VIA) Tool.

VIA is an image-annotating tool which specially used for defining regions of an image and which as well creates textual information about the selected images of those defined regions in an image. It is an open-source project development tool.

First step is that we need to annotate the images that we have in our dataset through VIA tool. As a result, we get the annotated images as JSON file. Mask R-CNN is implemented to this resulted JSON file. Next step is training which starts with 25 epoch and 100 steps per epoch. Once the model is ready, we can test it by giving an unknown image. Finally, we get an output for the test image and the model applies mask on the leaf affected area by displaying the accuracy of the prediction that the model has achieved. Figure 1 shows the steps that are involved in the methodology of the proposed model.

Convolutional neural network implemented in extracting features out of colored raw images of leaf. Then VGG annotating tool will be implemented for the features extracted from CNN. The feature extracted from CNN is then given as input to RPN (Regional proposal network) for identifying the objects present on leaf that is, it checks for healthy or diseased part on leaf.

Then the feature map generated from CNN and the RPN results will be given as input for ROI (region of interest) for max pooling. Once the images are completed with max pooling, one type of dimensions will be given to all images. These resized images are passed through fully connected layer, which uses ReLU as the activation function for identifying whether the leaf is diseased or healthy.

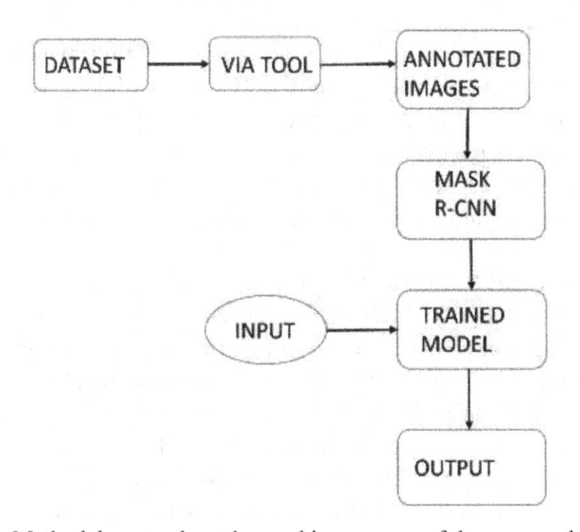

Fig. 1. Methodology to show the working process of the proposed Model.

Finally, the results will be given out with three divisions namely mask, regression, and solid max classifier. Figure 2 illustrated the architecture of Mask R-CNN. Here, mask refers to the part that has been masked by mask R - CNN, Regression refers to the plotting of images which means a boundary will be formed around the object and by solid max classifier we can find out if the leaf image is healthy or diseased. In total, 70% of dataset is used for training, 20% for valid and 10% for testing the model.

3.1 Experimental Results

Once the model has been trained properly, we can test the model. As an input an image from the test images dataset is given to the model, as an output we get the leaf images which is masked. The mask shows that the model has segmented the diseased part present in the leaf image, along with mask as well as the name of the particular disease is also predicted.

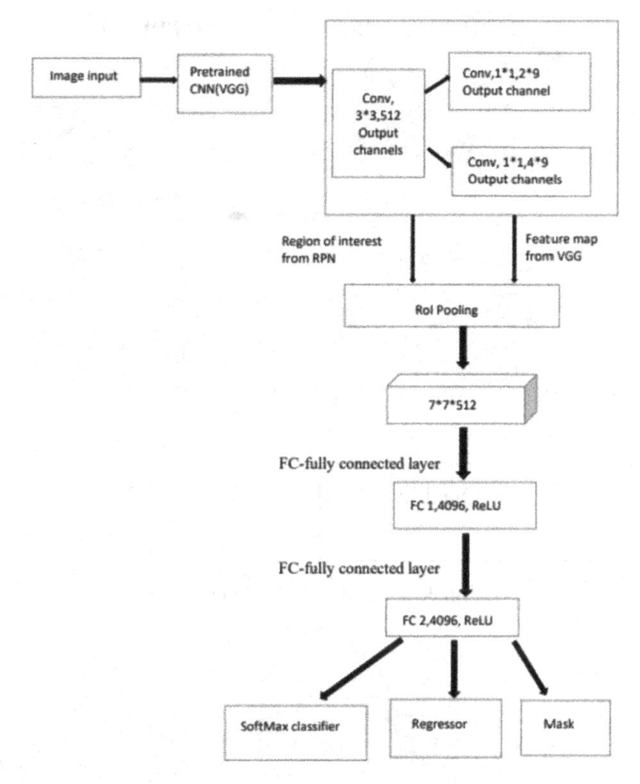

Fig. 2. Architecture of MASK R-CNN

Figure 3, Fig. 4 and Fig. 5 shows the images of some of the input images of various crops that have been tested along with the corresponding resultant output images. Leaf of healthy cherry is shown in Fig. 3, leaf of rot apple black is shown in Fig. 4, and leaf of black grapes having measles is shown in Fig. 5. Table 1 shows the accuracy gained by the existing and the proposed model. The existing model have obtained an accuracy of 96%, whereas our proposed model has gained an accuracy of 97.8% after using the techniques CNN and Mask R – CNN. Figure 6. Shows the comparative study of the works of various model along with the proposed model. Existing models have used combination of various techniques such as, SSD with Mask R – CNN, Mask R -CNN with ResNet, R – CNN with Mask R – CNN to name a few. The accuracy varies from 77.76 to 95.7.

Fig. 3. Leaf of Healthy Cherry

Fig. 4. Leaf of Rot Apple Black

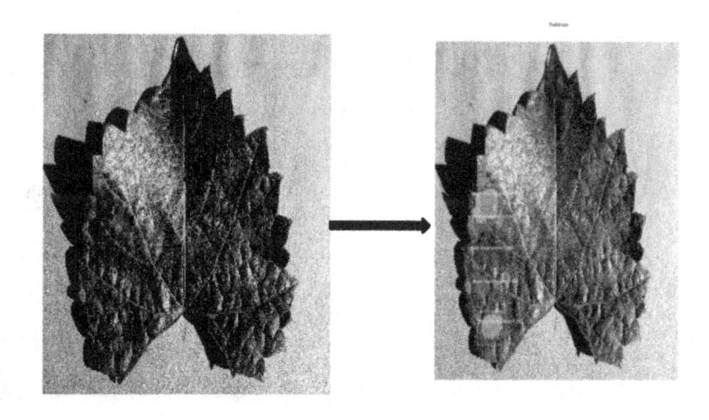

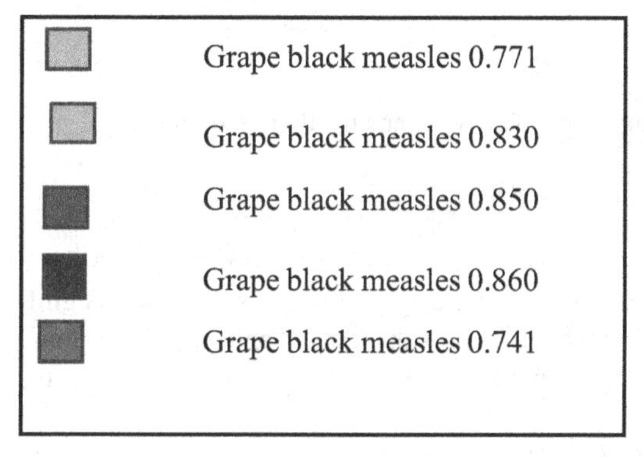

Fig. 5. Black Grape leaf having Measles.

Table 1. Accuracy obtained by the existing and the proposed model.

Models	Highest Accuracy Obtained
Existing model	96%
Proposed model	97.8%

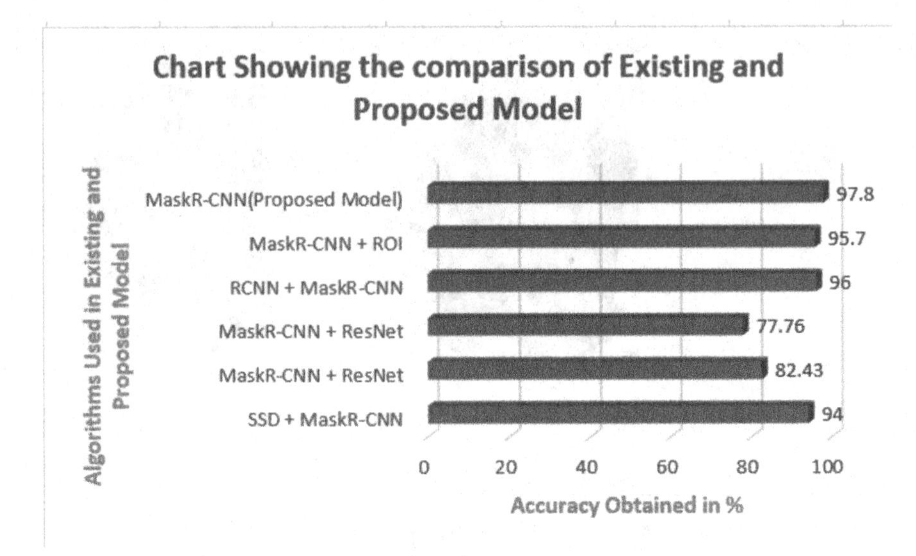

Fig. 6. Graph Showing the comparison of Existing and the Proposed Model

4 Conclusion

Crop disease causes a greater impact in damaging the agriculture, resulting in insignificant yield losses. Hence it is very important to prevent, detect and get rid of these diseases. The recent expansion of deep learning methods have provided a way in the detecting the disease in the crop.

A model is trained using Mask R-CNN by taking the Plant Village Dataset which has different variety of both healthy and unhealthy leaves. Various inspection was conducted by taking various hyperparameters of the Mask R-CNN architecture. Accuracy of results in the process of identification of disease showed that Mask R-CNN model is promising as well as can do a greater impact and is also efficient in identifying the disease and may have the potential in the detection of diseases in real-time agriculture system. The accuracy obtained is 97.8% for Apple Black Rot which is highest among all the crops. Future mainly concentrates on providing treatment for the identified type of pests that have affected the image of leaf and also concentrates on suggesting the treatment for the once that have been detected with disease in leaf images.

References

1. Udawant, P., Srinath, P.: Cotton leaf disease detection using instance segmentation. J. Cases Informat. Technol. (JCIT) **24**(4), 1–10 (2022)
2. Afzaal, U., Bhattarai, B., Pandeeya, Y.R., Lee, J.: An instance segmentation model for strawberry diseases based on mask R-CNN. Sensors **21**(19), 6565 (2021)
3. Su, W.-H., et al.: Automatic evaluation of wheat resistance to fusarium head blight using dual mask-RCNN deep learning frameworks in computer vision. Remote Sensing **13**(1), 26 (2020)

4. Anandhan, K., Singh, A.S.: Detection of paddy crops diseases and early diagnosis using faster regional convolutional neural networks. In: 2021 International Conference on Advance Computing and Innovative Technologies in Engineering (ICACITE), pp. 898–902. IEEE (2021)
5. Rehman, Z., et al.: Recognizing apple leaf diseases using a novel parallel real-time processing framework based on MASK RCNN and transfer learning: an application for smart agriculture. IET Image Process. **15**(10), 2157–2168 (2021)
6. Wang, S., Sun, G., Zheng, B., Yawen, D.: A crop image segmentation and extraction algorithm based on Mask RCNN. Entropy **23**(9), 1160 (2021)
7. Pallapothu, T., Singh, M., Sinha, R., Nangia, H., Udawant, P.: Cotton leaf disease detection using mask RCNN. AIP Conf. Proc. **2393**, 020114 (2022). https://doi.org/10.1063/5.0074814
8. Almazaydeh, L., Alsalameen, R., Elleithy, K.: Herbal leaf recognition using masl-region convolutional neural network (MASK R-CNN). J. Theoret. Appli. Informat. Technol. **100**(11) (2022)
9. Storey, G., Meng, Q., Li, B.: Leaf disease segmentation and detection in apple orchards for precise smart spraying in sustainable agriculture. Sustainability **14**(3), 1458 (2022)
10. Baireddy, S.: Leaf Tar Spot Detection Using RGB Images. arXiv preprint arXiv:2205.00952 (2022)
11. Yao, N., Ni, F., Wu, M., Wang, H., Li, G., Sung, W.-K.: Deep learning-based segmentation of Peach diseases using convolutional neural network. Front. Plant Sci. **13** (2022)
12. Harish, H., Sreenivasa Murthy, A.: Identification of lane lines using advanced machine learning. In: 2022 8th International Conference on Advanced Computing and Communication Systems (ICACCS), vol. 1. IEEE (2022)
13. Harish, H., Sreenivasa Murthy, A.: Identification of lane line using PSO segmentation. In: 2022 IEEE International Conference on Distributed Computing and Electrical Circuits and Electronics (ICDCECE). IEEE (2022)
14. Das, S., Roy, D., Das, P.: Disease feature extraction and disease detection from paddy crops using image processing and deep learning technique. In: Computational Intelligence in Pattern Recognition, pp. 443–449. Springer, Singapore (2020)
15. Kumar, D., Kukreja, V.: Image-based wheat mosaic virus detection with mask-RCNN model. In: 2022 International Conference on Decision Aid Sciences and Applications (DASA), pp. 178–182. IEEE (2022)
16. Felfeliyan, B., et al.: Self-Supervised-RCNN for Medical Image Segmentation with Limited Data Annotation. arXiv preprint arXiv:2207.11191 (2022)
17. Nandhini, M., Kala, K.U., Thangadarshini, M., Madhusudhana Verma, S.: Deep Learning model of sequential image classifier for crop disease detection in plantain tree cultivation. Comput. Elect. Agricult. 197, 106915 (2022). https://doi.org/10.1016/j.compag.2022.106915, ISSN 0168–1699
18. Harish, H., Sreenivasa Murthy, A.: Edge discerning using improved PSO and Canny algorithm. In: Communication, Network and Computing (CNC-2022). CCIS. Springer Nature. https://doi.org/10.1007/978-3-031-43140-1_17
19. Raj, N., Perumal, S., Singla, S., Sharma, G.K., Qamar, S., Chakkaravarthy, A.P.: Computer aided agriculture development for crop disease detection by segmentation and classification using deep learning architectures. Comput. Elect. Eng. **103**, 108357 (2022). ISSN 0045–7906, https://doi.org/10.1016/j.compeleceng.2022.108357. (https://www.sciencedirect.com/science/article/pii/S0045790622005742)

20. Kurmi, Y., Saxena, P., Kirar, B.S.: Deep CNN model for crops' diseases detection using leaf images. Multidim Syst Sign. Process **33**, 981–1000 (2022). https://doi.org/10.1007/s11045-022-00820-4
21. Harish, H., Bharathi, D.S., Pratibha, M., . Holla, D., Ashwini, K.B., Keerthana, K.R., Particle swarm optimization for predicting Breast Cancer. In: 2022 International Conference on Knowledge Engineering and Communication Systems (ICKES), Chickballapur, India, pp. 1–5 (2022). https://doi.org/10.1109/ICKECS56523.2022.10060690

An Improved Machine Learning Approach for Throughput Prediction in the Next Generation Wireless Networks

Aritri Debnath[1] and Barnali Dey[2(✉)]

[1] Electronics and Communication Dept. (VTU RC), CMR Institute of Technology, Bengaluru, India
aritri.d@cmrit.ac.in

[2] Department of Information Technology, Sikkim Manipal Institute of Technology, Majitar East, Sikkim, India
barnali.mou.dey@gmail.com

Abstract. Currently evolving 5G telecom networks require different intelligent learning and decision mechanisms to adapt to the varying network conditions. Further, considering additional requirements of low-latency and ultra-reliability, newer resource allocation schemes are to be explored to find the most effective way to predict the amount of resources required by a system. In this paper, a unique resource allocation scheme is devised for the 5G network using the properties of first packet transmission and the subsequent retransmissions. The proposed model is shown to accurately predict the system-level as well as user-level throughput for a set of mobile users, while ensuring lower consumption of network resources. The simulations show an accuracy of 85% for the user-level throughput that is acceptable for the dynamic resource planning in future networks.

Keywords: Resource Allocation · Decision Making algorithm · Machine learning · Regression Learning

1 Introduction

Communication and networking form an inseparable pair that is the fundamental technology to enable global connectivity and foster economic growth. It started with analog based communications of the first-generation networks in the 1980s. It was characterized by the advent of cellular technology where the land area was divided into cells and subsequently this resulted in increase in the usage of the spectrum which resulted in the accommodation of a bigger group of users. Then the second-generation communication enabled digital encryption. This ensured a reduction of the effects of interference, signal noise and distortion. Furthermore, the third generation had a greater frequency spectrum and faster transfer of data. The advent of 4G and then subsequently LTE (long term evolution) ushered a new era characterized by widespread availability of smart phones, global mobility support and personalized service. One significant aspect of 4G was that it used OFDM (Orthogonal Frequency Division Multiplexing) in place of FDMA or

P. Das et al. (Eds.): AMRIT 2023, CCIS 1953, pp. 23–30, 2024.
https://doi.org/10.1007/978-3-031-47224-4_3

TDMA [1]. It also brought about better user conveniences and sophisticated interfaces along with much faster data rates of transmission.

The currently evolving 5G network technologies are mainly characterized by immense amounts of data being transmitted over the wireless channels because of larger bandwidths and superior telecommunications technologies. While the earlier generations of wireless communications focused primarily on one-on-one communications and mobile services, the next generation of wireless communications ventures into newer fields of medicine, factories, industries, transportation and other societal fields which involve breakthroughs in other technologies like the internet of things(IoT) [2]. The International Telecommunications Union (ITU) defines the specific parameters and guidelines for every aspect of each of the objectives of 5G. Also, 5G is expected to be implemented for a variety of use cases and other features. The next generation wireless systems should be capable of providing fast (a low latency of 1ms) as well as highly efficient (reliability of 99.999% or in 5 nines) transmissions.

The most fundamental way of achieving these standards during transmission is to ensure an effective method of allocating radio resources during transmission [3]. Therefore, the most effective way of utilizing radio resources of time and frequency must be devised. Although expansion of the frequency spectrum is one way to go about doing this, other optimal ways of tackling these hurdles are also explored. To reach the required reliability goal, certain parts of the special domain spectrum must be re-utilized. The smallest unit of a resource is considered as a resource block. It consists of TTI (Transmission Time Interval) which is the amount of time taken for a transmission in the radio link and a subchannel which is an individual medium of transmission. The radio resources need to be allocated to the UE (User Equipment) as resource blocks. There could be mismatches or errors when the resource blocks or packets are transmitted therefore seamless resource allocation is crucial for effective and reliable retransmissions to take place.

Machine learning, particularly in recent years, has found its use in almost every major field. It provides optimal solutions by using previously obtained data to predict or suggest probable outcomes. Hence machine learning can be an extremely useful tool that can be used to fulfil the Ultra-Reliable-Low-Latency (URLLC) conditions of the next generation of wireless communications. It is expected that employing machine learning algorithms within the 5G framework will result in the system automatically being able to detect the levels of available spectrum and intelligently provide ways to allocate the radio resources accordingly. All this will ensure that resources are being assigned in an effective and optimal manner while still ensuring that the necessary stringent requirements of reliability and latency prescribed by the International Telecommunications Union (ITU) are met.

This paper proposes and evaluates an improved machine-based approach for throughput prediction using the characteristics of packet transmissions. The paper organization is as follows. Section 2 explores the literature in this field. The resource allocation model is presented in Sect. 3. The proposed regression training for throughput prediction is described in Sect. 4. Results obtained via simulation are provided in Sect. 5, and the paper concludes with a note on the possible future work in this field in Sect. 6.

2 Literature Survey

In the last decade, machine learning has gained pace due to a multiplicity of factors. Two main factors are the availability of high-capacity hardware components such as GPU/TPU for faster computation, and improvements in the software domain for the required low-latency calculations.

With the increasing usage of software as well as network virtualization techniques, estimation for User Equipment (UE) mapping to the correct gNB (gNodeB) has become a key requirement to sustain flow throughput. The resource estimation is then used for various use-cases such as resource allocation [4], capacity planning and network upgrade, for example. Techniques such as the Hidden Markov Model (HMM) [5], Star-garch [6], and fuzzy logic [7] are popular for estimation. However, these methods lack the dynamism and timeliness guarantee as required by the application like URLCC among others.

Using machine learning to predict throughput is therefore a natural choice among researchers that can not only provide the low-latency real-time output but also promises accuracy for a wide variety of applications. In [8], a vector regression based method is used for TCP throughput prediction. The technique provides promising outputs but cannot be directly applied to the access network's throughput estimation as it is based on the end-to-end paradigm. A mmWave throughput prediction technique is devised in [9] that focuses on the situation awareness. Several other research methods have been studied in [10–13] that focus on various aspects of wireless network throughput prediction.

We differentiate our work from the earlier studies in two ways. First, we used a regression based completely analytical technique that focuses on the error reduction for the estimated resource blocks to be allocated to the mobile UEs. Second, the proposed method creates a relationship between the packet transmission and retransmission thereby providing an inherent advantage of TCP-like performance for the access network. The next section describes the resource allocation model used in this work.

3 Proposed System

A) Resource Allocation.

This section presents our integrated approach of throughput prediction of access networks from the Internet Protocol (IP) packet characteristics. In this plan, we consider the utilization cases with deterministic bundle appearances where enough assets are saved for the transmission of the packets of every User Equipment (UE) [14]. It is to be noted that some of the packets might get lost due to the effect of erroneous radio conditions which are dealt with using retransmissions. This means that a pool of resources must be made available for this purpose. In the event that the measure of assets held for first transmissions is equivalent to that of the size of the pool, every single lost packet can be retransmitted. In this paper, an attempt is made to minimize the resource pool size with the goal that the general asset utilization is diminished while fulfilling the objective dependability [7].

A system with N UEs is considered to be indexed by i. As the latency limitation significantly affects the distribution of resources, it is guaranteed that the all-out time lapsed during the packet creation and the termination of the retransmission of the packets is not exactly the latency target. Along these lines, there's a limitation in the measuring of TTIs devoured for the continuous transmissions and hence in the measure of required range by pre-deciding the necessary measure of resources [16]

For a given "resource unit (RU)" that is equal to R resource blocks (RBs), in our system, each packet occupies one unit. We consider the number of RUs per TTI as M. The measure of accessible range per unit (T) is procured by separating the whole scope of the range W by the accessible measure of spectral resource per unit, i.e., M = bW/(Rω)c. In case of the first transmission, the number of the user equipment indicates how many resource units will be required. K is the amount of the resources to be designated for the retransmissions.

To assess how reliable the proposed resource allocation scheme is, the two possible outcomes that lead to a loss are examined. Firstly, if the number of resources that are required for retransmissions are greater than K, retransmission of certain lost packets cannot be done, which will result in loss. Secondly, even if there is sufficient room for the retransmissions of the packets, there is a possible chance for failure of retransmissions. We also need to see that when the space is not enough for all the retransmissions, random K packets are selected amongst the lost packets to be retransmitted.

Therefore, the equation for the error rate will be [16]:

$$e(K, \delta_1, \delta_2) = \sum\nolimits_{n=0}^{N-1} C_{N-1}^{n} \delta_1^{n+1} (1 - \delta_1)^{N-1-n}$$

$$\times (\delta_2 I_{n+1 \leq K} + \frac{\delta_2 K + n + 1 - K}{n+1} I_{n+1>K})$$

As we know the errors that occur will not be dependent on the possibility of losing n number of packets in the initial round is given by the law of binomial distribution. The summation represents the possible cases of losing n packets in the initial transmissions of (N − 1) UEs. The summation consists of the law of binomial distribution which is, multiplied by δ_1. To assume that the packets of the user under consideration has been lost, whereas $(1 - \delta_1)$ represents the retransmissions. In case there is sufficient room for every packet that is lost for its transmitted again, the possibility of error is represented as δ_2, giving the term $\delta_2 I_{n+1 \leq K}$. Or else, it will be 1 in case the packet's retransmissions are not regarded and in the case of retransmissions it will be δ2.

B) Regression Learning.

Machine learning is associated with various algorithms and learning models. One of the most common types is regression. Regression determines the relationship between the predictor variables. It describes the relationship between the dependent target variable and independent predictor variable and estimates the changes of dependent variables corresponding to an independent predictor variable when other variables are held constant. Regression helps us in determining the trends in the data and predict real or continuous values.

Regression learner app in MATLAB tool is used to predict the data model using machine learning by training the dataset, with 60% of training and 40% of testing data.

The steps involved in training the dataset model is given by:

- After choosing the information and approving the dataset model for specific regression problems, the dataset is taken for it to be trained according to regression learning algorithms.
- Suitable comparing and tuning options of the regression learning model have to be chosen.
- The effectiveness of the dataset model in regression learning is assessed by comparing model statistics and visualizing the obtained results.
- Simulation Set-up

We use the vehicular LTE eNB data collected at the University of Hertfordshire using ETSI API. The dataset consists of 178 users and their throughput values taken at different times. The UL Scrambling Code of 54, UL SIR Target 17.3, Min UL Channelization Code Length 8, DL Scrambling Code 1, DL channelization Code Number 15, Maximum DL Power 9.3 and Minimum DL Power 10.1 was used. The UEs move across the network on a predefined path. RSSI and throughput for a UE are measured at each time tick (along with logging the GPS coordinates).

The dataset is trained using the regression model. The dataset was imported to the workspace and trained via cross-validation. After the dataset model has been trained, various algorithms are checked and analysed to see which model has obtained the best overall results. The error rates are calculated for the various models. This result is used to determine the best model. The model with the lowest error rate is chosen. The selected model was trained using the fine tree algorithm.

4 Results

Figure 1 shows the prediction response plot for true vs. predicted values of throughput. Here, when the responses are plotted one can inspect the results of the obtained model. Once the dataset is trained by the means of regression modelling, the plot of the responses obtained are the predictions of the responses compared with the record number. The predicted values in contrast to the original responses are plotted. Prediction errors can be observed as one can see the disparities between the original responses (in blue) and the predicted ones(in yellow). The response variable here is the throughput and these responses are observed with respect to the users.

For all the UEs, at all the locations, every predicted throughput value is averaged to obtain the predicted response. Similarly, all the original throughput values are averaged to get the actual response. The error percentage is calculated as 100*(actual - predicted)/actual. The average error percentage of our system is 14.93% thus providing us with an accuracy of more than 85%.

Figure 2 provides the predicted vs. actual throughput performance using the proposed prediction model. The Predicted vs Actual plot can be used to ascertain whether the model is effective or not. This plot is used to determine how well the obtained model predicts values that are close to the true response. In this plot, one can observe the predictions in contrast to the true responses. Ideally if the predictions were to be 100% accurate, the predictions will be a diagonal line coinciding with the original response. In the figure, we can observe the error in the predictions by observing the perpendicular distance between the diagonal line to any one of the points. An acceptable model will be one where the predictions are spread across closely about the line.

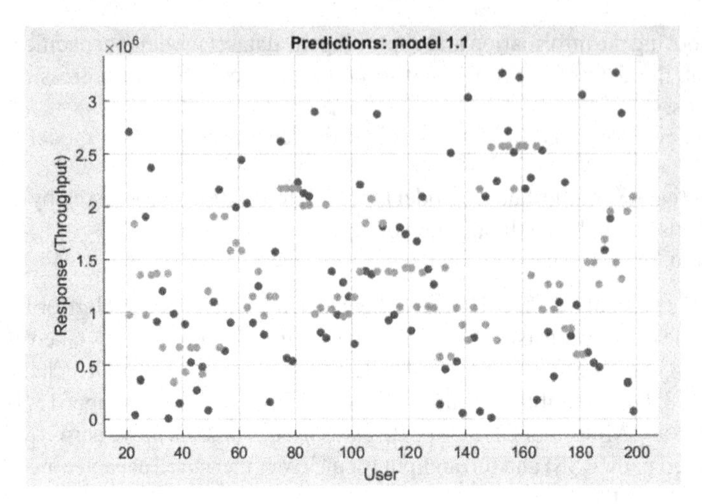

Fig. 1. Response Plot for ML Regression Model

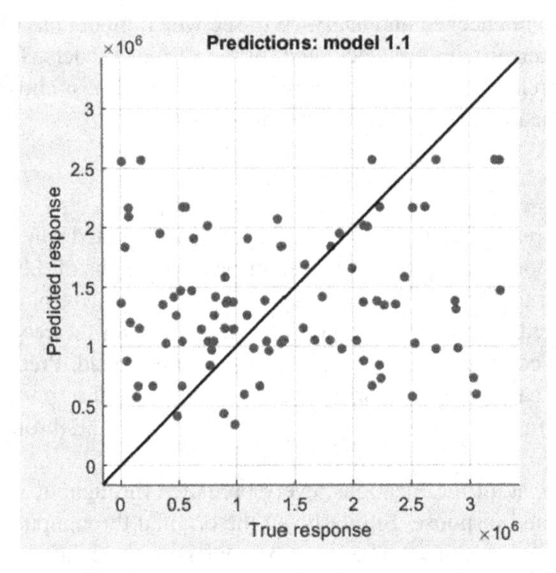

Fig. 2. Predicted vs. Actual Response Plot

5 Conclusion

Ensuring an effective way through which transmissions and subsequent retransmissions of resources take place is crucial for reducing latency. In this work, the error rate minimization mechanism is implemented that reduces the block error rates (BLER) for multiple transmissions and retransmissions. The approach takes into account various possible scenarios for the error probabilities on the basis of how many times the transmission of the same packet has taken place. It likewise considered the results relying upon whether there is sufficient space for all the packets that are lost to be transmitted again.

In the next generation of wireless communications, the gNodeB establishes the link between the mobile network and the user equipment (UE). The resource blocks are required to be allocated in an optimal manner to improve the system efficiency without compromising on the reliability. Based on a real-time dataset of vehicular traffic data in an LTE network, the work presented here showed that using regression analysis in machine learning, the average resources consumed by various users can be examined and trained to predict throughput with an accuracy of more than 85%.

6 Future Scope

A dynamic approach would broaden the scope of the data which would account for various situations under which the system would function. Therefore, to further improve upon accuracy of the predictions in the future, we would incorporate a multi parametric dynamic approach which would factor in several other parameters to ensure better prediction accuracy.

Acknowledgement. The work is carried out using licensed version of MATLAB 19 which is available at the signal processing centre of excellence. We would like to acknowledge the support and guidance provided by the members of the CoE.

References

1. Xiaodong, W.: OFDM and its application to 4G. In: 14th Annual International Conference on Wireless and Optical Communications (2005). WOCC 2005, Newark, NJ, USA, 2005, pp. 69-, doi: https://doi.org/10.1109/WOCC.2005.1553751
2. Olga, B.-L., Victor, L., Branka, J., Aditya, S., Ehsaneh, S., Bryson, P.: Microwave and Wearable Technologies for 5G (2015). https://doi.org/10.1109/TELSKS.2015.7357765
3. Chang, B., Zhang, L., Li, L., Zhao, G., Chen, Z.: Optimizing resource allocation in URLLC for real-time wireless control systems. IEEE Trans. Veh. Technol.Veh. Technol. **68**(9), 8916–8927 (2019). https://doi.org/10.1109/TVT.2019.2930153
4. Suganya, S., Maheshwari, S., Latha, Y.S., Ramesh, C.: Resource scheduling algorithms for LTE using weights. In: 2016 2nd International Conference on Applied and Theoretical Computing and Communication Technology (iCATccT), pp. 264–269. IEEE (2016)
5. Maheshwari, S., Mahapatra, S., Cheruvu, K.: Measurement and forecasting of next generation wireless internet traffic (No. 525). EasyChair (2018)
6. Kulkarni, P., Lewis, T., Fan, Z.: Simple traffic prediction mechanism and its applications in wireless networks. Wireless Pers. Commun.Commun. **59**(2), 261–274 (2011)
7. Vasu, K., Maheshwari, S., Mahapatra, S., Kumar, C.S.: QoS aware fuzzy rule based vertical handoff decision algorithm for wireless heterogeneous networks. In: 2011 National Conference on Communications (NCC), pp. 1–5. IEEE (2011)
8. Mirza, M., Sommers, J., Barford, P., Zhu, X.: A machine learning approach to TCP throughput prediction. ACM SIGMETRICS Perform. Eval. Rev. **35**(1), 97–108 (2007)
9. Wang, Y., Narasimha, M., Heath, R.W.: MmWave beam prediction with situational awareness: a machine learning approach. In: 2018 IEEE 19th International Workshop on Signal Processing Advances in Wireless Communications (SPAWC), pp. 1–5. IEEE (2018)

10. Alsheikh, M.A., Lin, S., Niyato, D., Tan, H.P.: Machine learning in wireless sensor networks: algorithms, strategies, and applications. IEEE Commun. Surv. Tutorials **16**(4), 1996–2018 (2014)
11. Chen, M., Challita, U., Saad, W., Yin, C. and Debbah, M.,. Machine learning for wireless networks with artificial intelligence: a tutorial on neural networks. arXiv preprint arXiv:1710. 02913, 9 2017
12. Jagannath, J., Polosky, N., Jagannath, A., Restuccia, F., Melodia, T.: Machine learning for wireless communications in the internet of things: a comprehensive survey. Ad Hoc Netw.Netw. **93**, 101913 (2019)
13. Jiang, C., Zhang, H., Ren, Y., Han, Z., Chen, K.C., Hanzo, L.: Machine learning paradigms for next-generation wireless networks. IEEE Wirel. Commun.Wirel. Commun. **24**(2), 98–105 (2016)
14. Jiang, D., Wang, H., Malkamaki, E., Tuomaala, E.: Principle and performance of semi-persistent scheduling for VoIP in LTE systems. In: International Conference on Wireless Principle and performance of semi-persistent scheduling for VoIP in LTE system. In 2007 International Conference on Wireless Communications, Networking and Mobile Computing, Sept 2007, pp. 2861–2864 (2007)
15. 3GPP: Study on latency reduction techniques for LTE. 3GPP TR 36.881 v14.0.0, Tech. Rep. (2016)
16. Elayoubi, S.E., Brown, P., Deghel, M., Galindo-Serrano, A., Elayoubi, S.: Radio resource allocation and retransmission schemes for URLLC over 5G networks. IEEE J. Sel. Areas Commun. Inst. Electr. Electron. Eng. **37**(4), 896–904 (2019). https://doi.org/10.1109/jsac. 2019.2898783. hal-02117082

Protein Secondary Structure Prediction Without Alignment Using Graph Neural Network

Tanvir Kayser and Pintu Chandra Shill(✉)

Department of Computer Science and Engineering, Khulna University of Engineering and Technology , Khulna 9203, Bangladesh
pintu@cse.kuet.ac.bd

Abstract. Complex molecules known as proteins carry out a number of essential tasks in the human body. Protein structure are interwoven and determine each protein's particular activity. Graph Neural Net (GNN) has developed as an effective deep learning method to extract information from protein structures, which may be represented by graphs of amino acid residues. This paper suggests utilizing a graph neural network to predict protein structure from amino acid sequences without alignment. In this instance, nodes (amino acids) and connecting edges (distances between amino acids) may be used to instantiate a protein in a network. To demonstrate the scalability of the suggested technique, many experiments are carried out using various benchmark datasets, including RS126 and RSCB PDB. The simulation results show that in the majority of cases for different data sets, the recommended strategy outperforms other comparable methods.

Keywords: Graph Neural Network · Amino Acid · Alignment · Secondary Structure Prediction · Window Padding

1 Introduction

Predicting protein structure from amino acid sequences is a challenging and complex task that requires identifying commonalities and variances among proteins. For the purpose of creating novel medications, the biological activity of various species is investigated using the protein secondary structure. Since the creation of protein sequences is now quick and inexpensive, determining the activities of newly found proteins is time-consuming and critical. The complexity of proteins' activities and the quick growth of their quantity in sequence databases make computational methods for automated function prediction difficult. The way that proteins fold into three-dimensional structures allows them to perform a wide range of tasks inside the cell. Furthermore, many functional regions of proteins are disordered, whereas most domains fold into a characteristic and organized three-dimensional shape. Protein structure determines binding selectivity, mechanical stability, and a host of other biological functions, such as transport and signal transduction. These many different protein activities are categorized using the Kyoto Encyclopedia of Genes and Genomes (KEGG), the Gene Ontology (GO) Consortium, Enzyme Commission (EC) numbers, and other widely used categorization techniques.

P. Das et al. (Eds.): AMRIT 2023, CCIS 1953, pp. 31–45, 2024.
https://doi.org/10.1007/978-3-031-47224-4_4

To illustrate different facets of protein functioning, GO, for example, groups protein classes into three unique ontologies: molecular function (MF), biological process (BP), and cellular component (CC).

Graph neural network (GNN) is a new and promising classification approach based on deep neural networks [1]. Because GNNs can extract hidden patterns from Euclidean data (pictures, videos, texts), their increasing popularity in recent years has been encouraging in a number of fields, including recommendation systems, knowledge graphs, protein interface prediction graph neural networks [2], social networks [3], bioinformatics [4], graph mining, and other research fields. As a novel paradigm, the use and investigation of GNN in bioinformatics are few. In order to research this cutting-edge theory and use it to advance humanity and the world, this study was motivated. Provide a new method for utilizing a graph neural network to predict a protein's secondary structure. Since each amino acid in the main sequence is connected to two of its neighbors, a graph illustrating their relationships may be made. A particular kind of neural network used to manage data in graph data structures is called a graph neural network. Evaluating graph data for deep learning and classic machine learning algorithms is still a difficult undertaking because of the variety of data kinds that are present in graphs and since these methods specialize in core data types. Create a graph using the dataset and main sequence (amino acid). The orthogonal encoding characteristic of the model gives each network node a distinct embedding. Then, iterate the complete network using the GNN method to combine the information from neighboring nodes.

There are an astounding number of combinatorial optimization techniques [5–9] that may be used to identify a protein's secondary structure, including SWISS-PROT [5], deep learning [6], and convolutional neural networks (CNN) [7]. The annotation information in SWISS-PROT is mined using a data mining method. Here, the unannotated proteins are subjected to 11306 rules, which produce the annotated proteins. Without relying on assumptions, the deep learning approach is used to retrieve the minute details from the protein sequence. The prediction of protein structure is used in drug discovery, antibody design, understanding protein-protein interactions, and molecular-molecule interactions [8]. These methods are often slower for large protein data sets and have poor prediction accuracy Based on the aforementioned examination of current techniques, including both conventional and combinatorial approaches, it is imperative and time-consuming to suggest a way that minimizes current shortcomings while forecasting the structure of proteins. The technique of finding commonalities or arrangements between biological sequences is known as sequence alignment, and it is this arrangement that is used to forecast the secondary structure of proteins [9]. These models' primary flaw is that they are not entirely accurate. These models are quicker, but they could miss certain crucial details about the secondary structure of a protein. Furthermore, these models do not ensure that the best protein sequence will be found. When protein databases are huge, this approach takes longer. Along with the database, the memory demand rises as well. They need a gap penalty to operate properly. The main flaw in these models is that they rely on propensity values for probability calculations. This approach may not always identify protein secondary structures accurately and may provide false positive results.

2 Related Works

Using structural alignment to improve the prediction of protein secondary structure accuracy Structure-based sequence alignments are a step in the secondary structure prediction process that Scott Montgomerie et al.'s technique incorporates. It is feasible to anticipate at least a portion of the secondary structure of the query protein by mapping the structure of a known homologue (sequence ID > 25%) onto the sequence of the query protein. It is feasible to get very high prediction accuracy by combining this structural alignment approach with traditional (sequence-based) secondary structure approaches and then combining it with a "jury-of-experts" system to produce a consensus conclusion. Applying this novel method to a sequence-unique test set of 1644 proteins from EVA yields an average Q3 score of 81.3%. [10] Protein Secondary Structure Prediction Utilized are deep convolutional neural fields Sheng Wang et al. provide Deep CNF (Deep Convolutional Neural Fields) for protein SS prediction. Conditional Neural Fields (CNF), which combine shallow neural networks with conditional random fields (CRF), are an extension of conditional neural fields (CNF) that uses deep learning. Deep CNF is significantly more effective than CNF because it can describe interdependency between neighboring SS labels in addition to sophisticated sequence- structure relationships using a deep hierarchical design. Experimental results show that Deep CNF performs substantially better than the most popular predictors, with 84% Q3 accuracy, 85% SOV score, and 72% Q8 accuracy on the CASP and CAMEO test proteins, respectively [11].

The neural network technique is predicated on the way synaptic connections function in brain neurons, where input is transformed into the final output after being processed several times. Using a training set of sequences with known structures, the values of the weights that affect signals are altered in neural networks during training. Typical neural network technique is demonstrated by Rost and Sander's [12] PHD program, which claims 72.2% accuracy after training on a test set of profiles from different sequences. When larger databases and PSI-BLAST [13] were utilized to build the training set, this accuracy climbed to 75% [14].

Kathuria et al [15] Random Forest classification method is used to estimate the protein's secondary structure. As a classification and regression tree (CART) method that may create several decision trees from a single set of training data, Breiman created Random Forest [16]. In this study, they simply distinguish between alpha (α) structure and non-alpha structure. All polypeptides with an alpha structure have their structure property set to alpha during the preprocessing stage, whereas the property for all other polypeptide structures is set to non- alpha. A ROC curve is used to evaluate their model and demonstrate that it yields the desired results. However, they restricted their investigation to binary categorization. The support vector machine (SVM), which employs frequent profiles and evolutionary information as an encoded input [17], was developed by Vapnik et al. [18] as a different approach to classification problems. Others were able to achieve 70% accuracy by using the PSI-BLAST PSSM profiles [19] as input vectors and combining the sliding window technique with SVM architecture [20]. SVM is gradually being applied to PSSP problems since it has occasionally beaten the bulk of other learning systems, including NNs [21].

With the use of various window widths and a basic distance classifier, Ghosh et al. [22] were able to attain an accuracy of about 60%. To obtain a Q3 = 68 percent accuracy,

Ahmed et al. used a genetic algorithm with a multiple window approach. A protein secondary structure predictor was developed by Spencer and colleagues [23] using deep learning network topologies that were swiftly trained on a collection of 198 proteins using the position-specific score matrix of BLAST.

Zhang et al. [24] describe the construction of a Graph Neural Network (GNN). Applying GNN to the process of predicting the secondary structure of the protein is astounding. As a consequence of this inspiration, GNN is utilized to forecast the secondary structure of a protein. Keeping all these reasons in mind, the use of a graph neural network in protein secondary structure has been considered.

This is how the rest of the paper is organized. The approach used to create the GNN model is presented in the major part that follows. Section 3 evaluates the performance and simulation results. Section 4 conclusion.

3 Proposed Method

3.1 Data Pre-possessing

Preprocessing the main and secondary structure of proteins by removing primary and secondary sequences from datasets and which dataset lengths more than 128 will first minimize time complexity, CPU use, and memory usage. Shorten protein primary and secondary structures and remove non-standard amino acids like B, Q, U, X, and Z during preprocessing to allay worries about sequence labeling.

3.2 Encoding

A unique binary vector known as the amino acid character's one-hot-encoding or orthogonal encoding is used to separately alter each of the twenty amino acids. This binary vector is then used for further computations, including sliding windows. We convert the eight secondary protein structures into integer values between 0 and 7 [25]. There are 20 amino acids, as can be observed, hence the feature has been set to 20. Each amino acid character thus has a distinct embedding length of precisely 20.

3.3 Graph Generation

The study's graph delegation, which details the protein's amino-acid linkages, was created using the predicted structure. Here, amino acid-level features—which are utilized to generate the nodes of the graph representation—were extracted using a language model, an architecture that usually learns from text (in this instance, protein sequence) [26]. Figure 1 demonstrates the construction of the protein's graph representation, where each vertex stores one-hot encodings of the current vertex's index.

Here, a GNN was trained to identify and then predict proteins based on their graph representation, as seen in Fig. 1. This GNN model feeds a set of neural networks a graph that describes a few proteins as input. These neural networks use message passing, a framework in which node information is sent to connecting nodes at each step, to make the graph representation more complicated in order to reveal potentially useful hidden

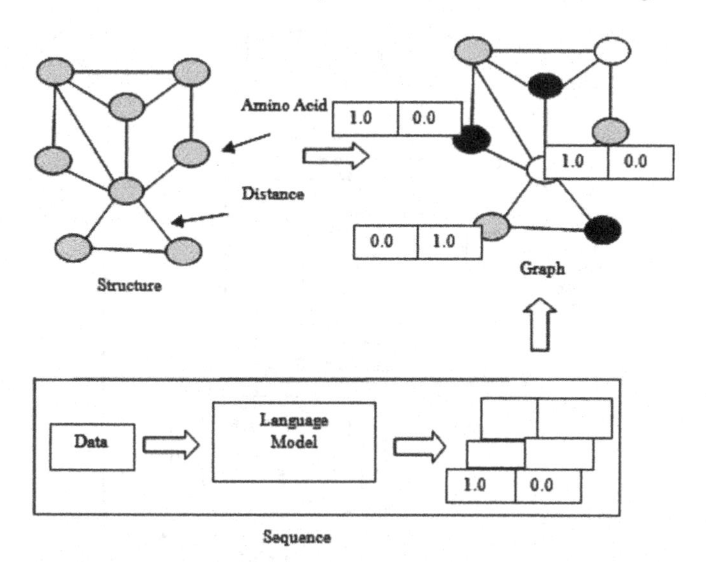

Fig. 1. Illustration of protein graph representation construction.

information about the protein. The information describing other amino acids is updated in this context using the sequencing data for each amino acid. Each node in the new, higher-level graph representation that the GNN creates has data on both itself and its neighbors in addition to information about itself. Subsequently, an activation function and a pooling layer reduce the graph's dimensionality to a single, usable score. The model generates an accurate suitable score by determining which node connections in the neural networks to provide higher priority for a more accurate prediction.

3.4 Window Padding

To prevent the predicted missing pattern of commencing and terminal amino acid sequences or a protein main sequence, zero paddings are used in the sliding window technique. The middle amino acid of the window, which foresees the amino acid's complementary secondary structure, is the goal value. Window padding mainly work through encoding where font and back padding works randomly another hand window size also depends on this padding process.

3.5 Neural Network

Lastly, use a neural network to determine the accuracy of protein secondary structure prediction. The hyperparameters of the kernel are:

C: C determines the classification of a data point in the penalty form. The better answer is produced by the smaller C value since it accepts a broader margin.

Gamma: The gamma symbol () denotes the decision region. The training data is over fit by gamma values that are bigger (Fig. 2).

Get better accuracy with $C = 0.1$, $\gamma = 1.5$, and kernel = rbf for the (rscb pdb) and rs126 datasets.

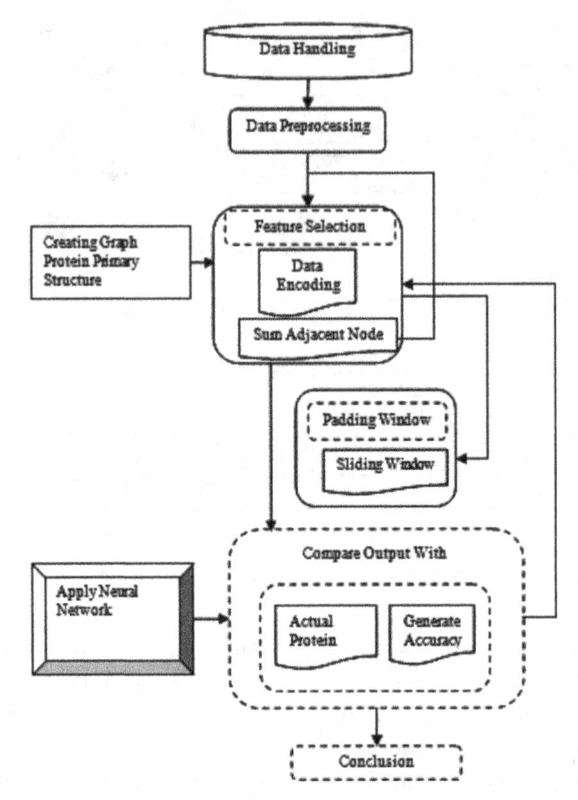

Fig. 2. Proposed Graph Neural Network Model.

4 Simulation Results

To The RS126 dataset and RSCBPDB dataset from related studies are used in a number of tests with various characteristics to see whether the suggested approach is correct [29].

4.1 RS126 Dataset

On the RS126 benchmark data set, the suggested method is evaluated. There are 152 total occurrences, each of which has two lines. The first line displays the protein's main sequence, while the second line displays the primary sequence's secondary structure. The average length of a protein sequence is 185 amino acids, and there are 22596 residues total. The beta strand contributes for 22%, the coil for 45%, and the alpha helix for 31% of RS126. Various more data files from past research or studies contain the RS126 dataset. Additionally, you might obtain it from a website like Protein Data Bank (PDB) [27]. It has 20 distinct main residue types and three secondary structural states (Table 1).

Table 1. Three States of PSS of RS126 Dataset

No	Symbol	Narration
I	E	Stand
II	H	Helix
III	C	Loops

4.2 RSCB PDB Dataset

The RSCB PDB dataset [34], which is available on relevant research, is one of the benchmark datasets for predicting protein secondary structure [29]. 200000 instances make up the dataset. The primary dataset includes a list of peptide sequences and secondary structures. It is a tabular representation of the file retrieved from the RSCB PDB.

A description of the dataset's columns:

pdb_id: The URL's

chain_code: The chain code is necessary to identify a particular peptide (chain) when a protein has numerous of them.

seq: The amino acid sequence of the peptide sst8: The secondary structure with 8state (Q8) sst3: The secondary structure with 3state (Q3) len: The peptide's total amount of amino acids

hasnonstdaa: Whether the peptide contains any unusual amino acids (B, O, U, X, or Z).

The nonstandard proteins B, O, U, X, and Z are all masked with the "*" character. For secondary structure prediction, it is customary to decrease the previously described eight states (Q8) to three (Q3) by combining (E, B) into E, (H, G, I) into E, and (C, S, T) into C.

4.3 Performance Metric Indicators

Ran numerous tests with varying compositions using benchmark datasets we acquired from various sources to evaluate the classification accuracy and computational efficiency of graph neural networks. Various error measurements are utilized for evaluation. True positive (TP) is properly classified as positive, whereas false negative (FN) is incorrectly classified (FN). True negative (TN) refers to an occurrence that is negative and accurately identified, whereas false- positive refers to the opposite (FP).

Accuracy

The quantity of similarity of assessments of a quantity to its true value is the accuracy of a measuring system.

$$\text{Accuracy} = \frac{TP + TN}{TP + TN + FP + FN} \tag{1}$$

Precision.

The ratio of True Positives (TP) to all Positives is known as precision. It displays the overall volume of classified data that has been correctly categorized.

$$\text{Precision} = \frac{\text{TP}}{\text{TP} + \text{FP}} \tag{2}$$

Recall

Recall is a metric for gauging a model's ability to identify True Positives. Out of all the data that has been categorized, it is the proportion of data that has been correctly categorized.

$$\text{Recall} = \frac{\text{TP}}{\text{TP} + \text{FN}} \tag{3}$$

F-score

A model's precision on a given dataset is measured by the F-score, often known as the F1-score. The F-score, defined as the harmonic mean of the accuracy and recall measures, is a way to integrate the two metrics for a model. You can change the F-score such that precision takes precedence over recall, or the other way around.

$$\text{F1} = 2 \times \frac{\text{Precision} \times \text{Recall}}{\text{Precision} + \text{Recall}} \tag{4}$$

Mean Absolute Error (MAE)

Mean absolute error is a measure of errors between the number of pairs reporting the same occurrence (MAE).

$$\text{MAE} = \frac{\sum^{n} i = 1\,\text{yi} - \text{xi}}{\text{n}} \tag{5}$$

Here,

yi = Predict value

xi = Actual value

n = Total number of object

Relative Absolute Error (RAE)

When a mean error (residual) is compared to errors produced by a simplistic or naive model, the Relative Absolute Error is measured as the ratio. A reasonable model (one that yields better results than a simplistic model) will yield a ratio of less than one.

$$RAE = \frac{\sum_{\text{i}=1}^{\text{n}} (\text{yi} - \text{xi})^2}{\sum_{\text{i}=1}^{\text{n}} (\text{xi})^2} \tag{6}$$

Here,

yi = Predict value

xi = Actual value

n = Total number of object

4.4 Results

Two distinct types of data sets, including RS126 and RSCB PDB, were used to execute the simulations. Proportionally split the dataset, using 75 percent for training and 25 percent for testing. The hyper parameter tinkering for the two datasets is displayed in Table 2. A sliding window technique is employed in the suggested GNN model, with different window sizes and types producing varying accuracy levels. The differences in accuracy as a result of sliding window size are shown in the Table 3. There are multiple window size.

Table 2. Hyper parameter tuning of GNN

Dataset	Kernel	C	λ
RS126	Rbf	100	0.01
RSCB PDB	Rbf	1.5	0.1

Table 3. Compare Datasets With Various Window Size

Window Size	RS126 Accuracy	RSCB PDB Accuracy
5	68.07%	85.78%
7	69.87%	86.92%
9	71.12%	88.18%
11	72.26%	89.7%
13	73.66%	89.10%
15	74.2%	88.46%
17	73.25%	86.93%
19	71.11%	85.20%
21	70.02%	84.55%

Table 3 shows that the highest accuracy of the RSCB PDB dataset is 89.7 percent and that of the RS126 dataset with window size15 is 74.2 percent. The suggested model is assessed using several performance assessments, which are evaluated and displayed in the table for those two datasets (Table 4).

Figure 3 shows how the accuracy varies with window size from a plot perspective. The illustrations below are meant to help you understand how the performance measuring factor works.

Cross Validation

A method of interpolation called cross-validation is used to assess machine learning algorithms on a small sample of data. Cross-validation is a technique used in learning algorithms to assess how effectively a machine learning model functions on unknown input. In other words, a tiny sample will be used to evaluate the model's performance

Table 4. Performance Evaluation

Dataset	Precision	Recall	F-score	MAE	RAE
RS126	.73	.74	.72	.31	.53
RSCB	.87	.80	.82	.31	.26
PDB					

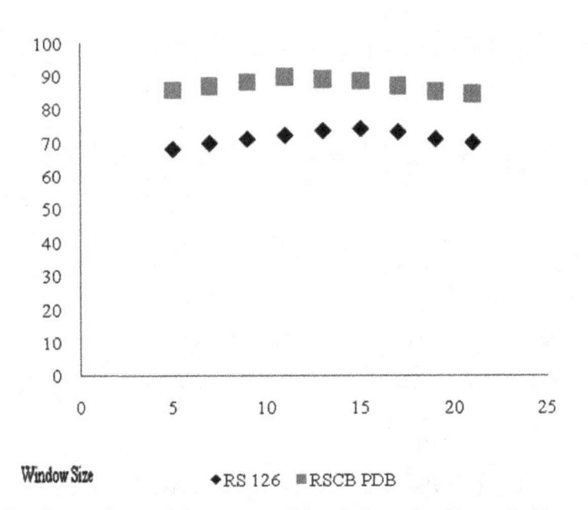

Fig. 3. Comparison of Accuracy with window size Between Datasets.

as a whole when predictions are made using data that was not used during the model's training.

One popular technique for evaluating machine learning methods on a dataset is K-fold cross-validation. From a limited dataset, the k-fold cross validation process generates k non- overlapping folds. While the remaining folds are merged to create a training dataset, each of the k folds has the opportunity to be utilized as a stand-alone test set. A total of k models are fitted, assessed, and their average performance is recorded on the k hold-out test sets. Cross- validation is used to determine whether the proposed model GNN is accurate. It is based on two datasets. The figure below displays the results of cross-validation using k-fold cross- validation (Figs. 4 and 5).

Comparative Analysis

A comparison of the graph neural network (GNN) with other special approaches in the Table 5's protein secondary structure prediction demonstrates that GNNs have developed to a standard level. There are several methods accuracy test rest given in the Table 5. But all those researches they find out results for only one data set. In this study worked for two different dataset which are RS126 and RSCB PDB. Where RS126 given 74.2 % and RSCB PDB given

89.7 % accuracy for the proposed Graph Neural Network model. Other hand takes a look on the Table 5 here highest accuracy is 73.3 % (Figs. 6 and 7).

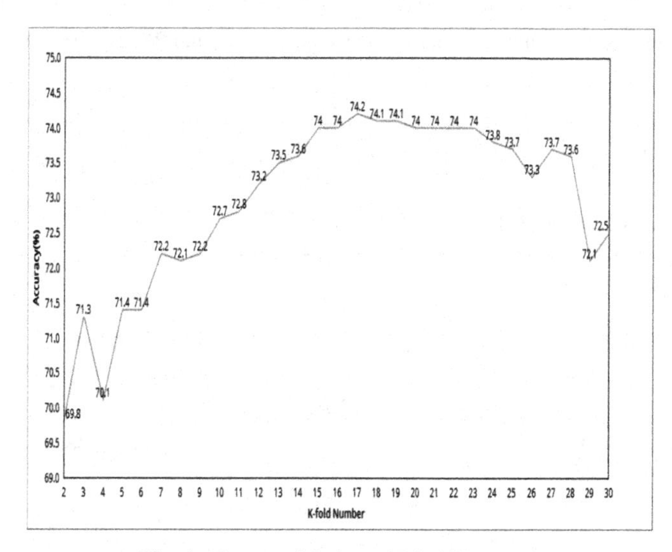

Fig. 4. Cross validation of RS126 Dataset.

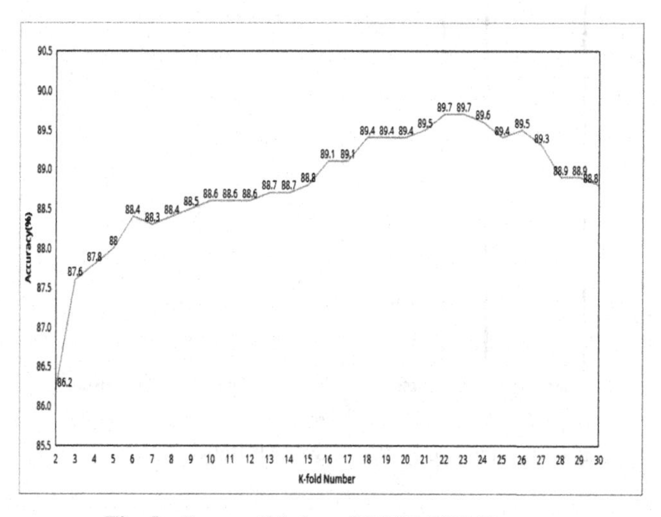

Fig. 5. Cross validation of RSCB PDB Dataset.

Table 5. Comparison with various model using RS126 and RSCB PDB datasets

Methods	Year	Accuracy
CNN [28]	2014	66.7
LSTM [29]	2014	67.4
Convolutional neural fields [11]	2016	68.3
biRNN [30]	2017	68.5
Multi-scale CNN	2018	70.3
[31] BiLSTM [32]	2019	70.4
SAINT [33]	2019	73.3
GNN (Proposed model)		74.2 & 89.7

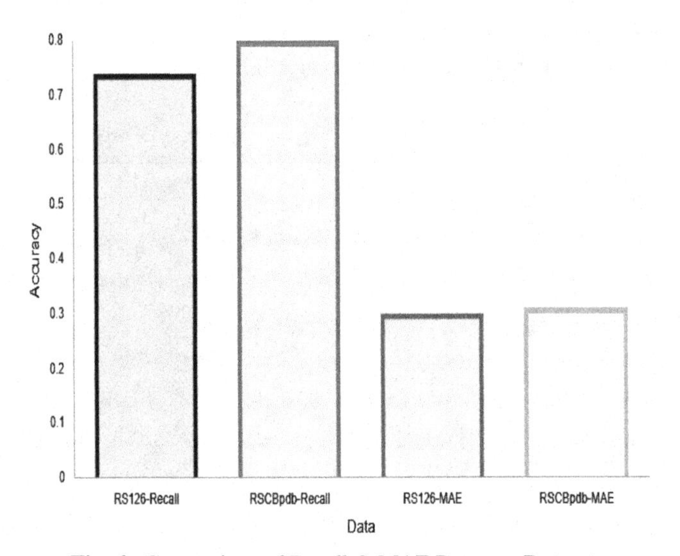

Fig. 6. Comparison of Recall & MAE Between Datasets.

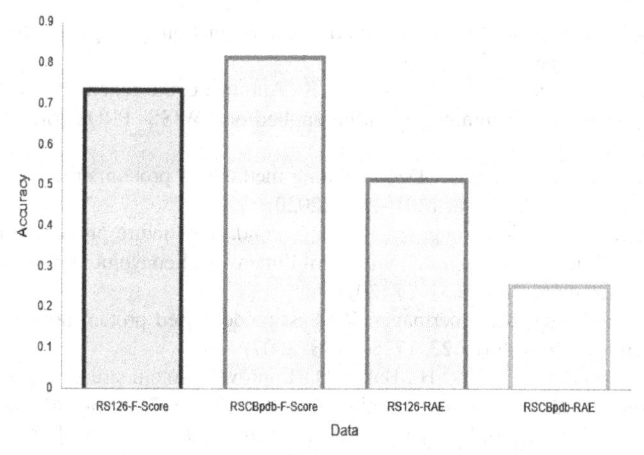

Fig. 7. Comparison of F-Score & RAE Between Datasets.

5 Conclusion

The big, complex molecules called proteins found in the human body are studied in this quest. GNN works well as a novel model for predicting protein secondary structure. This model deviates slightly from the conventional GNN in that it is composed of many components that join together to form the real model. The data is first graphed, and then the graph is processed and a neural network is employed in the following two processes. Because the amino acid sequence is linear, a linear graph will be produced. The node will add its feature to those of the neighboring connected nodes in order to analyze the graph. For even more optimization, window padding and orthogonal encoding are used. Different widths of window padding are used to get the best outcome. The accuracy of the RS126 dataset was 74.2 percent after hyper parameter adjustment, while the accuracy of the RSCB PDB dataset was 89.7 percent. The absence of substantial data and large processing power is one of the suggested model's drawbacks.

Future work:

More accuracy can be achieved by combining neural networks with other models. Other optimization techniques can be used to improve accuracy.

Incorporate 3D graphing of secondary structure into this proposed model to improve it.

References

1. Strokach, A., Becerra, D., Corbi-Verge, C., Perez-Riba, A.: Fast and flexible protein design using deep graph neural networks. Cell Syst. **11**, pp. 402–411 (2020)
2. Alex, F., Byrd, J., Shariat, B., Ben-Hur, A.: Protein interface prediction using graph convolutional networks. In: Advances in Neural Information Processing Systems. NeurIPS, pp. 1–10 (2017)
3. Wu, Y., Lian, D., Xu, Y., Wu, L., Chen, E.: Graph convolutional networks with mark over random field reasoning for social spammer detection. In: Proceedings of the AAAI Conference on Artificial Intelligence, vol. 34, pp. 1054–1061. IEEE (2020)

4. Zitnik, M., Leskovec, J.: Predicting multi-cellular function through multi- layer tissue networks. Bioinformatics **33**, 190–198 (2017)
5. Kretschmann, E., Fleischmann, W., Apweiler, R.: Automatic rule generation for protein annotation with the C4.5 data mining algorithm applied on SWISS- PROT. Bioinformatics **17**, 920–926 (2001)
6. Torrisi, M., Pollastri, G., Le, Q.: Deep learning methods in protein structure prediction. J. Comput. Struct. Biotechnol. **18**, 1301–1310 (2020)
7. Guo, Y., Wang, B., Li, W., Yang, B.: Protein secondary structure prediction improved by recurrent neural networks integrated with two- dimensional convolutional neural networks. J. Bioinform. Comput. Biol. **16**, 1–17 (2018)
8. Hochreiter, S., Heusel, M., Obermayer, K.: Fast model-based protein homology detection without alignment. Bioinform. **23**, 1728–1736 (2007)
9. Yang, J., Anishchenko, I., Park, H., Baker, D.: Improved protein structure prediction using predicted interresidue orientations. Biophys. Comput. Biol. **117**, 1496–1503 (2020)
10. Scott, M., Shan, S., Warren, J.G., David, S.W.: Improving the accuracy of protein secondary structure prediction using structural alignment. BMC Bioinform. **7**, 1–13 (2006)
11. Wang, S., Peng, J., Ma, J., Xu, J.: Protein secondary structure prediction using deep convolutional neural fields. Sci. Rep. **6**, 1–11 (2016)
12. Rost, B., Sander, C.: Improved prediction of protein secondary structure by use of sequence profiles and neural networks. In: Proceedings of the National Academy of Sciences, PNAS, pp. 7558–7562 (1993)
13. Jones, D.T.: Protein secondary structure prediction based on position- specifics coring matrices. J. Mol. Boil. **292**, 195–202 (1999)
14. Salamov, A.A., Solovyev, V.V.: Protein secondary structure prediction using local alignments. J. Mol. Boil. **268**, 31–36 (1997)
15. Kathuria, C., Mehrotra, D., Misra, N.K.: Predicting the protein structure using random forest approach. Procedia Comput. Sci. **132**, 1654–1662 (2018)
16. Breiman, L.: Random forests. Mach. Learn. **45**, 5–32 (2001)
17. Guo, J., Chen, H., Sun, Z., Lin, Y.: A novel method for protein secondary structure prediction using dual-layer SVM and profiles. Struct. Funct. Bioinform. **54**, 738–743 (2004)
18. Cortes, C., Vapnik, V.: Support-vector networks. Mach. learn. **20**, 273–297 (1995). https://doi.org/10.1007/BF00994018
19. Soheilifard, R., Toussi, C.A.: On the contribution of normal modes of elastic network models in prediction of conformational changes. In: 23rd Iranian Conference on Biomedical Engineering, 1st International Iranian Conference on Biomedical Engineering, pp. 263– 266. IEEE (2016)
20. Hu, H.J., Pan, Y., Harrison, R., Tai, P.C.: Improved protein secondary structure prediction using support vector machine with a new encoding scheme and an advanced tertiary classifier. IEEE Trans. NanoBiosci. **3**, 265–271 (2004)
21. Hossain, A., Faisal Zaman, M., Nasser, M.M., Islam,: Comparison of GARCH, neural network and support vector machine in financial time series prediction. In: Chaudhury, S., Sushmita Mitra, C.A., Murthy, P.S., Sastry, S.K., Pal, (eds.) Pattern Recognition and Machine Intelligence, pp. 597–602. Springer, Heidelberg (2009). https://doi.org/10.1007/978-3-642-11164-8_97
22. Reyaz-Ahmed, A.B.: Protein Secondary Structure Prediction Using Support Vector Machines, Nueral Networks and Genetic Algorithms, pp. 1–80 (2007)
23. Spencer, M., Eickholt, J., Cheng, J.: A deep learning network approach to ab initio protein secondary structure prediction. In: IEEE/ACM Transactions on Computational Biology and Bioinformatics, vol. 12, pp. 103–112. IEEE (2014)
24. Zhang, Z., Leng, J., Ma, L., Miao, Y., Li, C., Guo, M.: Architectural implications of graph neural networks. Comput. Architect. Lett. **19**, 59–62 (2020)

25. Lin, K., May, A.C., Taylor, W.R.: Amino acid encoding schemes from protein structure alignments: Multi-dimensional vectors to describe residue types. J. Theor. Boil. **216**, 361–365 (2002)
26. Gligorijević, V., Renfrew, P.D., Kosciolek, T.: Structure-based protein function prediction using graph convolutional networks. Nat. Commun. **12**, 3168 (2021)
27. Chin, Y.F., Hassan, R., Mohamad, M.S.: Optimized local protein structure with support vector machine to predict protein secondary structure. In: Lukose, D., Ahmad, A.R., Suliman, A. (eds.) KTW 2011. CCIS, vol. 295, pp. 333–342. Springer, Heidelberg (2012). https://doi.org/10.1007/978-3-642-32826-8_34
28. Zhou, J., Troyanskaya, O.: Deep supervised and convolutional generative stochastic network for protein secondary structure prediction. In: 31st International Conference on Machine Learning, vol. 32, pp. 745–753. ICML (2014)
29. Sønderby, S., Winther, O.: Protein secondary structure prediction with longshort term memory networks, ar Xiv preprint ar Xiv, Cornell University (2014)
30. Johansen, A., Sønderby, C., Sønderby, S., Winther, O.: Deep recurrent conditional random field network for protein secondary prediction. In: Proceedings of the 8th ACM International Conference on Bioinformatics, pp. 73–78 (2017)
31. Zhou, J., Wang, H., Zhao, Z., Xu, R., Lu, Q.: BMC Bioinformatics: Protein8 - class secondary structure prediction by convolutional neural network with highway. CNNHPSS **19**, 1–10 (2018)
32. Asgari, E., Poerner, N., McHardy, A., Mofrad, M.: BioRxiv. Deep Prime 2Sec: deep learning for protein secondary structure prediction from the primary sequences. CSH, 1–8 (2019)
33. Uddin, M., Mahbub, S., Rahman, M., Bayzid, M.: Self-attention augmented inception-inside-inception network improves protein secondary structure prediction. Bioinformatics **36**, 4599–4608 (2019)
34. Protein data bank. https://www.kaggle.com/datasets/alfrandom/proteinsecondary-structure
35. Zhuang, C., Ma, Q.: Dual graph convolutional networks for graph-based semi- supervised classification. In: Proceedings of the World Wide Web Conference on World Wide Web, WWW, pp. 1–10 (2018)

A Lexicon-Based Approach for Sentiment Analysis of Bodo Language

Jaya Rani Mushahary(✉), Bipul Roy, and Mandwip Baruah

National Institute of Electronics and Information Technology, BTR, Kokrajhar EC 783370, Assam, India
jayaranimhy@gmail.com

Abstract. With the proliferation of digital content, it is increasingly important to understand the sentiment of the information that is being shared. Sentiment Analysis, the procedure of identifying the emotional tone of a given text, has become a crucial area of research in Natural Language Processing (NLP). However, performing sentiment analysis on Bodo language texts is challenging due to inadequate labeled data and the complexity of the language. The amount of resources and research available for Bodo language is quite limited compared to more widely spoken languages, such as English or Hindi. This study proposes a lexicon-based approach for sentiment analysis of Bodo language texts, which is a low-resource language spoken in India. For our purpose, we have considered sentences and words from various social sites and then created a Bodo sentiment lexicon through manual labeling of Bodo language words and sentences with their sentiment polarity (positive, negative, or neutral). Results from evaluating the approach on a dataset of Bodo language texts show an accuracy of 70%. This lexicon-based approach is a simple and effective method for sentiment analysis of Bodo language texts and can be used to enhance the performance of other NLP tasks such as text classification and opinion mining.

Keywords: Sentiment Analysis · Natural Language Processing · Bodo Language

1 Introduction

The surging trend of social media platforms such as Instagram, Facebook and Twitter has led to an increase in the amount of user-generated content online. This data is used in many areas such as marketing strategies, government policies, and individual decision making. With the rise of social media, people are expressing their opinions in their native languages, which has led to an increase in the amount of data in Indian languages on the internet. Sentiment analysis aims to determine the emotional tone of a text and is a crucial task in the field of Natural Language Processing (NLP). It is widely used in various fields such as business, marketing, social media, and politics to understand public opinion and consumer behavior.

P. Das et al. (Eds.): AMRIT 2023, CCIS 1953, pp. 46–56, 2024.
https://doi.org/10.1007/978-3-031-47224-4_5

While there is significant research and resources available for sentiment analysis in English, the same cannot be said for other languages, particularly low-resource languages like Bodo. Bodo is a Sino-Tibetan language spoken by the Bodo people, an indigenous ethnic group primarily found in the northeastern Indian state of Assam, as well as parts of Arunachal Pradesh, West Bengal, and Nagaland. It is one of the officially recognized languages in India and is used in education, government and media. Bodo is written using the Devanagari script, and has its own distinct grammar and vocabulary. It has a rich literary tradition, and various literary works such as novels, poetry, and plays have been written in Bodo. With over a million speakers, as per the 2011 census [3], the number of Bodo speakers is on the rise as shown in Fig. 1. The number of Bodo speakers totals to 1,482,929, out of which 1,454,547 are native speakers [4]. Bodo is considered a relatively new language in terms of written literature in comparison to other scheduled languages with standard scripts. Although the written literature of Bodo has grown in recent decades, this growth is impeded by several challenges, such as the absence of up-to-date dictionaries, difficulties in handling spelling variations [1], and the necessity to incorporate new words [2]. Despite the growth, performing sentiment analysis on low-resource languages like Bodo is challenging due to the shortage of labeled data and the language's complexity.

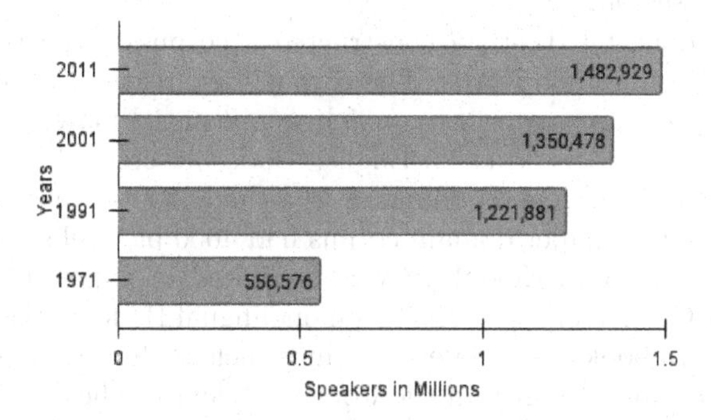

Fig. 1. Graph depicting the number of Bodo speakers over the years. The census for Bodo was not conducted in 1981.

The objective of sentiment analysis for Bodo language is to accurately recognize and extract subjective information from Bodo text, such as opinions, evaluations, appraisals, and emotions. It can help in understanding public opinion on various issues, products, and services, and can be used to improve decision-making in various fields. This work aims to address this challenge by proposing a lexicon-based approach for sentiment analysis of Bodo language. The main contribution of this work are summarized as follows:

a. Decreasing the gap in resources for sentiment analysis and other NLP works for Bodo language by creating a Bodo text corpus through collection of sentences, phrases and words from social media and dictionary sites.

b. Development of sentiment lexicon for Bodo language by manually labeling the collected data to their sentiment polarity (positive, negative, or neutral)
c. Proposal of a lexicon-based approach for sentiment analysis in Bodo language, which uses the sentiment lexicon to classify Bodo text into positive, negative, and neutral sentiments.
d. Contribution to the development of NLP tools for Bodo language and to promote further research in the field of sentiment analysis for low-resource languages.

The rest of the paper is structured in the following way: Sect. 2 provides a review of related work, Sect. 3 describes our proposed approach and methodology, and Sect. 4 presents the analysis of our results. We conclude the paper in Sect. 5 with a summary of our research contribution, limitations, and future research directions.

2 Related Work

The field of NLP research in the Bodo language is in its infancy, and few studies have been conducted to date. In particular, no previous work has been done on sentiment analysis in the Bodo language, making this an important and challenging area of research.

In the work done by [1], they constructed a corpus of the Bodo language that consisted of 1.5 million words. The corpus was created by collecting Bodo language texts from three categories, which were learned materials, media, and literature. In [5], a parallel corpus for English-Bodo was created from general and newspaper domains. The general domain corpus had 6500 sentences from various sources, while the newspaper domain corpus had 4000 parallel sentences related to news and events. A methodology was presented in paper [6] that utilized Google Keep for OCR in order to create a monolingual Bodo corpus from various books, including 13 books from different genres such as children's, fiction, novels, plays, biographies, and dramas. In the work [7], a lexicon dictionary-based approach was introduced by the authors for detecting polarity in Bengali language text data. They compared their proposed system with Machine Learning(ML) classifiers and concluded that lexicon dictionary-based approach functions as a more accurate model with 92% accuracy for Bengali text polarity detection. In the paper [8], by utilizing lexical features on the news domain, ML classifiers were employed to develop a sentiment polarity classification model for Assamese language. They collected a total of 18806 sentences from the local daily newspapers' news articles - Ganaadhikar and Pratidin. Paper [9] performs sentiment analysis on Manipuri articles using a manually modified verb lexicon and POS tagging. The system achieved a F-measure of 75.00%, a recall of 72.10%, and a precision of 78.14%. In the paper [10] a technique was suggested to create G-SWN, a Gujarati SentiWordNet, by leveraging the synonym relationships of H-SWN, a Hindi SentiWordNet, and IWN, an IndoWordNet. The accuracy of Gujrati G-SWN was evaluated achieving 52.95% and 52.72% accuracy for simple scoring and unigram presence classifiers respectively.

3 Methodology

The steps of the proposed approach for sentiment analysis of Bodo language is illustrated in Fig. 2. The first step involves the collection of Bodo language texts from various sources. The collected data are then cleaned to remove unwanted characters, duplicates, and other noise. After cleaning, the data are preprocessed by tokenizing the text into words, removing stop words, etc. Finally, the preprocessed text is used for sentiment analysis using a lexicon-based approach. Each of these steps will be described in detail in the following subsections.

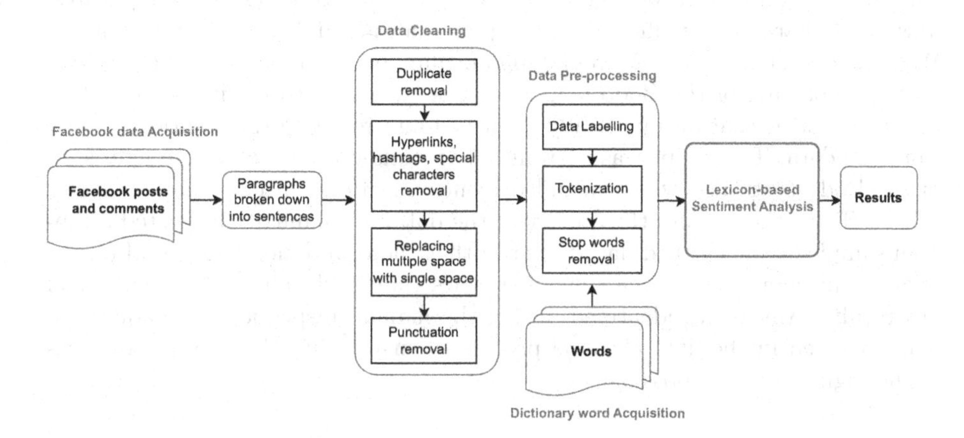

Fig. 2. Proposed system architecture

3.1 Data Collection

Dataset acquisition is a crucial and laborious task. In this research, a dataset of Bodo language was prepared by manually collecting Bodo posts and comments from Facebook news pages. The collected data included posts and comments in paragraphs, and the text dataset was obtained from these paragraphs. We also collected a separate dataset of Bodo words from dictionary website[1] available on Google using a Python web scraping script developed using the Beautiful Soup and Requests libraries. The Facebook data collected was from the period of October to December 2022, and the resulting dataset comprises 3000 sentences and 9,366 words. The statistics of the Bodo corpus created are presented in Table 1.

[1] https://www.bsarkari.com/bodo/dictionary.

Table 1. Statistics of data collected

Source	Total number of data
Facebook news pages	3000 sentences
Dictionary sites	9366 words

3.2 Data Cleaning

The collected data were subjected to a cleaning process to eliminate duplicates, unwanted characters, multiple spaces, punctuations and other forms of noise. We removed URLs, email addresses, phone numbers, and any special characters that are not part of the Bodo language. The cleaning process involved writing a Python script that automatically identified and removed the mentioned noise from the data. The script was customized to handle the specific characteristics of the Bodo language, which included identifying Bodo-specific characters and tokens. To further ensure the quality of the data, we manually inspected a random sample of the cleaned data to verify that there were no instances of errors, missing information, or other inconsistencies that could affect the accuracy of our results. Any issues identified during the manual inspection were addressed and corrected in the data cleaning process to ensure that the final dataset was of the highest quality possible.

3.3 Data Pre-processing

After the data cleaning process, the next step is to preprocess the Bodo text corpus. The preprocessing involves converting the text into a format that can be used for sentiment analysis. In this step, we perform the following operations:

Data Labeling: In the data labeling process, the dataset has been manually labeled into three different categories: positive, negative, and neutral. These categories were represented by the numbers '1', '−1', and '0' respectively. The sentences and words data were saved in separate files. Data that convey emotions like joy, appreciation, admiration, motivation, and encouragement have been categorized as positive. Similarly, data that convey negative emotions like loathing, disregard, sorrow, offend, discord, or resistance have been tagged as negative. Neutral was the tag assigned to data that did not express a powerful emotion, such as basic information, figures, or declarations. Examples of the labeled sentences and labeled words are provided in Fig. 3 and Fig. 4 for reference.

Tokenization: The process of tokenization involves breaking down text into individual units, typically words or subwords, which are called tokens. It is a crucial step in the data preprocessing phase. The aim of tokenization is to transform a document or a sentence into a list of tokens, which can then be used for text processing and analysis.

Stop Words Removal: Words that are frequently used in a language and do not contribute any significant meaning to the text are known as stop words. These words should be eliminated to enhance the efficiency and accuracy of subsequent NLP tasks such as sentiment analysis, text classification, and information retrieval.

The output of the preprocessing step is a cleaned and preprocessed text corpus, which is then used for sentiment analysis using a lexicon-based approach.

Sentiment	Sentence
Positive	जोबोद मोजां हाबाफारि। (Excellent work)
Negative	ग'साइगावआव गिख्रंथाव जाब्रथाइ जानानै सासे बर' आइजोनि जिउ खहा । (A tragic accident in Gossaigaon resulted in the death of a Bodo woman)
Neutral	दिनै बिजया दशमि एबा दुर्गा फुजानि जोबथा सान। (Today is the last day of Vijayadashami or Durga Puja)

Fig. 3. Labeled sentence data with their corresponding English meanings

Sentiment	Word
Positive	बाख्नायथाव (Appreciable)
Negative	दुखुनांथाव (Saddening)
Neutral	सावगारि (Picture)

Fig. 4. Labeled word data with their corresponding English meanings

3.4 Lexicon-Based Sentiment Analysis

Sentiment analysis employs two main approaches, one of which involves using a lexicon to compute the sentiment by evaluating the polarity of phrases and words present in a given text. This method requires a dictionary of positive, neutral, and negative words, and a sentiment value must be assigned to each word. Various methods for developing these dictionaries are available, including manual and automatic methods. Lexicon-based methods generally depict a text message as a collection of words known as a "bag of words". Then the sentiment values corresponding to all positive, neutral, and negative words or phrases in the text are allocated based on the lexicon or sentiment dictionary. To determine the overall sentiment of the text, a final prediction is made by applying a function, like sum or average. When determining the sentiment of a word, its local context, like negation, is also taken into account.

In this work, we chose to utilize a lexicon-based approach to eliminate the requirement for creating a labeled training dataset. The primary limitation of models using ML is their dependence on labeled data, which can be challenging to obtain in sufficient quantities and with appropriate labeling in case of languages like Bodo. Moreover, the ease with which this approach can be grasped and modifiable by humans is a substantial benefit for our work.

Algorithm 1: Algorithm used for sentiment classification

```
Input: text: A piece of Bodo text
       positive_words: A set of positive words
       negative_words: A set of negative words
       negation_words: A set of negation words
Output: sentiment polarity score (1 for positive, -1 for negative, 0 for
        neutral)
1  Initialize sentiment_score variable to 0.
2  Initialize negate_sentiment variable to False.
3  for word in text do
4      if word in negation_words then
5          negate_sentiment = not negate_sentiment
6      else if word in positive_words then
7          if negate_sentiment = False then
8              sentiment_score += 1
9          else
10             sentiment_score -= 1
11         end
12         negate_sentiment = False.
13     else if word in negative_words then
14         if negate_sentiment = False then
15             sentiment_score -= 1
16         else
17             sentiment_score += 1
18         end
19         negate_sentiment = False.
20     end
21 end
   /* Determine the sentiment based on the sentiment score */
22 if sentiment_score > 0 then
23     return 1
24 else if sentiment_score < 0 then
25     return -1
26 else
27     return 0
28 end
```

Negation. In our work, we have collected and utilized negation words. In a lexicon-based approach, the most commonly used approach for dealing with negation is to reverse the polarity of the lexicon item located adjacent to the negator in a text. For instance, "good" would have a polarity of 1, while "not good" would have a polarity of −1. However, in the Bodo language, the negating word appears after the word, such that "mwjang" has a polarity of 1 and "mwjang nonga," where "nonga" is a negating word, has a polarity of −1. To address this issue, we have reversed the order of the words in the text, enabling us to traverse the text in the opposite direction. Moreover, it is noteworthy that Bodo is also written in the Roman script for online communication, and "mwjang nonga" is a Bodo word written in Roman script.

Sentiment Classification. The sentiment classification process involves iterating through each word in the Bodo text and checking its sentiment from the word sentiment lexicon (or sentiment dictionary) created in this study. If the word is a negation word, the sentiment of the following word is reversed. If it is a positive word, the sentiment score is incremented by 1, or decremented by 1 if it is negated. Similarly, if it is a negative word, the sentiment score is decremented by 1, or incremented by 1 if it is negated.

After iterating through all the words in the text and calculating the sentiment score, the sentiment is determined based on the value of the sentiment score. Sentiment scores greater than 0 are considered positive, whereas scores less than 0 are classified as negative. A sentiment score of 0 is classified as neutral. The algorithm used for sentiment classification is shown in Algorithm 1.

4 Results

Our dataset contains a total of 3000 sentences and 9366 words. The words consist of 527 positive words, 780 negative words, and 8059 neutral words. Our analysis achieved an accuracy of 70% on the dataset of 3000 sentences. Accuracy is defined as the percentage of text lines or sentences in the test set that the classifier identifies correctly. We calculated the accuracy of our proposed system using Eq. 1 given below.

$$Accuracy = \frac{C}{T} * 100 \tag{1}$$

where,
C = number of correct prediction, and T = number of total prediction

Bodo text sentiment analysis system heavily relies on the Bodo word sentiment lexicon. Inadequate sentiment words listed in the dictionary may lead to changes in the results. By increasing number of correct sentiment words in the sentiment dictionary, we can improve the accuracy of our analyzer and obtain better results. The comparison cloud of the most frequent words, depicted in Fig. 5, provides an overview of the words that have the greatest impact on the overall sentiment in the dataset. This figure displays the most common words, with the size of the word indicating the frequency of its occurrence. By analyzing this figure, we can identify the words that are most closely associated with

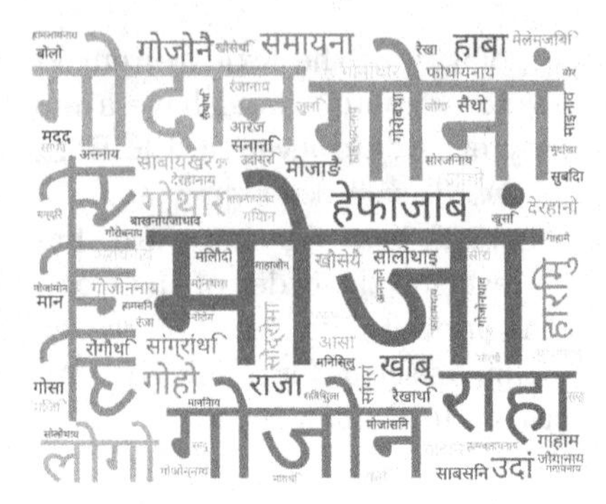

Fig. 5. Comparision cloud of most frequent words

positive, negative, or neutral sentiments, and gain insight into the factors that influence the sentiment of the Bodo language.

The bar chart illustrated in Fig. 6 provides a visual representation of the sentiment distribution of the predicted text data. The x-axis represents the three sentiment categories: positive, negative, and neutral, and the y-axis shows the frequency count of each category.

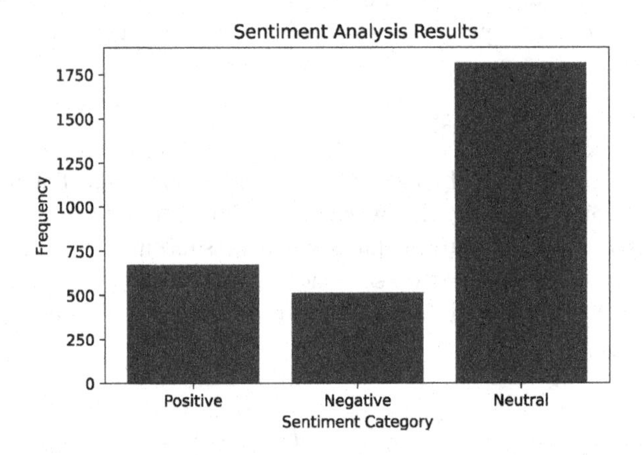

Fig. 6. Sentiment distribution of predicted text data

The chart shows that the majority of the predicted text belongs to the neutral category with a frequency count of around 1750, followed by the positive category with a frequency count of around 700, and the negative category with a frequency count of 500. The chart suggests that the predicted text data has a higher frequency of neutral sentiments, followed by positive sentiments and then negative sentiments.

5 Conclusion

In this work, we developed a Bodo text sentiment analysis system using a lexicon-based approach. We created a Bodo sentiment lexicon by manually collecting and labeling a dataset, which was a time-consuming task. Our analysis indicated that the system attained an accuracy of 70% on a dataset of 3000 sentences. We also identified that the system's performance was mostly dependent on the sentiment word dictionary, and the shortage of sentiment words in the lexicon could impact the results. Increasing the sentiment words in the dictionary and proper labeling of the dataset could improve the accuracy of the system. The dataset we have created is a valuable resource for researchers working on sentiment analysis of Bodo language and other NLP tasks. With the availability of labeled datasets, researchers can develop and evaluate new models for Sentiment Analysis and other NLP tasks for Bodo language. For future work, we intend to add a greater number of sentences and words to the dataset and evaluate the performance of the system. Additionally, we can explore other ML techniques, such as deep learning, to improve the accuracy of the system. Finally, we believe that this work can pave the way for developing sentiment analysis tools for other under-resourced languages, which can greatly benefit researchers and businesses working in these areas.

References

1. Brahma, B., Barman, A., Sarma, S.K., Boro, B.: Corpus building of literary lesser rich language-Bodo: insights and challenges. In: Proceedings of the 10th Workshop on Asian Language Resources, pp. 29–34 (2012)
2. Narzary, M., Muchahary, G., Brahma, M., Narzary, S., Singh, P.K., Senapati, A.: Bodo resources for NLP-an overview of existing primary resources for Bodo. In: AIJR Proceedings, pp. 96–101 (2021)
3. Census. Abstract of speakers' strength of languages and mother tongues. Census 2011 (2011). https://censusindia.gov.in/2011Census/C-16_25062018_NEW.pdf
4. Census. Comparative speakers' strength of languages and mother tongues - 1971, 1981, 1991, 2001 and 2011. Census of India 2011 (2011). https://censusindia.gov.in/2011Census/C-16_25062018_NEW.pdf
5. Islam, S., Paul, A., Purkayastha, B.S., Hussain, I.: Construction of English-Bodo parallel text corpus for statistical machine translation. Int. J. Nat. Lang. Comput. (IJNLC) **7**, 93–104 (2018)
6. Narzary, S., et al.: Generating monolingual dataset for low resource language Bodo from old books using google keep. In: Proceedings of the Thirteenth Language Resources and Evaluation Conference, pp. 6563–6570 (2022)
7. Dey, R.C., Sarker O.: Sentiment analysis on Bengali text using lexicon based approach. In: 2019 22nd International Conference on Computer and Information Technology (ICCIT), pp. 1–5. IEEE (2019)
8. Das, R., Singh, T.D.: A step towards sentiment analysis of Assamese news articles using lexical features. In: Maji, A.K., Saha, G., Das, S., Basu, S., Tavares, J.M.R.S. (eds.) Proceedings of the International Conference on Computing and Communication Systems. LNNS, vol. 170, pp. 15–23. Springer, Singapore (2021). https://doi.org/10.1007/978-981-33-4084-8_2

9. Nongmeikapam, K., Khangembam, D., Hemkumar, W., Khuraijam, S., Bandyopadhyay, S.: Verb based Manipuri sentiment analysis. Int. J. Nat. Lang. Comput. **3**(3), 113–118 (2014)
10. Gohil, L., Patel, D.: A sentiment analysis of Gujarati text using Gujarati senti word net. Int. J. Innovative Technol. Exploring Eng. (IJITEE) **8**(9), 2290–2293 (2019)

An Osprey Optimization Based Efficient Controlling of Nuclear Energy-Based Power System

Prince Kumar[1](✉), Kunal Kumar[2], Aashish Kumar Bohre[3], and Nabanita Adhikary[1]

[1] The Department of Electrical Engineering, NIT Silchar, Silchar, Assam, India
princemuz95@gmail.com, nabanita@ee.nits.ac.in
[2] The Department of Electrical Engineering, NIT Rourkela, Rourkela, Odisha, India
[3] The Department of Electrical Engineering, NIT Durgapur, Durgapur, WB, India
aashishkumar.bohre@ee.nitdgp.ac.in

Abstract. Increased utilization of automated machine to reduce human effort has enormously increased demand of power units to be produced. This increased demand cannot be realized with conventional sources of energy and hence researchers have tried exploring possibility of fulfilling this heavy demand with non-conventional or green energy sources. This paper presents the energy exploitation model of nuclear power system. Nuclear energy is highly sensitive to extract and transmit to end consumers. In this paper, a combined frequency and automatic voltage restoration is achieved efficiently with the help of Osprey Optimization Technique (OOT) tuned TID-PID controller under condition of sudden loadings or disturbances in power network. The proposed work is processed and simulated with the help of MATLAB/SIMULINK tool.

Keywords: Nuclear Power · TID-PID · OOT · PSO · ITMWAE · Generation Control Challenges

1 Introduction

Nuclear power plant has potential to satisfy enormous increased in the power demand but this type of power needs to be delicately handled and hence a suitable controller for generation and voltage control for multi loading has been planned and processed. A nuclear power system refers to the technology used to generate electricity through the controlled use of nuclear reactions. Nuclear power plants use nuclear reactors to produce heat, which is then used to generate steam that drives turbines and produces electricity. The heart of a nuclear power plant is the nuclear reactor, where a controlled nuclear chain reaction takes place. The reactor uses fuel rods made of uranium or plutonium to sustain the chain reaction. The heat produced by the nuclear reaction is used to generate steam, which drives turbines and produces electricity. Nuclear power plants are known for their high energy density and reliability, as they can operate for long periods of time without interruption. However, nuclear power plants also pose potential risks, such as accidents

P. Das et al. (Eds.): AMRIT 2023, CCIS 1953, pp. 57–72, 2024.
https://doi.org/10.1007/978-3-031-47224-4_6

that could result in the release of radioactive materials into the environment. To ensure the safe operation of nuclear power plants, strict regulations and safety measures are in place to prevent accidents and minimize their impact if they do occur. Additionally, the proper storage and disposal of nuclear waste is also a critical issue in the operation of nuclear power plants. Load frequency control (LFC) is a mechanism used in power systems to maintain the balance between the supply and demand of electricity. LFC ensures that the power output from generators matches the power demand from consumers, in order to maintain a stable frequency and voltage level in the power grid. In an electric power system, the load and generation are constantly changing due to factors such as weather, consumer demand, and equipment failures. If the power generation and demand are not balanced, it can cause a change in frequency and voltage, which can result in equipment damage or power outages. LFC works by adjusting the power output of generators in response to changes in load demand. LFC controllers monitor the frequency and adjust the power output of the generators accordingly, to maintain a stable frequency level. The controllers can also adjust the power output in response to changes in voltage levels, to ensure that the voltage remains within the acceptable range. LFC is an important aspect of power system stability, and is typically implemented through the use of advanced control algorithms and automated systems. These systems can help to improve the efficiency and reliability of power generation, while also minimizing the risk of equipment damage and power outages.

Disturbances in nuclear power network is inevitable and these disturbances can cause synchronism loss. To get rid of this situation, several model of frequency control in different sources of power model is given by authors. Transfer function model of different sources of power such as thermal, hydro, wind and nuclear power system has been given in [1] by authors. For thermal power system, automatic voltage regulating loop and load frequency loop has been shown by authors in [2]. Tunning of parameters values of controllers such as TID, PID and FOPID controllers with the help of soft computing technique has been given in [3] by authors. In multi area power systems, frequency of power system is successfully controlled with the help of hybrid controllers by authors of [4]. Designing and optimization of fitness or cost function is given by authors in [5, 6]. A Meta heuristic technique, PSO (Particle Swarm Optimization) is used by authors of [5, 6] for optimization problem. Implementation of fractional order controller for inverted pendulum case is given by authors in [7]. Also tunning of parameters for controller has been estimated using PSO in [7]. Damping of fluctuations in frequency with several communication delays is controlled using fractional order controller in [8] by authors. Load frequency problem in presence of distributed generation is solved appreciably using fractional order PID controller with the help of optimization technique by authors in [9]. Frequency oscillation in presence of load variation in deregulated environment is addressed by authors of [10]. Automatic voltage regulation with the help of controller with Henry Gas Solubility Optimization(HGSO) is given by authors of [11]. Also, comparative study in presence of other optimization techniques with HGSO is shown in [11]. 3-area system automatic generation control issue is addressed by authors using Antlion Optimizer Algorithm(ALO) in [12]. Fuzzy controllers are low cost controller and it is used in [13] for controlling frequency oscillations in 2-area connected system. Also, fuzzy-PID controller is compared among PI, PID, fuzzy PI and fuzzy PID.

After comparing fuzzy PID controller is found to be more efficient among all shown in [13] by authors. In [14], authors focused on future scenario of power exploitation from renewable sources in Egypt which is set to increase share of renewable sources of power to 42%. Hence, generation control studies on multi renewable sources of power is done in [14] and its control by utilizing suitable controller is proposed in the paper. Tunning of fractional order controller for utilization in system fitted with energy storage system is given in [15] by authors. Also, Krill Herd algorithm is used for optimization of fitness function considered in the paper [15]. A multi area system is considered in [16] for improved stability in transient response when disturbance is subjected. Author in [16] has introduced TLBO (Teaching Learning Based Optimization) for tunning controller for better stability in multi area system.

After going through research survey, controlling of frequency and voltage in case of nuclear power model need improvement and hence in this paper, combined frequency and voltage fluctuations in transmission of power from nuclear power plant in presence of disturbances in power network has been addressed with suitable fractional order controller tuned with robust optimization technique Osprey optimization technique (OOT). A detailed comparison chart for osprey optimization technique is given in [17] by authors. The performance of OOT is found to be superior as compared to other algorithm in optimizing or solving problem [17].

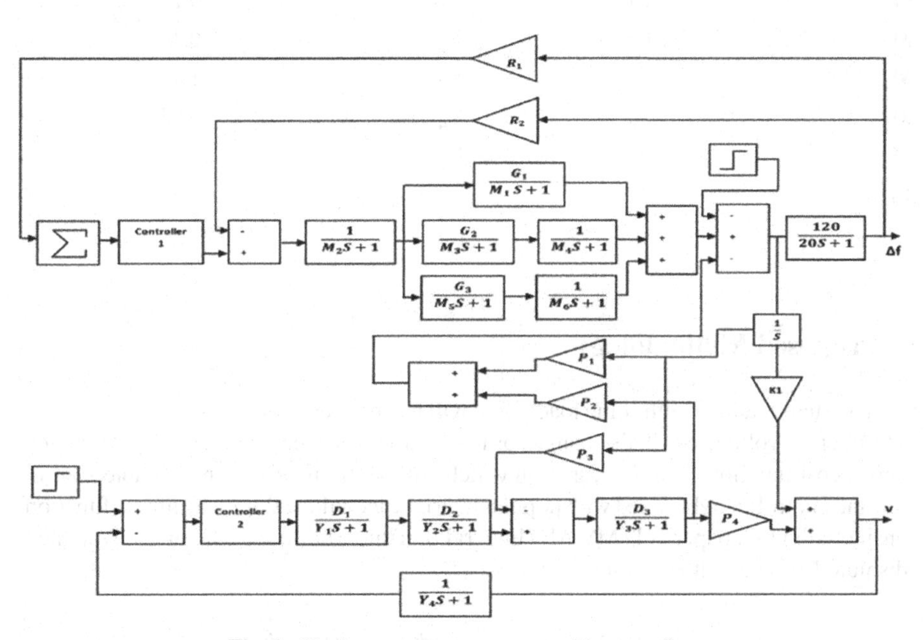

Fig. 1. Nuclear power test system with controller

2 Nuclear Power Test System

Nuclear power test system consists of two parts namely frequency control loop and automatic voltage regulation loop. Frequency control loop consists of controller, governor, turbines and power system. Voltage regulator loop consists of controller, exciter, generator field and sensor. A detailed transfer model of both loop is shown in Fig. 1. Values of constraint used in nuclear test system is given in Table 1.

Table 1. Values of constraints considered in nuclear power test system.

PARAMETERS	VALUE	PARAMETERS	VALUE
R_1	0.425	P_2	0.2
R_2	1/2.4	P_3	1.4
G_1	0.2	P_4	0.5
G_2	0.3	D_1	10
G_3	5	D_2	10
M_1	0.2	D_3	0.8
M_2	0.08	Y_1	0.1
M_3	0.5	Y_2	0.4
M_4	7	Y_3	1.4
M_5	10	Y_4	0.05
M_6	9	K_1	−0.1
P_1	1.5		

3 Proposed Methodology

When system is added with some load, there will be changes in different parameters such as frequency, voltage etc. This changes or fluctuation in system parameter is formulated using a cost function or fitness function which will be minimized using suitable control loop and controller hybridized with popular heuristic technique PSO. The fitness function considered in this paper is ITMWAE (Integral of Time Multiplied Magnified Weighted Absolute Error) and it is given by Eq. (1):

$$\text{ITMWAE} = m_1 \int_0^t |\Delta f| dt + m_2 \int_0^t |\Delta v| dt \tag{1}$$

$$m_1 = w_1 * M \tag{2}$$

$$m_2 = w_2 * M \tag{3}$$

Here, M = 10

Where, M is magnification factor;

$w_1 = 0.7$, $w_2 = 0.3$

w_1 and w_2 are priority based weighted values attached to frequency and voltage changes;

Where, m_1, m_2 is priority based magnified weighting factor and the values are $m_1 = 7$, $m_2 = 3$. $|\Delta f|$ is absolute frequency change; $|\Delta v|$ is absolute voltage change; m_1 is magnified weighted value attached to frequency; m_2 is magnified weighted value attached to voltage (Fig. 3).

TID controller is fractional order controller and its detailed model is shown in Fig. 2

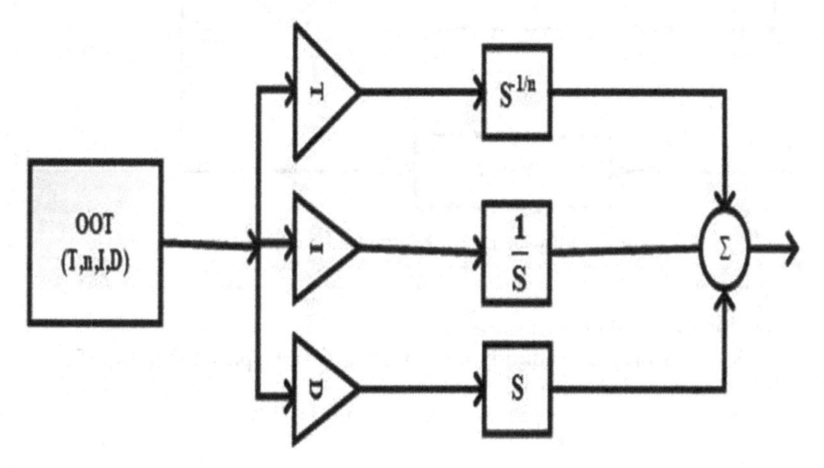

Fig. 2. Detailed model of TID controller

Transfer function equation for TID controller is as shown below:-

$$TID(S) = T * S^{-\frac{1}{n}} + I * \frac{1}{S} + D * S \tag{4}$$

3.1 Optimization Technique

3.1.1 Osprey Optimization Technique (OOT)

OOT is developed on the basis of behavior of bird namely osprey. Osprey is also called fish hawk. The strategy of osprey in catching fish from a river or sea and then take it at suitable place to eat it comfortably is natural intelligent strategy and hence this strategy can be utilized by human to solve real life solution. In this paper, fluctuations in power network due to sudden disturbances is solved using the intelligent strategy of catching fish by osprey. This intelligent strategy is mathematically modelled to optimize the proposed fitness function considered in this paper.

Stepwise implementation of OOT:-

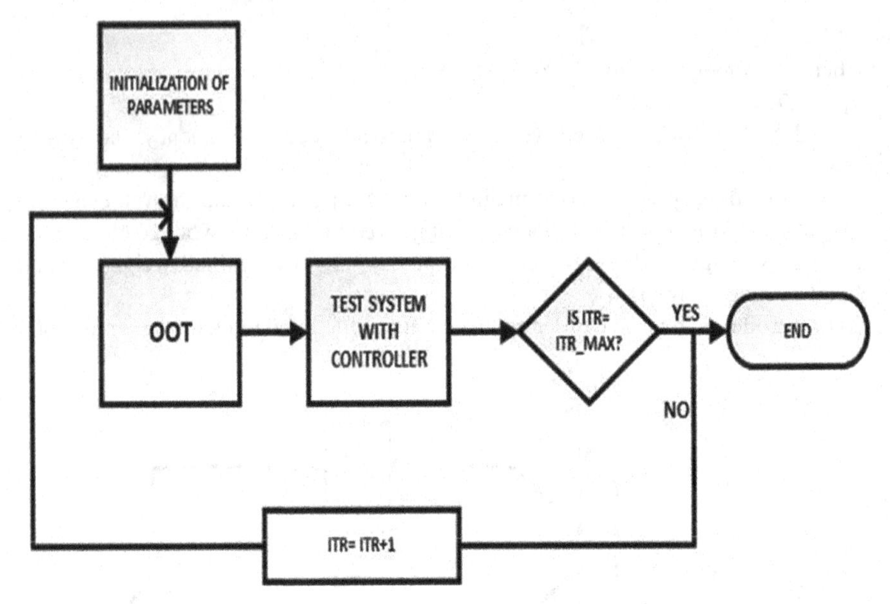

Fig. 3. Flow chart of the proposed methodology

i. Initialization of population
   Position of osprey is randomly initialized in search space using Eq. (5)

$$p_{ij} = lb_j + rand * (ub_j - lb_j) \tag{5}$$

   where, i is the number of osprey(solution) and j is the number of problem variable.
ii. Evaluation of osprey's fitness using fitness function or objective function.
iii. Phase 1(PH1): Location identification for hunting of fish by osprey

   The set of fish for each osprey is given by Eq. (6)

$$FP_i = \{P_k | k \in \{1, 2, 3, \ldots..N\} \wedge F_k < F_i\} \cup \{P_{best}\} \tag{6}$$

where FP_i is set of fish position for ith osprey and P_{best} is the best osprey or solution
   Updated position of osprey is calculated using Eq. (7)

$$p_{ij}^{PH1} = p_{ij} + rand * (SF_{ij} - I_{ij} * p_{ij}) \tag{7}$$

$$P_i = \begin{cases} P_i, F_i^{PH1} < F_i \\ P_i, else \end{cases} \tag{8}$$

where PH1 is phase 1 i.e. identification of fish location by ith osprey

iv. Phase 2(PH2):-Carrying out fish to suitable position for eating.
   Suitable position for eating fish can be calculated using Eq. (9)

$$p_{ij}^{PH2} = p_{ij} + \frac{lb_j + rand * (ub_j - lb_j)}{t} \quad (9)$$

Updated position of osprey can be given by Eq. (10)

$$P_i = \begin{cases} P_i, F_i^{PH2} < F_i \\ P_i, else \end{cases} \quad (10)$$

v. Repetition of phase 1 and phase 2 till maximum iteration is attained

3.1.2 PSO

PSO is optimization technique and can be used as maximization or minimization. In this paper, minimization of fitness function is going to be done. Particle velocity of the swarm is updated using equation as shown in Eq. (11).

$$v_i^{q+1} = l * v_i^q + u_1 * rand * (p_{best} - x_i^q) + u_2 * rand * (g_{best} - x_i^q) \quad (11)$$

where, v_i^q is the current velocity; v_i^{q+1} is the updated particle velocity; u_1, u_2 is the constriction factor; l is the weighting factor of the PSO algorithm and x_i^q is the current position of the particle of the swarm.

Position is being updated using Eq. (12) as follows:-

$$x_i^{q+1} = x_i^q + v_i^{q+1} \quad (12)$$

PSO is optimization technique and can be used as maximization or minimization. In this paper, minimization of fitness function is going to be done. Particle velocity of the swarm is updated using equation as shown in Eq. (13).

$$v_i^{q+1} = l * v_i^q + u_1 * rand * (p_{best} - x_i^q) + u_2 * rand * (g_{best} - x_i^q) \quad (13)$$

where, v_i^q is the current velocity; v_i^{q+1} is the updated particle velocity; u_1, u_2 is the constriction factor; l is the weighting factor of the PSO algorithm and x_i^q is the current position of the particle of the swarm.

Position is being updated using Eq. (14) as follows:-

$$x_i^{q+1} = x_i^q + v_i^{q+1} \quad (14)$$

4 Result and Discussion

Case-a:- Nuclear power plant with TID-PID-OOT controller
Case-b:- Nuclear power plant with TID-PID-PSO controller.
Case-c:- Nuclear power plant with TID-PID controller.

4.1 Case-1

A nuclear power plant model is considered and 150 MW loading has been provided for the testing of proposed controllers. As load is increased by 150 MW, system frequency will get dipped and during restoration of this dipped frequency to normal point, oscillations or fluctuations in frequency will occur. In the proposed model, a TID-PID-OOT controller is proposed to mitigate these fluctuations in order to avoid the case of synchronism loss of system. The proposed controller is very handy in efficiently controlling of both voltage and frequency at a time. The proposed model has been compared with two other controllers namely TID-PID-PSO and TID-PID for comparison under similar condition. Tunning parameter of TID-PID has been shown in Table 3 after simulation using OOT and PSO.

Table 2. Cost function value

	Case-a	Case-b	Case-c
Cost function value	5.2535859	9.140079	719.657794

Fitness value shown in Table 2 with TID-PID-OOT controller is less in comparison to others controllers fitted nuclear power system. Fitness function is indication of performance index of test system with different controller. If fitness value is less that means system's performance is better.

Table 3. TID-PID controller parameters value for OOT and PSO

	OOT	PSO
T(1)	73.678	40
n(1)	69.085093	2.69883
I(1)	77.52646	40
D(1)	13.854777	1.1
P(2)	13.080378	23.997565
I(2)	42.321611	19.270925
D(2)	5.0913163	4.2241364

Table 4 shows the load frequency control of nuclear power test system and from the data obtained and as per shown frequency analysis in Fig. 4, it can be inferred that TID-PID-OOT controller is more efficient in comparison to other controllers fitted test system.

As per Table 5 and Fig. 5, it can be observed that settling time for voltage is reduced for case-a although peak overshot is larger for case-a.

Graph given in Fig. 6 shows the robustness of OOT in comparison to PSO in optimizing fitness function considered in this paper.

Table 4. Frequency analysis for case-1 with case-A

Parameters	Case-a	Case-b	Case-c
Rise Time	1.5677e-06	1.6873e-06	0.0397
Settling Time	5.2169	8.6302	48.2815
Settling Min	44.4868	39.4732	38.9109
Settling Max	61.5222	63.2739	71.3588
Overshoot	2.5372	5.4567	28.8363
Undershoot	0	0	0
Peak	61.5222	63.2739	71.3588
Peak Time	3.4468	1.8030	11.8361

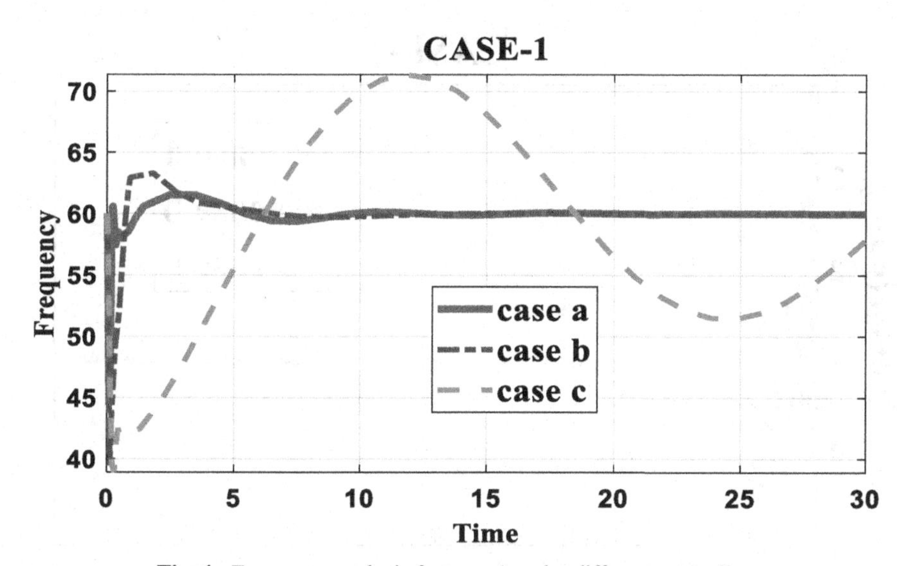

Fig. 4. Frequency analysis for case-1 under different controllers

4.2 Case 2

Under this section, 200 MW loading has been considered for creating fluctuations in the nuclear test system and corresponding frequency and voltage response of the test system with different controllers are analyzed (Table 6 and 7).

Fitness value for TID-PID-OOT controller fitted system is less than other controllers fitted system as per Table 10.

Frequency-time response data for TID-PID-OOT, TID-PID-PSO and TID-PID controllers fitted system is shown in Table 8. From the Table 8 and as per graph shown in Fig. 7, it can be concluded that TID-PID-OOT fitted nuclear test system is more efficient in damping fluctuations due to loading of 200 MW.

Table 5. Voltage analysis for case-1 with case-A

Parameters	Case-a	Case-b	Case-c
Rise Time	0.0061	0.0218	0.0568
Settling Time	0.2981	0.5881	1.2181
Settling Min	−0.2095	0.2446	0.7443
Settling Max	3.9744	2.3349	1.5902
Overshoot	297.4363	133.4917	58.9791
Undershoot	20.9498	0	0
Peak	3.9744	2.3349	1.5902
Peak Time	0.0481	0.0684	0.2003

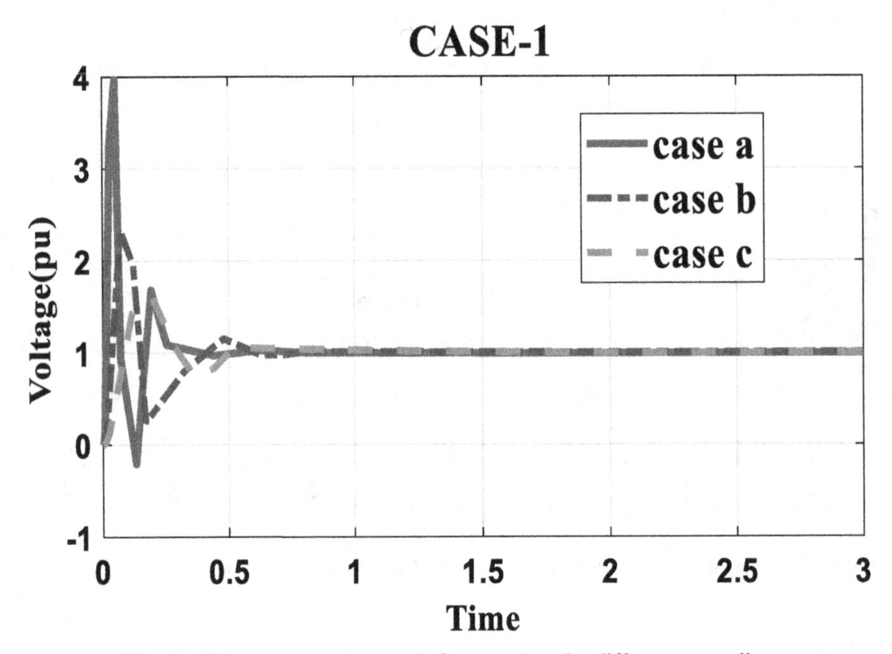

Fig. 5. Voltage restoration graph for case-1 under different controllers

Voltage-time response data obtained from simulated result in MATLAB environment is shown in Table 9. Looking into Table 9 data and graph given in Fig. 8, it can be inferred that TID-PID-OOT controller fitted test system is more efficient in curbing fluctuations in voltage after loading of 200 MW on the test system. Settling time for case-a is smaller in comparison to other cases considered.

Optimization of fitness function using OOT and PSO for 200 MW loading can be observed from Fig. 9. OOT is found to better in minimizing fitness function as compared to PSO.

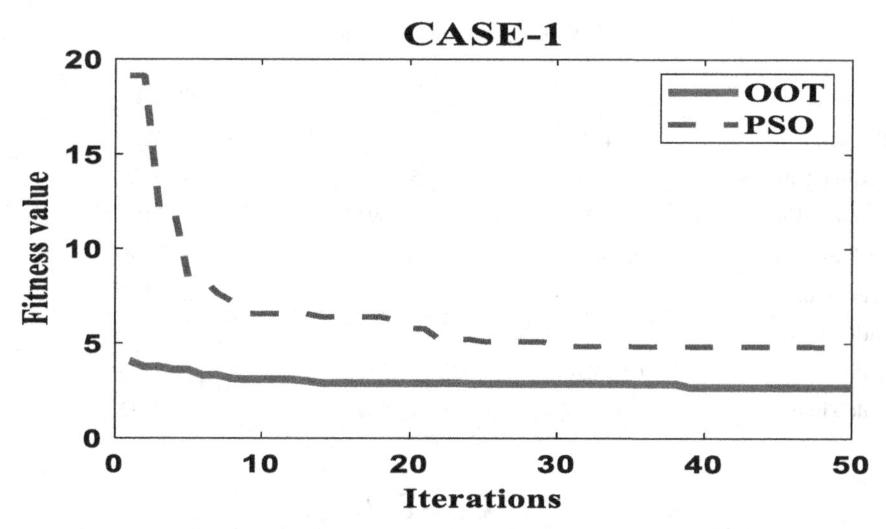

Fig. 6. Convergence graph for OOT and PSO for case-1

Table 6. Cost function value under case-2

	Case-a	Case-b	Case-c
Cost function value	3.99706	5.387137	544.585831

Table 7. TIID-PID controller parameters value for case 2

	OOT	PSO
T(1)	97.58441	40
n(1)	33.6963	2.906837
I(1)	92.2061	40
D(1)	1.910261	1.1
P(2)	95.02589	29.79661
I(2)	40.87436	22.45812
D(2)	29.14233	1.376326

Table 8. Frequency analysis for case 2

Parameters	Case-a	Case-b	Case-c
Rise Time	1.6424e-09	1.1096e-06	0.0229
Settling Time	5.0575	5.3894	46.6521
Settling Min	44.1384	36.6937	38.2101
Settling Max	63.4424	64.2762	70.7209
Overshoot	5.7373	7.1271	23.2371
Undershoot	0	0	0
Peak	63.4424	64.2762	70.7209
Peak Time	0.5885	1.0867	11.2392

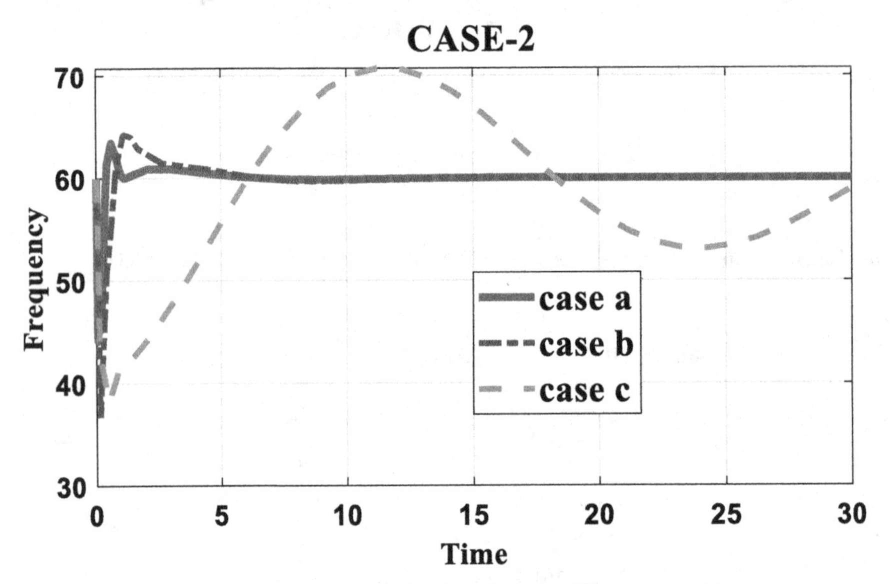

Fig. 7. Frequency analysis for case-2 under different controllers

Table 9. Voltage analysis for case 2 with case A

Parameters	Case-a	Case-b	Case-c
Rise Time	0.0067	0.0291	0.0479
Settling Time	0.4893	0.8216	1.2101
Settling Min	0.5508	0.5238	0.9048
Settling Max	2.9003	2.0839	1.8736
Overshoot	190.0320	108.3878	87.3418
Undershoot	0	0	0
Peak	2.9003	2.0839	1.8736
Peak Time	0.0275	0.0850	0.1239

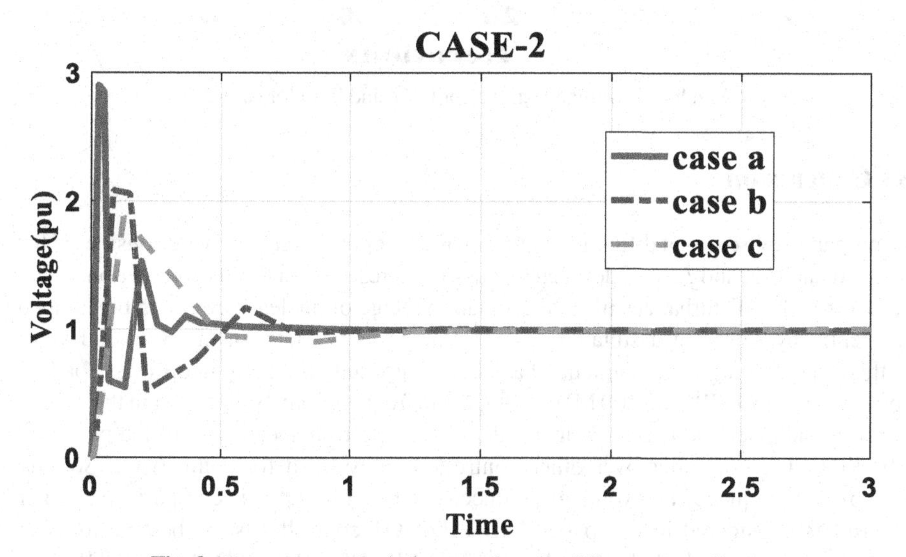

Fig. 8. Frequency analysis for case-2 under different controllers.

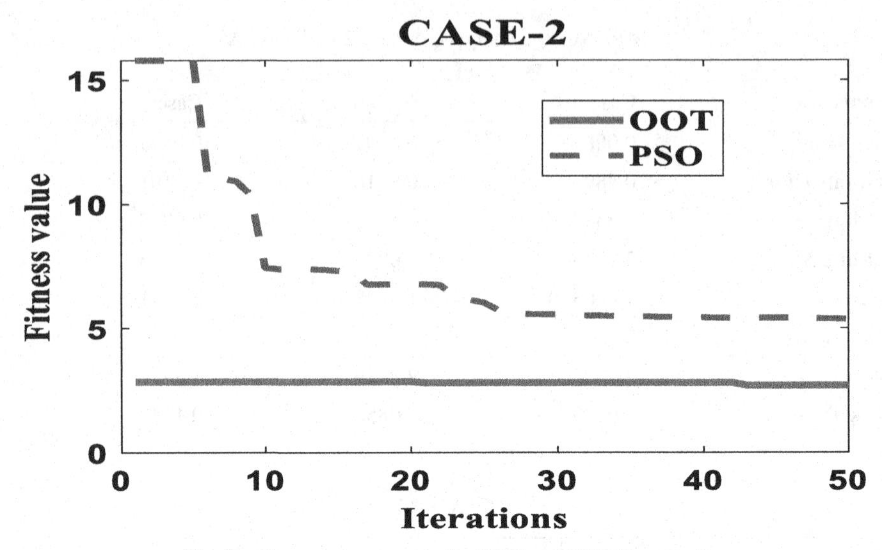

Fig. 9. Convergence graph for OOT and PSO for case-2

5 Conclusion

In this paper, two cases of different loading on nuclear power network were considered for detailed analysis and After a detailed discussion in results and discussion section, it can be inferred that disturbances in frequency and voltage of nuclear power can be damped efficiently by selecting a suitable combination of controllers. In the proposed work in this paper, voltage and frequency has been controlled efficiently in different loading conditions of 150 MW and 200 MW. Voltage and frequency analysis shown in the figures in result and discussion section clearly depicts the performance report of the proposed TID-PID-OOT controller over other controllers considered for comparison. Settling time of both frequency and voltage is reduced for TID-PID-OOT in comparison to other controllers considered in this paper. TID-PID-OOT controller is the best performing controller among 3 selected controllers i.e. TID-PID-OOT, TID-PID-PSO and TID-PID for comparing the results of controller fitted in the nuclear power test system. Robustness of meta heuristic algorithm OOT in optimizing the considered fitness or cost function ITMWAE in the paper is also shown with the help of graph in the results and discussion section. It is observed that OOT is superior in comparison to PSO as per Fig. 9 and Fig. 6.

Conflict of Interest. On behalf of all authors, the corresponding author states that there is no conflict of interest.

References

1. Kachhwaha, Pandey, S.K., Dubey, A.K., Gupta, S.: Interconnected multi unit two-area automatic generation control using optimal tuning of fractional order PID controller along with electrical vehicle loading. In: 2016 IEEE 1st International Conference on Power Electronics, Intelligent Control and Energy Systems (ICPEICES), pp. 1–5 (2016). https://doi.org/10.1109/ICPEICES.2016.7853167
2. Prakash, A., Parida, S.K.: Combined frequency and voltage stabilization of thermal-thermal system with UPFC and RFB. In: 2020 IEEE 9th Power India International Conference (PIICON), pp. 1–6 (2020). https://doi.org/10.1109/PIICON49524.2020.9113034
3. El-Dabah, M.A., Kamel, S., Khamies, M., Shahinzadeh, H., Gharehpetian, G.B.: Artificial gorilla troops optimizer for optimum tuning of TID based power system stabilizer. In: 2022 9th Iranian Joint Congress on Fuzzy and Intelligent Systems (CFIS), pp. 1–5 (2022). https://doi.org/10.1109/CFIS54774.2022.9756463
4. Mohamed, E.A., Ahmed, E.M., Elmelegi, A., Aly, M., Elbaksawi, O., Mohamed, A.-A.A.: An optimized hybrid fractional order controller for frequency regulation in multi-area power systems. IEEE Access **8**, 213899–213915 (2020). https://doi.org/10.1109/ACCESS.2020.3040620
5. Kumar, P. Bohre, A.K.: Efficient planning of solar-PV and STATCOM using optimization under contingency. In: 2021 International Conference on Computational Performance Evaluation (ComPE), pp. 109–1142021). https://doi.org/10.1109/ComPE53109.2021.9751899
6. Kumar, P., Bohre, A.K.: Optimal allocation of solar-PV and STATCOM Using PSO with multi-objective approach to improve the overall system performance. In: Gupta, O.H., Sood, V.K., Malik, O.P. (eds.) Recent Advances in Power Systems. Lecture Notes in Electrical Engineering, vol. 812. Springer, Singapore (2022). https://doi.org/10.1007/978-981-16-6970-5_49
7. Baruah, A., Buragohain, M.: Design and implementation of FOPID and modified FOPID for inverted pendulum using particle swarm optimization algorithm. In: 2018 2nd International Conference on Power, Energy and Environment: Towards Smart Technology (ICEPE), pp. 1–6 (2018). https://doi.org/10.1109/EPETSG.2018.8659047
8. Ahuja, A., Narayan, S., Kumar, J.: Robust FOPID controller for load frequency control using Particle Swarm Optimization. In: 2014 6th IEEE Power India International Conference (PIICON), pp. 1–6 (2014). https://doi.org/10.1109/POWERI.2014.7117663
9. Shayeghi, H., Molaee, A., Valipour, K., Ghasemi, A.: Multi-source power system FOPID based load frequency control with high-penetration of distributed generations. In: 2016 21st Conference on Electrical Power Distribution Networks Conference (EPDC), pp. 131–136 (2016). https://doi.org/10.1109/EPDC.2016.7514796
10. Pandey, R.K., Gupta, D.K., Dei, G.: Hybrid intelligent optimization technique (HIOT) driven FOPID controller for load frequency control of deregulated power system. In: 2022 IEEE Global Conference on Computing, Power and Communication Technologies (GlobConPT), pp. 1–6 (2022). https://doi.org/10.1109/GlobConPT57482.2022.9938247
11. Ekinci, S., Izci, D., Hekimoğlu, B.: Henry gas solubility optimization algorithm based FOPID controller design for automatic voltage regulator. In: 2020 International Conference on Electrical, Communication, and Computer Engineering (ICECCE), pp. 1–6 (2020). https://doi.org/10.1109/ICECCE49384.2020.9179406
12. Dei, G., Sahoo, S., Kumar Sahu, B.: Performance analysis of ALO tuned FOPID controller for AGC of a three area power system. In: 2018 International Conference on Recent Innovations in Electrical, Electronics & Communication Engineering (ICRIEECE), pp. 3116–3120 (2018). https://doi.org/10.1109/ICRIEECE44171.2018.9008646

13. Behera, A.R., Mallick, R.K., Agrawal, R., Nayak, P., Sinha, S.: Performance analysis of Fuzzy-PID controller in nuclear, dish-stirling solar thermal and reheat thermal energy integrated deregulated power system. In: 2022 International Conference on Intelligent Controller and Computing for Smart Power (ICICCSP), pp. 1–6 (2022). https://doi.org/10.1109/ICICCSP53532.2022.9862365
14. Nour, M., Magdy, G., Chaves-Ávila, J.P., Sánchez-Miralles, Á., Petlenkov, E.: Automatic generation control of a future multisource power system considering high renewables penetration and electric vehicles: Egyptian power system in 2035. IEEE Access **10**, 51662–51681 (2022). https://doi.org/10.1109/ACCESS.2022.3174080
15. Mohamed, R., Boudy, B., Gabbar, H.A.: Fractional PID controller tuning using krill herd for renewable power systems control. In: 2021 IEEE 9th International Conference on Smart Energy Grid Engineering (SEGE), pp. 153–157 (2021). https://doi.org/10.1109/SEGE52446.2021.9534982
16. Kumar, P., Kumar, K., Bohre, A.K., Adhikary, N.: Intelligent priority based generation control for multi area system. Smart Sci. (2023). https://doi.org/10.1080/23080477.2023.2189628
17. Dehghani, M., Trojovsky, P.: Osprey optimization algorithm: a new bio-inspired metaheuristic algorithm for solving engineering optimization problems. Front. Mech. Eng. **8**, 1126450 (2023). https://doi.org/10.3389/fmech.2022.1126450

Fuzzy Association Rule Mining Techniques and Applications

Rajkamal Sarma and Pankaj Kumar Deva Sarma(✉)

Department of Computer Science, Assam University, Silchar, Assam, India
pankajgr@rediffmail.com

Abstract. Fuzzy Association Rule Mining (FARM) is considered as a significant part of data mining approach that uses fuzzy logic to extract meaningful patterns and relationship from large dataset. Fuzzy set are used to represent the membership degree of data in a particular set, which allows for more flexible and accurate modelling of data. The application of fuzzy association rule mining provides valuable insights and knowledge for decision-making in different domains. In this paper, basic concept of FARM and how to generate meaningful rules based on FARM algorithm is discussed. The issue of sharp boundary found in classical Association Rule Mining (ARM) technique can be solved using FARM technique. Different Membership Function can be used to convert crisp data into fuzzy data. Applying fuzzification method quantitative dataset is converted into fuzzy category dataset and binary category dataset. Then both dataset is implemented through different algorithms to generate rules and analyzed those rules.

Keywords: Fuzzy Association Rule Mining · Data Mining · Fuzzy Logic · Membership Degree · FTDA

1 Introduction

Different Techniques have been applied to find Fuzzy Association Rules under data mining process to discover useful information in recent time. To find association among different data and mining those association in a standard framework to generate some useful rules, is an important activity of data mining approach. FARM plays a key role in KDD (knowledge Discovery in Database) [1] process, which includes data identification, pre-processing, transformation, pattern generation, knowledge representation and knowledge utilization etc. Classical rule mining, widely known as ARM is a Technique used to discover interesting relationship among different attributes in a large dataset [2]. However, some problems involved in this technique and one of the major problems is identified as sharp boundary problem. At this point, FARM technique provides a solution named as smooth boundary. Unlike traditional rule mining approach which is operated on binary dataset or categorical dataset, FARM technique is used to operate on Fuzzy dataset, where data are expressed in terms of membership degree represented by the range [0, 1] rather than 0 or 1. Fuzzy sets are effective means to design imprecise concept and relations as human understandable form used for communication with more flexibility [3]. Due to its close alignment with human knowledge representation, the fuzzy technique has been considered as a key component of mining approach [4].

P. Das et al. (Eds.): AMRIT 2023, CCIS 1953, pp. 73–87, 2024.
https://doi.org/10.1007/978-3-031-47224-4_7

Fuzzy Association Rule Mining Technique has become increasingly popular due to their ability to handle uncertainty and imprecision of data. Fuzzy association Rule is typically used to express in the form of X⇨Y, suggesting there occurs a relation between two set of items, X and Y, showing if X occurs then Y occurs as well. This can be presented as "if X, then Y." This involves the discovery of linguistic "if-then" rules that can be used to make predictions and draw inferences from data. These rules are based on fuzzy logic, a framework that allows the representation of ambiguity and vagueness in data. Fuzzy Rule Mining Technique have been applied to different fields, including finance, medical diagnosis, engineering, healthcare etc.

In this paper, an overview of some fuzzy Association Rule Mining Techniques that have been developed based on some key algorithms and their application in different domains is presented. Practical implementation of Fuzzy association Rule mining techniques in a dataset and experimental results are discussed. Finally, analysis of output is done and find relevance for our future work.

2 Related Work

Fuzzy set theory was introduced in 1965 by L.A. Zadeh [5] to provide an approximate and yet effective way for describing the characteristics of a complex system to admit precise mathematical analysis [6]. Fuzzy sets are regarded as an expansion of the classical crisp sets. Since the introduction of fuzzy concept in Association rule mining approach in last three decade of last century, there have been various works found related to FARM. Initially, some algorithms had been proposed by different researchers. Those algorithms were implemented in different type of dataset in different domains. Some of those algorithms are such as F-APACS [7], FTDA [8, 12], FQARM [9], CFARM [10], FWARM [11] FAR-Miner [15] etc. However, later, some other algorithms such as IFWARM [13], E-FWARM [14] etc. have been modified and used to find rules with more accuracy and efficiently. Besides this algorithm modification, FARM Techniques are used in different application domain for extraction of meaningful information.

3 Farm Techniques

Fuzzy Association Rule, represented as $X \rightarrow Y$, gives association between two fuzzy sets. But for the generation of meaningful and useful rule there is a standard framework. All the generated rules have to be inserted in the framework, where those rules will be tested under different criteria, which is known as interestingness measures. Interestingness measures play a decisive role in generating fuzzy association rules [7]. Two such measures are Fuzzy support (FS) and Fuzzy confidence (FC) which are user defined. These measures are calculated as given below :

$$FS(X) = \frac{\text{sum of votes satisfying X}}{\text{Number of records in T}}$$

where, X is a set ant T is Transaction of the dataset.

Fuzzy Confidence (FC) of a rule X → Y can be calculated like classical association rule as follows:

$$\mathrm{FC}(\mathrm{X} \rightarrow \mathrm{Y}) = \frac{FS(A, B)}{FS(A)}$$

Depending on these two thresholds, two key information can be observed-firstly, about the association between attributes generated in the form of rules and secondly, how strongly those generated rules are associated.

In the process of generating rules, one of the key steps is conversion of crisp value into fuzzy value, which can be termed as fuzzification. In fuzzification, some key terms are used to define, firstly, in which category a crisp value will be fuzzified and, secondly, what will be the participation values in that particular category. These categories are called linguistic term and the participation value are called membership degree, for this, some standard membership functions are used. Two types of often used in fuzzy logic are triangular and trapezoidal membership function]. However, selection of membership function, classification of linguistic term depends the context of dataset and specific problem addressed (Fig. 1).

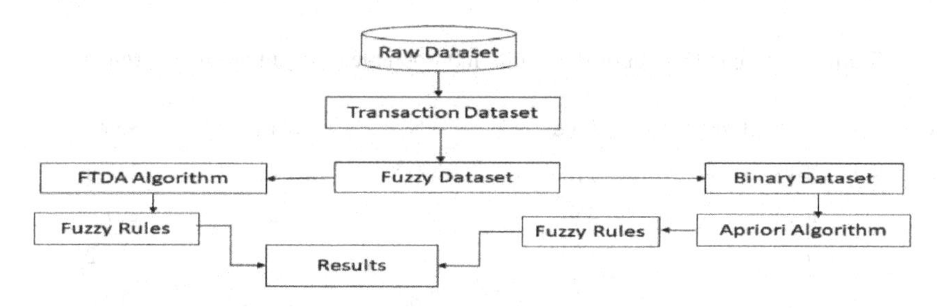

Fig. 1. Proposed Model

As shown in the proposed model, we have taken a meteorological dataset as raw data. This raw dataset is pre-processed and converted to transaction dataset with number of attributes. Transaction data is fuzzified to fuzzy dataset using trapezoidal membership function. The fuzzy dataset is classified into five linguistic terms as very low, low, average, high and very high. Taking this fuzzy dataset as input dataset we followed two approaches. Firstly, continuing the flow, a fuzzy rule-based algorithm is implemented and found some rules. Secondly, the fuzzy dataset is converted into categorical dataset with binary values. The conversion process (defuzzification) is done considering higher value as 1 and lower value as 0, where membership degree overlaps in two linguistic terms. Other values taken as it was in the fuzzy dataset. So, we have got a binary valued dataset. Taking this new binary dataset as input, classical ARM based algorithm is applied and found some rules. Finally, generated rules from two approaches have been compared to find common rules.

4 FTDA Approach

In this section, the FTDA (Fuzzy Transaction Data-Mining Algorithm) is used in rainfall data to generate association rules. This data is about the amount of rainfall occurred in monsoon season, which includes months of June to September of year 2011 to 2020. Since, in monsoon season, rainfall occurs heavily and it is also the season of cultivation of crops as well as associated with hazards like flood etc., hence, we consider this season to get more effective and useful rules. The sample dataset is shown in Table 1, that includes 10 transactions. Each transaction comprises a total rainfall in four months i.e. June to September. The dataset is a meteorological data of north bank of Brahmaputra valley and collected from BNCA, AAU, Jorhat. In mining process each month is considered as an attribute. In this example, each attribute is classified into five regions: VL for Very Low, L for Low, AV for Average, H for High and VH for Very High. So, five membership degrees are calculated for each month as per trapezoidal membership function. Here, amount of rainfall in the raw dataset is diversity and classification of linguistic terms are more than three, hence in this context trapezoidal membership function is used for fuzzification process. Representation of membership function and the fuzzification with different linguistic terms can be seen in Fig. 2.

Table 1. Sample Dataset of Rainfall in monsoon measured in millimeter (mm)

TID	Year/ Month	June	July	Aug	Sept
1	2011	172.4	460.5	478.1	247.6
2	2012	444.2	300.4	324.1	297.6
3	2013	156.7	532.6	230.3	169.2
4	2014	335.4	386.9	331.5	397.3
5	2015	473.2	391.8	546.3	269.1
6	2016	309.8	372.4	268.8	259.8
7	2017	470.8	321.1	329.7	287.2
8	2018	307.0	404.2	349.3	195.6
9	2019	183.6	317.9	166.4	289.2
10	2020	424.7	285.6	175.6	241.8

Step1: At first, transaction wise quantitative data are converted into fuzzy data. For example, the rainfall in the month of June as case1 is considered. The score "172.4" is converted into fuzzy set (0.56/Very Low + 0.44/Low + 0.0/Average + 0.0/High + 0.0/Very High) using membership function.

Step 2: The scalar cardinality of each attribute is computed for each linguistic term in the transaction. Taking the attribute of June. VL as a case, the scalar cardinality = $(0.56 + 0.0 + 0.86 + \ldots + 0.32 + 0.0) = 1.74$. Same calculation method can be applied for other attributes too, which is shown Table 2.

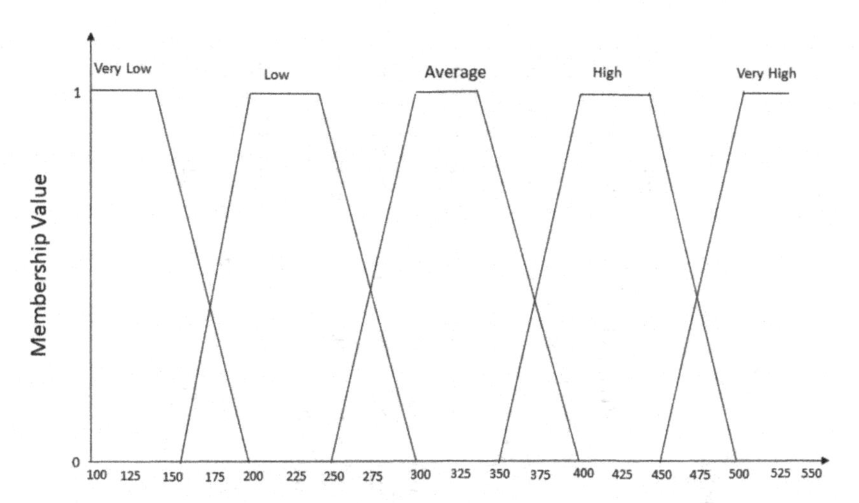

Fig. 2. Dataset expressed using Trapezoidal Membership Function

Step 3: Now the highest count among the five possible attributes for each linguistic term is found. In the case of June, the count is 1.74 for Very Low, 1.26 for Low, 3.0 for Average, 3.12 for High and 0.88 for Very High. Since the High is the largest among them Jun. is considered for June. For mining, this step is repeated for other attributes and thus four regions are being selected. These are given in the following Table 3 with their support count.

Step 4: Now the support counts of different regions are checked with the given threshold which is 2.0 in this example and since all the selected regions have more support count than 2.0, therefore, all the four regions will remain as the same for next step of the mining process.

Step 5: It is assumed r = 1.

Step 6: Generating the candidate set C_{r+1} from $L_{r:}$ C_2 is found from L_1 as given below:

(June.H,July.AV), (June.H,Aug.AV), (June.H,Sept.L), (July.AV,Aug.AV), (July.AV,Sept.L), (Aug.AV,Sept.L).

Step 7: Here, the sub steps for each candidate item set are as follows:

(a) The fuzzy membership degree for individual transaction datum is calculated by using intersection. Here, for example, (June.H, July.AV) is considered. For case 1, the derived membership degree is computed as min (0.0,0.0) = 0.0. Same procedure can be used for the other 2-itemsets.
(b) The counts are compared with pre-defined minimum support value 3.0. This results in two Itemsets (June.H, Aug.AV), (July.AV,Aug.AV) and (July.AV, Sept.L) as L_2 as in Table 11.

Step 8: If Lr +1 is null, then it goes to next step, otherwise r is set as r = r + 1 and the step 6–8 is repeated. Since, in this case L2 is null, it cannot be processed further.

Step 9: Using the following sub steps association rules can be constructed for each large item sets:

Table 2. Fuzzy Dataset with Scalar Cardinality of each attribute

TID	June					July					Aug					Sept				
	VL	L	AV	H	VH	VL	L	AV	H	VH	VL	L	AV	H	VH	VL	L	AV	H	VH
1	0.56	0.44	0.0	0.0	0.0	0.0	0.0	0.0	0.78	0.22	0.0	0.0	0.0	0.44	0.56	0.0	1.0	0.0	0.0	0.0
2	0.0	0.0	0.0	1.0	0.0	0.0	0.0	1.0	0.0	0.0	0.0	0.0	1.0	0.0	0.0	0.0	0.04	0.96	0.0	0.0
3	0.86	0.14	0.0	0.0	0.0	0.0	0.0	0.0	0.0	1.0	0.0	1.0	0.0	0.0	0.0	0.62	0.38	0.0	0.0	0.0
4	0.0	0.0	1.0	0.0	0.0	0.0	0.0	0.26	0.74	0.0	0.0	0.0	1.0	0.0	0.0	0.0	0.06	0.94	0.0	0.0
5	0.0	0.0	0.0	0.54	0.46	00	0.0	0.16	0.84	0.0	0.0	0.0	0.0	0.0	1.0	0.0	0.62	0.38	0.0	0.0
6	0.0	0.0	1.0	0.0	0.0	0.0	0.0	0.56	0.44	0.0	0.0	0.62	0.38	0.0	0.0	0.0	0.80	0.20	0.0	0.0
7	0.0	0.0	0.0	0.58	0.2	0.0	0.0	1.0	0.0	0.0	0.0	0.0	1.0	0.0	0.0	0.0	0.26	0.74	0.0	0.0
8	0.0	0.0	1.0	0.0	0.0	0.0	0.0	0.0	1.0	0.0	0.0	0.0	1.0	0.0	0.0	0.08	0.92	0.0	0.0	0.0
9	0.32	0.68	0.0	0.0	0.0	0.0	0.0	1.0	0.0	0.0	0.68	0.0	0.0	0.0	0.0	0.0	0.20	0.80	0.0	0.0
10	0.0	0.0	0.0	1.0	0.0	0.0	0.28	0.72	0.0	0.0	0.48	0.0	0.0	0.0	0.0	0.0	1.0	0.0	0.0	0.0
Total	1.74	1.26	3.0	312	0.88	0.0	0.28	4.70	3.80	1.22	1.16	2.46	4.38	0.44	1.56	0.70	5.28	4.02	0.0	0.0

Table 3. Highest Support Count

Item set	Support
June.H	3.12
July.AV	4.70
Aug.AV	4.38
Sept.L	5.28

(a) Among all generated rules, following four associations can be found:

If June = High, then August = Average. If August = Average, then June = High.
If July = Average, then August = Average. If August = Average, then July = Average.
If July = Average, then September = Low
If September = Low, then July = Average.

(b) The confidence values are calculated for the extracted rules and the confidence threshold denoted by λ is assumed as 0.5. The first association rule above is considered as example, where, the fuzzy value of June.H $\cap$ Aug.AV is calculated in the Table 5 which is 2.38. For the rule "If June = High, then August = Average", the confidence is given by

$\sum$ (June. H $\cap$ Aug. AV)/ $\sum^{10}$(June. H) = 0.76

(See Tables 4, 6, 7, 8, 9 and 10).
Outcome of remaining association rules can be outlined as follows:
"If August=Average, then June=High" with confidence value of 0.54
"If July=Average, then August=Average" with confidence value of 0.56
"If August=Average, then July=Average" with confidence value of 0.60
"If July=Average, then September=Low" with confidence value of 0.42
"If September=Low, then July=Average" with confidence value of 0.37

Step 10: After checking the given confidence threshold which is 0.5 with the above confidence factor the following three association rules will be considered:
"If June=High, then August=Average" with a confidence of 0.76
"If August=Average, then June=High" with a confidence of 0.54
"If July = Average, then August = Average" with a confidence of 0.56
"If August = Average, then July = Average" with a confidence of 0.60

But, in real life backward predication is not possible, therefore following two valid rules may be considered:
"If June = High, then August = Average".
"If July = Average, then August = Average".

Thus, the First rule suggests that if rainfall in the month of June is high then rainfall in the month of August is Average. Second rule suggest if the rainfall in the month of July is Average, then rainfall in the month of August is to be average as well.

Table 4. Membership degree of June.H ∩ July.AV

SL No.	June.H	July.AV	June.H ∩ July.AV
1	0.0	0.0	0.0
2	1.0	1.0	1.0
3	0.0	0.0	0.0
4	1.0	0.26	0.26
5	0.0	0.16	0.0
6	1.0	0.56	0.56
7	0.0	1.0	0.0
8	1.0	0.0	0.0
9	0.0	1.0	0.0
10	0.0	0.72	0.0
Total	3.0	4.70	1.82

Table 5. Membership degree of June.H ∩ Aug. AV

SL No.	June.H	Aug.AV	June.H ∩ Aug.AV
1	0.0	0.0	0.0
2	0.0	1.0	0.0
3	0.0	0.0	0.0
4	1.0	1.0	1.0
5	0.0	0.0	0.0
6	1.0	0.38	0.38
7	0.0	1.0	0.0
8	1.0	1.0	1.0
9	0.0	0.0	0.0
10	0.0	0.0	0.0
Total	3.0	4.38	2.38

Table 6. Membership degree of June.H ∩ Sept.L

SL No.	June.H	Sept.L	June.H ∩ Sept.L
1	0.0	1.0	0.0
2	0.0	0.04	0.04
3	0.0	0.38	0.0
4	1.0	0.06	0.06
5	0.0	0.62	0.0
6	1.0	0.80	0.80
7	0.0	0.26	0.0
8	1.0	0.92	0.92
9	0.0	0.20	0.0
10	0.0	1.0	0.0
Total	3.0	5.28	1.82

Table.7. Membership degree of July.AV ∩ Aug.AV

SL No.	July.AV	Aug.AV	July.AV ∩ Aug.AV
1	0.0	0.0	0.0
2	1.0	1.0	1.0
3	0.0	0.0	0.0
4	0.26	1.0	0.26
5	0.16	0.0	0.0
6	0.56	0.38	0.38
7	1.0	1.0	1.0
8	0.0	1.0	0.0
9	1.0	0.0	0.0
10	0.72	0.0	0.0
Total	4.70	4.38	2.64

Table 8. Membership degree of July.AV ∩ Sept.L

SL No.	July.AV	Sept.L	July.AV ∩ Sept.L
1	0.0	1.0	0.0
2	1.0	0.04	0.04
3	0.0	0.38	0.0
4	0.26	0.06	0.06
5	0.16	0.62	0.16
6	0.56	0.80	0.56
7	1.0	0.26	0.26
8	0.0	0.92	0.0
9	1.0	0.20	0.20
10	0.72	1.0	0.72
Total	4.70	5.28	2.0

Table 9. Membership degree of Aug.AV ∩ Sept.L

SL No.	Aug.AV	Sept.L	Aug.AV ∩ Sept.L
1	0.0	1.0	0.0
2	1.0	0.04	0.04
3	0.0	0.38	0.0
4	1.0	0.06	0.06
5	0.0	0.62	0.0
6	0.38	0.80	0.38
7	1.0	0.26	0.26
8	1.0	0.92	0.92
9	0.0	0.20	0.0
10	0.0	1.0	0.0
Total	4.38	5.28	1.66

Table 10. The Fuzzy Count of the Item sets in C2

Itemsets	Count
(June.H,July.AV)	1.82
(June.H,Aug.AV)	2.38
(June.H,Aug.AV)	1.82
(July.AV, Aug.AV)	2.64
(July.AV,Sept.L)	2.0
(Aug.AV, Sept.L)	1.66

Table 11. The Itemset and their Fuzzy counts in L2

Itemset	Count
(June.H,Aug.AV)	2.38
(July.AV, Aug.AV)	2.64
(July.AV,Sept.L)	2.0

4.1 Apriori Algorithm Based Approach

Apriori Algorithm is well known algorithm, proposed by Agarwal and his co-researcher [2] used to find association among different items. The algorithm is implemented and experimented in different application since it was introduced, specially to generate association rules. Here, before implementation of the algorithm the fuzzy dataset (shown on the Table 2) is converted to a binary dataset based on the maximum-membership method. Using this method, all the attribute having higher membership degree is assigned as 1 and others value as 0. The converted binary dataset is shown in the Table 12.

For the convenience, the attributes of months (June, July, Aug, Sept) and linguistic terms (VL, L, AV, H, VH) are represented by allopathically (A, B, C, D) and Numerically (1, 2, 3, 4, 5) respectively. For example, the attribute June.VL is represented as A1, July.L

Table 12. Binary Dataset converted from Fuzzy Dataset

TId	June					July					Aug					Sept				
	VL	L	AV	H	VH	VL	L	AV	H	VH	VL	L	AV	H	VH	VL	L	AV	H	VH
1	1	0	0	0	0	0	0	0	1	0	0	0	0	0	1	0	1	0	0	0
2	0	0	0	1	0	0	0	1	0	0	0	0	1	0	0	0	0	1	0	0
3	1	0	0	0	0	0	0	0	0	1	0	1	0	0	0	1	0	0	0	0
4	0	0	1	0	0	0	0	0	1	0	0	0	1	0	0	0	0	1	0	0
5	0	0	0	1	0	0	0	0	1	0	0	0	0	0	1	0	1	0	0	0
6	0	0	1	0	0	0	0	1	0	0	0	1	0	0	0	0	1	0	0	0
7	0	0	0	1	0	0	0	1	0	0	0	0	1	0	0	0	0	1	0	0
8	0	0	1	0	0	0	0	0	1	0	0	0	1	0	0	0	1	0	0	0
9	0	1	0	0	0	0	0	1	0	0	1	0	0	0	0	0	0	1	0	0
10	0	0	0	1	0	0	0	1	0	0	0	1	0	0	0	0	1	0	0	0

as B2, Aug.AV as C3, Sept.H as D4 and so on. Now, the Apriori Algorithm is applied on this binary dataset to find possible association among the attributes, often called as items. As input, minimum support and confidence is predefined as 2.0 and 0.5 respectively.

Step 1: In this step, the binary dataset is converted to a training dataset based on the occurrence in the transaction, which can be identified by 1 or 0. That is to say, "1" indicates present and "0" indicates absent of a particular item in a transaction. Based on this condition, a training dataset can be achieved as shown in the Table 13.

Table 13. Training dataset

TID	Items
1	A1, B4, C5, D2
2	A4, B3, C3, D3
3	A1, B5, C2, D1
4	A3, B4, C3, D3
5	A4, B4, C5, D2
6	A3, B3, C2, D2
7	A4, B3, C3, D3
8	A3, B4, C3, D2
9	A2, B3, C1, D3
10	A4, B3, C2, D2

Step 2: In this step, frequency of occurrences for each item is counted and compared with minimum support count, which is initialized as 2.0 for this example. Then, the highest frequent item from each of the attribute (A, B, C, D) is selected for next iteration. The Table 14 shows the items and its frequency.

Table 14. Highest frequent item with frequency

Items	Support count
A4	4
B3	5
C3	4
D2	5

Step 3. In this step, frequent items are associated and checked their occurrences per transaction. If the frequent itemset qualify the minimum support count, then those item sets are considered for next iteration. In this example, six possible item sets can be found. Among those six associations, five associations have more or equal support count compared to pre-defined support count. Those item sets are shown in the Table 15.

Table 15. Item sets with support count

Item sets	Support count
{A4, B3}	3
{A4, C3}	2
{A4, D2}	2
{B3, C3}	2
{B3, D2}	2

Step 4. In this step, frequent item sets are joined and 10 possible association can be formed. Again, using pruning method based on predefined support count, three frequent item set can be found. Here, only one item set is selected, i.e. {A4, B3, C3} with support count 2, as shown in the next Table 16.

Table 16. Frequent set with support count

Item set	Support count
{A4, B3, C3}	2

Step 5: From the frequent Item set (shown in the Table 15) some rules can be formed. But, for this example, we have considered the rainfall data during monsoon seasons. So, backward rules can be prescribed. Hence, following two valid and possible rules can be found.

$$\mathrm{A4} \rightarrow \{\mathrm{B3}, \mathrm{C3}\}$$

$$\{\mathrm{A4}, \mathrm{B3}\} \rightarrow \mathrm{C3}$$

Calculating the confidessnce measure of those rules and comparing the pre-defined confidence final rules can be generated.

confidence of measure of the "A4→{B3, C3}" can be calculated using equation.

$$\mathrm{Conf.}\ \{\mathrm{A4} \rightarrow \{\mathrm{B3}, \mathrm{C3}\}\} = \frac{Support(A4 \cap (B3, C3))}{Support(\mathrm{A4})} = 0.5$$

confidence of measure of the "{A4, B3}→C3" is 0.66

Now, converting these items into its original attributes, following rules can be found:

"If June = High, then July = Average and August = Average."

"If June=High and July=Average," then August= Average."

Thus, rainfall prediction can be done as follows

"If rainfall is high in the month of June, there may occurs average rainfall in the month of July and August."

"If rainfall in June and July is high and Average respectively, there may occurs Average rainfall in August."

5 Experimental Analysis

In the experiments, the minimum support and minimum confidence are assumed as 2.0 and 0.5 respectively. Then experiments are performed with different support count and confidence values and different results are obtained. After changing the threshold, it is noticed that when support and confidence value is smaller more itemset are generated and when the threshold value is larger than a smaller number of itemset are generated as shown below in the Fig. 3.

From the experimental results, as shown below, the Fig. 4 shows the relationship between number of association rules and minimum support values for different confidence values. From this figure it can be observed that when support and confidence are low more rules can be generated. But if support and confidence values are high or even later one low, a smaller number of rules are mined.

Similarly, when the minimum confidence values and support values are increases, number of association rules decreases as shown below in the Fig. 5.

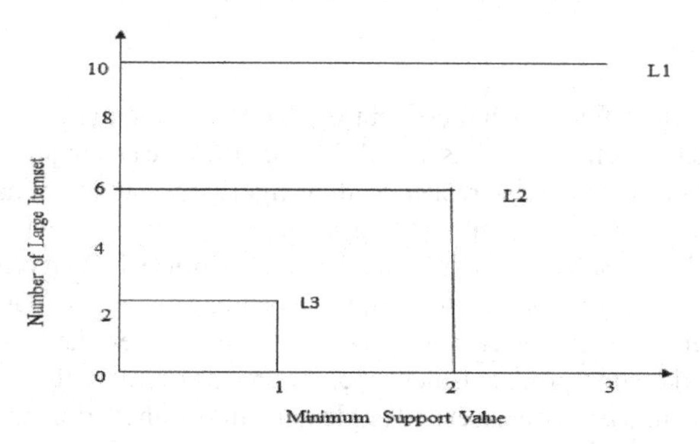

Fig. 3. Relationship between number of large items sets and minimum support values

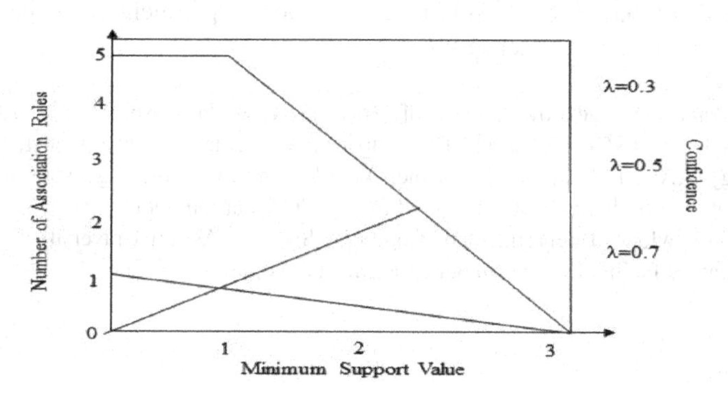

Fig. 4. Relationship between number of association rules with minimum support

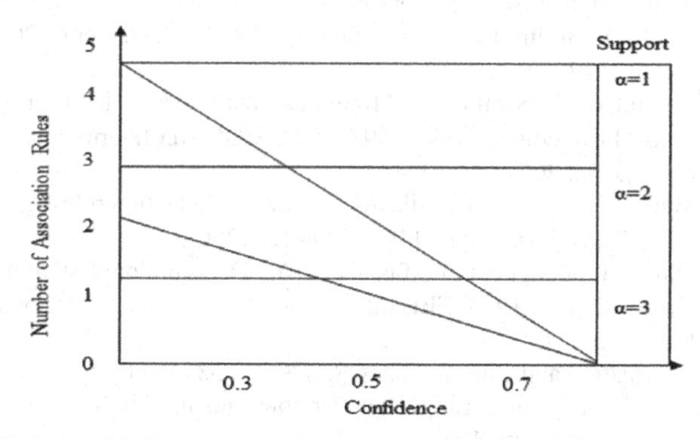

Fig. 5. Relationship between number of association rules and confidence values

6 Conclusion

This paper deals with fundamental concept of Fuzzy Association rule mining and its application-based work have been discussed. The significance and requirement of Fuzzy logic in ARM is presented. Some related leading algorithms and issues have been mentioned. Two Fuzzy association rule mining algorithm are experimented on a sample rainfall data. FTDA approach is flexible and based on Apriori TID algorithm. Classical Apriori algorithm is easy and simple technique. Depending on range of dataset both algorithms are useful. Different experiment is performed in different level with a standard framework. Fuzzification process is described to pre- process the fuzzy data. Conversion of fuzzy data into binary dataset and implementation with Apriori approach may be considered as an improvised technique. Fuzzy rule mining techniques are rapidly developing and used in practical field. Our work is a small step of modification and developing different algorithm-based activity. In future, with more experiment and implementation shall be carried out to find better results.

Acknowledgement. We acknowledged Prof. Prasanta Neog, Professor & HOD (Ag. Meteorology), PI (GKMS)BNCA, AAU and Dr. Arup Kumar Sarma, Senior Scientist, Department of Entomology, AAU, Jorhat (Assam) for their kind help and valuable suggestion in collection and analysis of meteorological data, collected from BNCA Station under AAU, Jorhat (Assam). Finally, we acknowledged department of Computer Science, Assam University for providing required laboratory facility to perform our experimental work.

References

1. Fayyad, U.: Knowledge discovery in databases: an overview. In: Inductive Logic Programming: 7th International Workshop, ILP-97 Prague, Czech Republic September 17–20 1997 Proceedings, pp. 1–16. Berlin, Heidelberg: Springer Berlin Heidelberg. (2005). https://doi.org/10.1007/3540635149_30
2. Agrawal, R., Imieliński, T., Swami, A.: Mining association rules between sets of items in large databases. In: Proceedings of the 1993 ACM SIGMOD International Conference on Management of Data, pp. 207–216 (1993)
3. Delgado, M., Marín, N., Sánchez, D., Vila, M.A.: Fuzzy association rules: general model and applications. IEEE Trans. Fuzzy Syst. **11**(2), 214–225 (2003)
4. Maeda, A., Ashida, H., Taniguchi, Y., Takahashi, Y.: Data mining system using fuzzy rule induction. In: Proceedings of 1995 IEEE International Conference on Fuzzy Systems, vol. 5, pp. 45–46. IEEE (1995)
5. Zadeh, L.A.: Information and control. Fuzzy Sets **8**(3), 338–353 (1965)
6. Zadeh, L.A.: The Concept of a Linguistic Variable and its Application to Approximate Reasoning. Inf. Sci. **8**, 199–249 (1975)
7. Chan, K.C., Au, W.H.: Mining fuzzy association rules. In: Proceedings of the Sixth International Conference on Information and Knowledge Management, pp. 209–215 (1997)
8. Hong, T.P., Kuo, C.S., Chi, S.C.: Mining association rules from quantitative data. Intelli. Data Anal. **3**(5), 363–376 (1999)
9. Gyenesei, A., Teuhola, J.: Interestingness measures for fuzzy association rules. In: De Raedt, L., Siebes, A. (eds.) Principles of Data Mining and Knowledge Discovery: 5th European Conference, PKDD 2001, Freiburg, Germany, 3–5 September 2001 Proceedings 5, vol. 2168, pp. 152–164. Springer, Berlin, Heidelberg (2001). https://doi.org/10.1007/3-540-44794-6_13

10. Muyeba, M., Khan, M.S., Coenen, F.: Fuzzy weighted association rule mining with weighted support and confidence frame-work. In: Chawla, S., eds. et al. New Frontiers in Applied Data Mining. PAKDD 2008. LNCS (), vol. 5433, pp. 49–61. Springer, Berlin, Heidelberg (2008). https://doi.org/10.1007/978-3-642-00399-8_5
11. Khan, M.S., Muyeba, M., Coenen, F.: Weighted association rule mining from binary and fuzzy data. In: Perner, P. (eds.) Industrial Conference on Data Mining, vol. 5077, pp. 200–212. Springer, Berlin, Heidelberg (2008). https://doi.org/10.1007/978-3-540-70720-2_16
12. Hong, T.P., Lee, Y.C.: An overview of mining fuzzy association rules. In: Bustince, H., Herrera, F., Montero, J. (eds.) Fuzzy Sets and Their Extensions: Representation, Aggregation and Models. Studies in Fuzziness and Soft Computing, vol. 220, pp. 397–410. Springer, Berlin, Heidelberg (2008). https://doi.org/10.1007/978-3-540-73723-0_20
13. Yang, T., Li, C.: A study of fuzzy quantitative items based on weighted association rules mining. In: 2nd International Conference on Intelligent Computing and Cognitive Informatics (ICICCI 2015), pp. 42–46. Atlantis Press (2015)
14. Mangayarkkarasi, K., Chidambaram, M.: E-FWARM: enhanced fuzzy-based weighted association rule mining algorithm. J. Theor. Appl. Inform. Technol. **96**(2), 322–332 (2018)
15. Mangalampalli, A., Pudi, V.: FAR-miner: a fast and efficient algorithm for fuzzy association rule mining. Int. J. Bus. Intell. Data Min. **7**(4), 288–317 (2012)
16. Princy, S., Dhenakaran, S.S.: Comparison of triangular and trapezoidal fuzzy membership function. J. Comput. Sci. Eng. **2**(8), 46–51 (2016)

Brain Tumor Detection Using VGG-16

Taniya Nandy, Laishram Munglemkhombi Devi(✉), and Ishita Chakraborty

Department of Computer Science and Engineering, Royal Global University, Guwahati, Assam, India
munglemkhombi123@gmail.com

Abstract. Brain Tumour exposes serious medical conditions and threat to human life. Early and accurate detection of these tumours are crucial for effective treatment. In the recent years, deep learning methods are used in medical imaging techniques that have shown promising results in detecting brain tumours. In this study, we proposed to use the Visual Geometry Group (VGG-16) that is based on deep neural network architecture for detecting brain tumours using magnetic resonance images (MRI). The proposed model showed improved result in terms of accuracy. In this work were able. To achieve 96% and 90% accuracy on the training and the testing set respectively with a precision of 96% accuracy on the training set and90% accuracy on the testing set. The recall of the model is also found to be reasonable with a recall of 92% on training set and 88% on the testing set.

Keywords: Brain Tumor · CNN · MRI · VGG-16 · Transfer Learning

1 Introduction

Brain tumours are a serious medical condition that can greatly impact a person's quality of living and may lead to death even. Early and accurate detection of these tumours is crucial for effective treatment, nevertheless, given the complexity and variety of these malignancies, it is still a difficult undertaking. MRI is frequently used for the diagnosis of brain tumours as it provides detailed information about the structure of the brain and its functionalities [1]. Brain tumors can be categorized into malignant tumors and benign or non-malignant.The tumors that develop itself within the brain is called as primary. The secondary tumors are usually caused due to the trespassing of the other tumors situated elsewhere.

Machine learning has emerged as a promising tool in the field of medical imaging, particularly in the detection and diagnosis of brain tumors. Brain tumors are a complex and heterogenous group of neoplasms that can present with a wide range of symptoms and imaging features [2]. The traditional approach to brain tumor detection involves visual inspection and interpretation of medical images by trained radiologists and clinicians which can be time consuming [3].

Deep Learning algorithms have proved itself to be very potential in various medical imaging analysis tasks, that includes early detection of tumours and other such ail-

P. Das et al. (Eds.): AMRIT 2023, CCIS 1953, pp. 88–95, 2024.
https://doi.org/10.1007/978-3-031-47224-4_8

ments. Convolutional Neural Networks (CNNs), are often used here due to its ability to represent data hierarchically [4]. In this study, we have used the VGG architecture, which in a well-established deep CNN, for detecting brain tumors [5].

1.1 Convolutional Neural Network

Convolution Neural Networks (CNNs) are a class of deep learning algorithms that have been extensively used in computer vision applications such as semantic segmentation, object detection and image categorization. The primary component of a CNN consists of a convolution layer, a pooling layer and a fully connected layers [6].

Convolutional layers are the main feature extraction component of a CNN, where the input image is processed with filters to discover regional patterns and linkages [7]. Pooling layers reduces the size of the feature maps produced by the convolutional layers, helping to reduce computational complexity and improve spatial invariance. The fully connected layers perform the final classification task based on the features learned by the convolutional and pooling layers [8]. All these combined forms the architecture of Convolutional Neural Network.A typical CNN architecture is showed in Fig. 1.

In recent years, CNNs have accomplished state-of-the-art outcomes in different fields of computer vision tasks and have widely adopted field in medical imaging autonomous vehicles, and robotics.

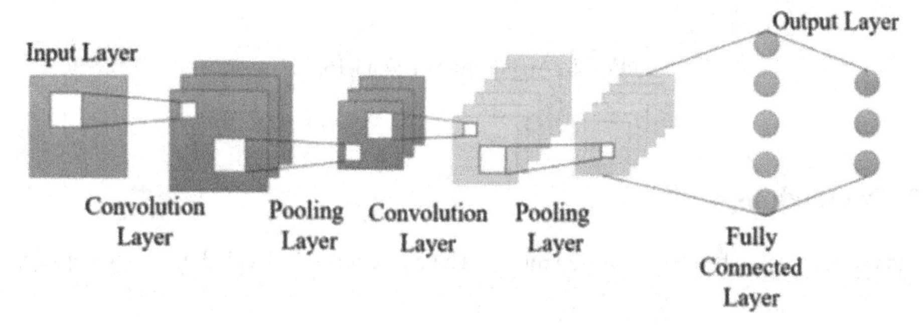

Fig. 1. Architecture of Convolutional Neural Network

1.2 Vgg-16

VGG-16 is an architecture of a deep convolutional neural network (DCNN) that was proposed by Simonyan and Zisserman in 2014. The network is named after the Visual Geometry Group (VGG) at the University of Oxford, where it was developed. The VGG-16 architecture is characterized by its deep and narrow nature, with 16 weight layers and a large number of filters [9].

Computer vision tasks including image classification, object recognition, and semantic segmentation have all seen widespread adoption of the VGG-16 architecture. VGG-16 architecture includes several convolutional layers, a pooling layer, and a fully linked layers with a large number of 3x3 filters in its convolutional layers, which in turn allows it to learn rich and complex features of the image [10]. One of the main strengths of the VGG-16 architecture is its ability to learn robust features that are invariant to translations and rotations in the image. Additionally, the VGG-16 architecture is trained on a large dataset allowing it to generalize well to new images. The visual representation of the VGG-16 is shown in Fig. 2.

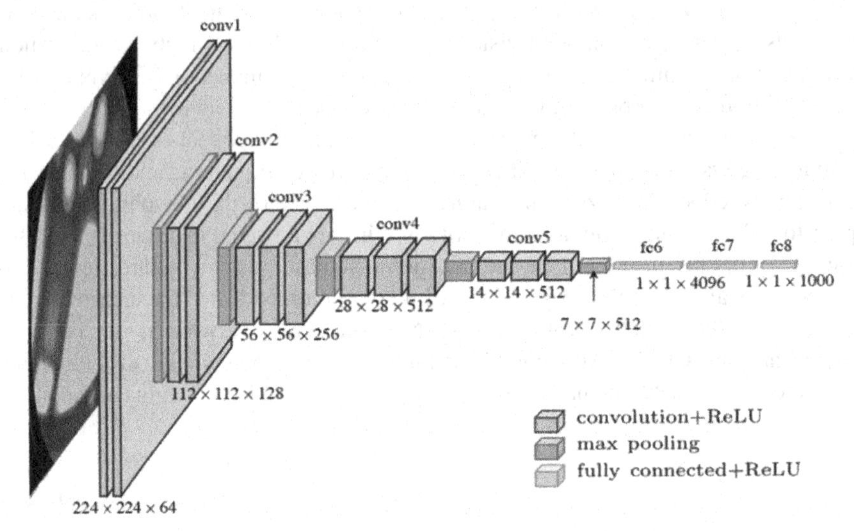

Fig. 2. Architecture of VGG-16.

2 Methodology

This section describes the proposed model. The entire work is divided into four steps as follows:

1. Data acquisition
2. Data augmentation
3. Image pre-processing
4. Brain tumour classification using VGG-16.

2.1 Data Acquisition

Data acquisition in machine learning refers to the process of collecting, importing, and preparing data that are used to train a machine learning model. The quality and relevance of the data has a substantial influence on the performance and accuracy of the resulting model, so it's important to carefully consider the sources of the data and how it is collected and processed. In this step we have collected our data i.e. Images of MRI scans from Kaggle. The name of the dataset used in this study is "Brain. MRI Images for Brain Tumour Detection". There are 253 brain MRI images in the dataset. The dataset consists

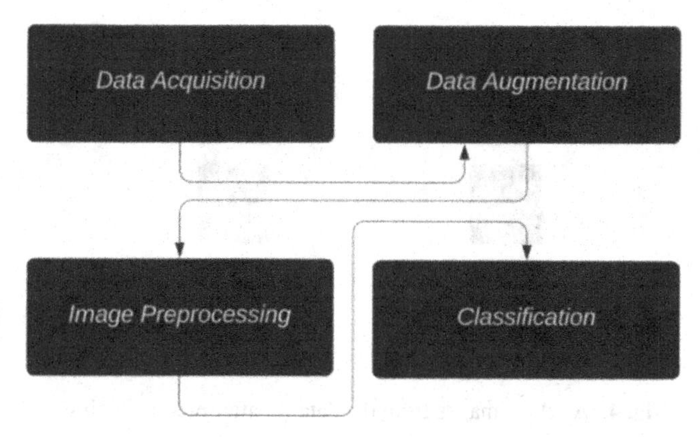

Fig. 3. Flow of methodology.

of two folders, namely yes and no. The no folder contains 98 (benign) Brain MRI Images that are not tumorous, while the yes folder consists of 155 tumorous (malignant) Brain MRI Images (Fig. 3).

2.2 Data Augmentation

Data Augmentation approach is used by experts to broaden the variety of data that is available for training models without having to actually collect new data [11]. Due to the restrictions on the size of the dataset the augmented data was used for training the neural network. This also solved the data imbalance issue [12]. After augmentation we generated 1078 positive (malignant tumours) and 960 negative images (benign tumours) totalling to 2038 images.

2.3 Data Preprocessing

Data pre-processing is a very important step for image classification [13]. In this step, the image containing the section of the brain is cropped using the open CV cropping technique to locate the brain's most remote views like top, bottom, left and right points. Then the images that are required as input for the VGG-16 model is pre-processed and resized into the data set as shown in Fig. 4 to have a shape of (224,224,3) = (image_width, image_height, number of channels).This reshaping is done because every image comes in different sizes. The neural network only accepts it as an input provided the images are of the same shape. The data is also normalized to scale pixel values ranging from 0 to 1.

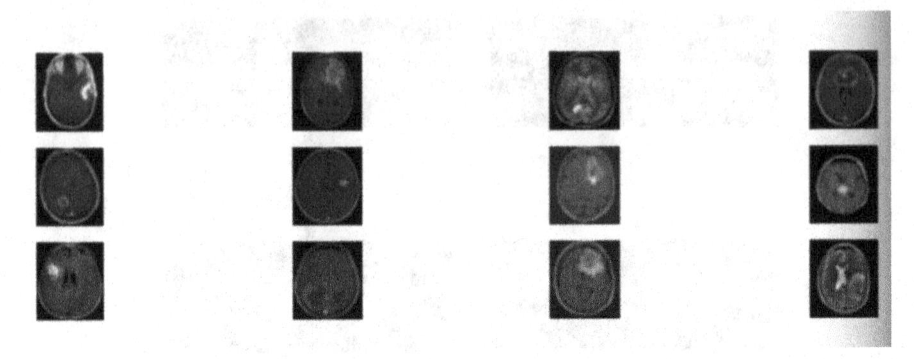

Fig. 4. A set of images from the data set after pre-processing.

2.4 Brain Tumor Classification Using VGG-16

In this step, for the classification of images we have applied transfer learning using VGG-16. The last layer was replaced with a sigmoidal function for representing the output unit and the rest of the parameters of the other layers were frozen [14-16].

3 Performance Evaluation

The goal of this study was to create a model that fits well while avoiding problems with underfitting and overfitting. We deduced from the result that our model didn't have an overfitting or underfitting issue. On the training set, we achieved an accuracy of 96% as shown in Fig. 5 and on the test set, we reached an accuracy of 90%. This suggests that the model is able to generalize well on the unseen data and is able to accurately identify brain tumors in a larger dataset. In terms of performance metrics, the model has demonstrated an excellent precision of 96% on the training set and a precision of 90% on the validation set. The model is evaluated by the metrics such as: recall of 92% on the train data and 88% on the test data. Considering the above results, the accuracy of the model can be improved further.

Table 1 below shows a comparative analysis of the proposed model with existing models.

Table 1. Comparative analysis of proposed analysis with existing models.

Reference	Year	Methodology	Result	Drawbacks
[3]	2021	CNN	97.79 accuracy on test set	No mention of other important evaluation metrics such as precision, recall, or F1-Score
[4]	2021	Depthwise Seperable CNN	92% accuracy	Lack of comparison to existing models and limited evaluation metrics
[5]	2022	U-Net & VGG-Net	0.92 accuracy	Some significant aspects are not addressed
Proposed Model	2023	VGG-16		High accuracy on training set

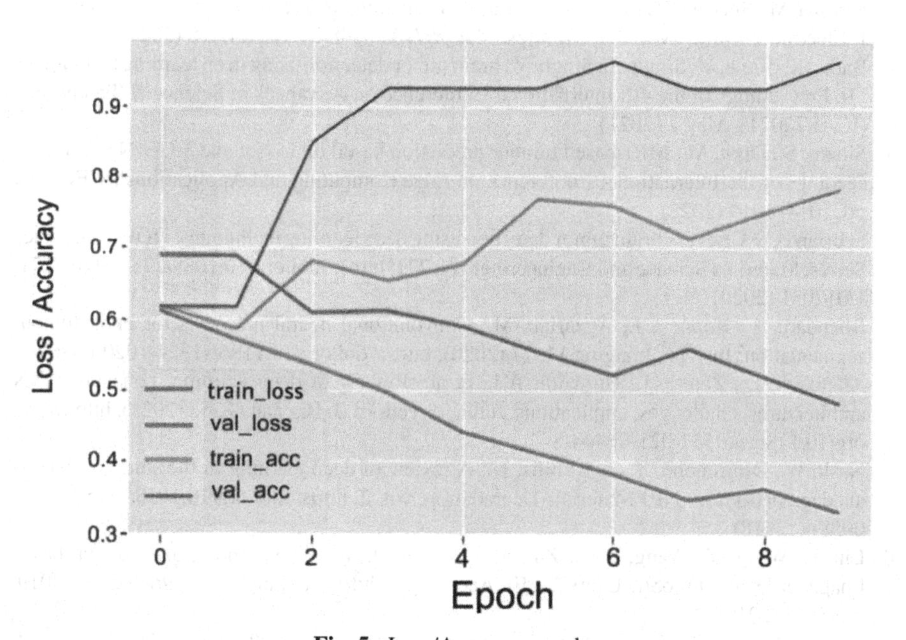

Fig. 5. Loss/Accuracy graph

4 Conclusion

To effectively intervene with patients, brain tumour detection is crucial. Since, medical images comes in different variations, it is essential to pre-process the medical images. The likelihood of a patient surviving is greatly increased with the automatic brain tumour detection approach. This approach makes the diagnosis simple and intuitive. In order to increase the precision of tumour identification and classification, CNN models have been used. In this paper, the MRI images of the brain from different patients were collected to train a DCNN for determining automatically whether a patient has brain tumour or not. In this work, VGG-16 was used for detecting brain tumor. The model was able to accurately classify tumours in MRI scans with higher accuracy. This opens the door for further research in medical image analysis, particularly in the area of brain tumor detection. Future studies involves improving the accuracy as well as exploring alternative architectures to further improve the model.

References

1. Deepa and Singh, A.: Review of brain tumor detection from MRI images. In: Proceedings of the 3rd International Conference on Computing for Sustainable Global Development (INDIACom), New Delhi, India, pp. 3997–4000 (2016)
2. Dhole, N.V., Dixit, V.V.: Review of brain tumor detection from MRI images with hybrid approaches. Multimedia Tools Appl. **81**, 10189–10220 (2022). https://doi.org/10.1007/s11042-022-12162-1
3. Sharma, M., Sharma, V., Mittal, R., Gupta, K.: Brain tumour detection using machine learning. J. Electron. Inform. , vol. **3**, DOI: https://doi.org/10.36548/jei.2021.4.005 (2021)
4. Bathe, K., Rana, V., Singh, S., Singh, V.: brain tumor detection using deep learning techniques. : In Proceedings of the 4th International Conference on Advances in Science & Technology (ICAST2021), May 7 (2021)
5. Sihare, S., Dixit, M.: MRI-based tumour prediction based on U-Net and VGG-Net. : In Proceedings of the International Conference on Edge Computing and Applications (ICECAA), pp. 1014–1019 (2022)
6. Febrianto, D.C., et al.: brain tumor detection using deep learning techniques. : IOP Conference Series: Materials Science and Engineering, vol. **771**: https://doi.org/10.1088/1757-899X/771/1/012031 (2020)
7. Bhandari, A., Koppen, J., Agzarian, M.: Convolutional neural networks for brain tumour segmentation. Insights Imaging **11**, 77 (2020). https://doi.org/10.1186/s13244-020-00869-4
8. Alzubaidi, L., Zhang, J., Humaidi, A.J., et al.: Review of deep learning: concepts, CNN architectures, challenges, applications, future directions. J. Big Data **8**, 53 (2021). https://doi.org/10.1186/s40537-021-00444-8
9. Nash, W., Drummond, T., & Birbilis, N.: A review of deep learning in the study of materials degradation. In: NPJ Materials Degradation, vol. **2**, https://doi.org/10.1038/s41529-018-0058-x (2018)
10. Liu, F., Wang, Y., Wang, F.-C., Zhang, Y., & Lin, J.: intelligent and secure content-based image retrieval for mobile Users," IEEE Access, p. 1, https://doi.org/10.1109/ACCESS.2019.2935222 (2019)
11. Shorten, C., Khoshgoftaar, T.M.: A survey on image data augmentation for deep learning. J. Big Data **6**, 60 (2019). https://doi.org/10.1186/s40537-019-0197-0
12. Mikołajczyk, A., & Grochowski, M.: Data augmentation for improving deep learning in image classification problem. In: Int. Interdisc. PhD Workshop (IIPhDW), Świnouście, Poland, pp. 117–122. https://doi.org/10.1109/IIPHDW.2018.8388338 (2018)
13. Pal, K., & Sudeep, K.: Preprocessing for image classification by convolutional neural networks. In: Proceedings of the International Conference on Recent Trends in Electronics, Information and Communication Technology (RTEICT), pp. 1778–1781. https://doi.org/10.1109/RTEICT.2016.7808140 (2016)

14. Tammina, S.: Transfer learning using VGG-16 with deep convolutional neural network for classifying images. In: Int. J. Sci. Res. Publ. (IJSRP), vol. 9, p. 9420, https://doi.org/10.29322/IJSRP.9.10.2019.p9420 (2019)
15. Tao, J., Gu, Y., Sun, J., Bie, Y., Wang, H.: Research on vgg16 convolutional neural network feature classification algorithm based on transfer learning. In: Proceedings of the 2nd China International SAR Symposium (CISS), Shanghai, China, pp. 1–3. https://doi.org/10.23919/CISS51089.2021.9652277 (2021)
16. Sihare, S., Dixit, M.: MRI-based tumour prediction based on U-Net and VGG-Net. In: Proceedings of the International Conference on Electronics, Computer and Computation (ICEC), pp. 1014–1019. https://doi.org/10.1109/ICECAA55415.2022.9936184 (2022)

Deep Learning Model for Fish Copiousness Detection to Maintain the Ecological Balance Between Marine Food Resources and Fishermen

O. M. Divya(✉), M. Ranjitha, and K. Aruna Devi

Department of Computer Science (PG), Kristu Jayanti College (Autonomous), Bangalore, India
divyammo@gmail.com

Abstract. Fish copiousness detection is crucial for the monitoring and management of aquatic ecosystems. Deep learning models, such as Mask R-CNN, have shown great potential in accurately detecting and segmenting fish in underwater images. This study explores the effectiveness of Mask R-CNN in fish detection and presents a detailed analysis of its performance. The dataset used for training and testing consists of a large number of underwater images of various fish species. The results show that the Mask R-CNN model can accurately detect and segment fish in complex underwater environments. This study also compares the performance of the Mask R-CNN model with other popular deep learning models and demonstrates the superiority of the Mask R-CNN model in terms of accuracy and efficiency. Overall, the study highlights the potential of deep learning models in fish detection and their usefulness in managing aquatic resources.

Keywords: Fish Copiousness · Deep Learning model · Mask R-CNN · Object Detection · Image Segmentation

1 Introduction

Fish copiousness detection from the ocean is an important task for fisheries management, as it helps to ensure that fish populations are sustainably managed, and that fishing practices are not depleting fish stocks to unsustainable levels. There are many different methods for detecting fish abundance from the ocean, including direct sampling methods such as trawling, seining, and acoustic surveys. These methods involve physically capturing or observing fish in the ocean, and then using statistical models to estimate the total population size. Other methods for detecting fish abundance from the ocean include remote sensing techniques, such as satellite imagery and sonar systems. These methods allow researchers to detect and track schools of fish over large areas of the ocean, and can provide valuable information on the distribution and abundance of fish populations. Fish copiousness detection from the ocean is a complex and challenging task, requiring careful planning, data collection, and analysis. However, it is a critical component of fisheries management, and is essential for ensuring the long-term sustainability of our oceans and the communities that depend on them.

P. Das et al. (Eds.): AMRIT 2023, CCIS 1953, pp. 96–104, 2024.
https://doi.org/10.1007/978-3-031-47224-4_9

Deep learning models are a subset of machine learning models that are designed to mimic the way the human brain works. These models use multiple layers of artificial neurons to learn and make predictions based on large datasets. Deep learning models are used in a wide variety of applications, including computer vision, natural language processing, speech recognition, and predictive analytics. Some common deep learning models include convolutional neural net.

Works (CNNs), recurrent neural networks (RNNs), and deep belief networks (DBNs).CNNs are commonly used in image and video recognition tasks, while RNNs are used in natural language processing tasks such as language translation and speech recognition. DBNs are often used in unsupervised learning tasks, such as data clustering and anomaly detection. Deep learning models have revolutionized many industries and are used in everything from self-driving cars to virtual assistants. They have also enabled breakthroughs in fields such as medical diagnosis and drug discovery.

Mask R-CNN is a deep learning model that combines object detection with instance segmentation, meaning it can not only detect objects in an image but also identify the pixels that belong to each instance of the object. It is an extension of the popular Faster R-CNN model, which uses a region proposal network (RPN) to propose potential object locations, and then applies a convolutional neural network (CNN) to classify and refine those proposals.

Mask R-CNN builds on this architecture by adding a branch to the network that produces a binary mask for each detected object instance, in addition to the bounding box coordinates and class labels. This branch is a fully convolutional network that takes as input the region proposals generated by the RPN and produces a pixel-level segmentation mask for each object.

Mask R-CNN is widely used in computer vision tasks, such as instance segmentation, object recognition, and tracking in images and videos. It has achieved state-of-the-art performance on various benchmark datasets, such as COCO (Common Objects in Context) and Pascal VOC (Visual Object Classes).

Mask R-CNN can work with various image features and formats, including:

1. RGB images: Mask R-CNN is designed to work with RGB images, which are the most common format for images. RGB images are composed of three color channels: red, green, and blue, and are represented as three-dimensional arrays.
2. Grayscale images: Mask R-CNN can also work with grayscale images, which contain a single channel representing the intensity of each pixel. Grayscale images are often used in medical imaging and other applications where color is not necessary.
3. High-resolution images: Mask R-CNN can handle high-resolution images with a large number of pixels. However, processing such images may require a lot of computational resources and can be slow.
4. Image patches: Mask R-CNN can process image patches or sub-images extracted from a larger image. This approach can be useful for processing large images that do not fit into memory or for detecting small objects that are hard to detect in the full image.
5. Augmented images: Mask R-CNN can work with images that have been augmented with various transformations such as rotation, scaling, flipping, and color jittering. Augmented images can help improve the generalization and robustness of the model.

6. Preprocessed images: Mask R-CNN can work with preprocessed images, such as those that have been resized, cropped, or normalized, to fit the input size and format required by the model. Preprocessing the images can help improve the performance and speed of the model.

2 Literature Review

Identifying fishes from the ocean is a challenging task, as there are a large number of different fish species, and many of them have similar features. To solve this problem, researchers have been exploring the use of deep learning models to identify fishes from the ocean.

"Comparison of three sampling methods for small-bodied fish in lentic nearshore and open water habitats" by Merz et al. (2021) compares the effectiveness of three different fish sampling methods in lentic (still water) nearshore and open water habitats. The study aims to provide guidance on the most appropriate method for sampling small-bodied fish in these types of habitats. The three methods tested were boat electrofishing, nearshore fyke nets, and open water gill nets. The study was conducted in 20 small lakes in Minnesota, USA, over two summers. The researchers compared the catch rates, species richness, and species composition of fish caught using each method. The results showed that the catch rates and species richness were highest using the boat electrofishing method, followed by the nearshore fyke nets and open water gill nets. However, the species composition varied between the methods and habitats, with some species being caught more frequently in one method or habitat than in others [1].

Electrofishing is a common technique used to estimate fish populations in freshwater streams and rivers. In this study, Saunders et al. (2011) evaluates the accuracy of nighttime removal electrofishing in estimating salmonid abundance in small streams, using marked fish as a reference. The study was conducted in three small streams in British Columbia, Canada, over a two-year period. The authors marked a subset of the fish population using visible implant elastomer tags, which allowed them to determine the accuracy of the electrofishing estimates. The results showed that nighttime removal electrofishing was an effective method for estimating salmonid abundance in small streams. The electrofishing estimates were highly correlated with the actual population size, and the accuracy of the estimates improved with increasing sample size [2]. There are underwater visual surveys for assessing fish abundance and community structure in a tropical estuary in Australia.

Several studies have explored the use of deep learning models for underwater fish species recognition.

Abangan et al. (2019) [3] developed Deep-Fish, a system composed of a fish detection module using YOLOv3 and a fish recognition module based on ResNet-50. Their system achieved a 94% accuracy on a dataset of 47 fish species.

Hridayami et al. (2019) [4] used transfer learning to fine-tune a pre-trained VGG16 CNN model on a dataset of 13 fish species and reported an accuracy of 89.47% for fish recognition.

Tamou et al. (2020) [5] employed a pre-trained Inception-V3 CNN model and fine-tuned it on a dataset of 20 fish species, achieving an accuracy of 87.47% for fish classification.

Shen et al. (2021) [6] utilized a ResNet-50 CNN model and data augmentation techniques on a dataset of 30 fish species to attain an accuracy of 97.6% for fish recognition, surpassing traditional machine learning models.

The use of transfer learning and data augmentation techniques can further improve the accuracy of these models. The proposed systems can have important applications in the monitoring and conservation of marine biodiversity. Identifying the location of fish abundance is crucial for fishermen to optimize their fishing efforts and improve their catch. Deep learning models have been used to identify the location of fish abundance in the ocean.

Allken et al. [15] in their study aimed to detect specific types of fish, such as blue whiting, Atlantic herring, Atlantic mackerel, and mesopelagic fishes, from images taken in the Norwegian sea. Due to the need for a large amount of annotated data to train the models, a combination of real and synthetic images were used. The results showed a mean average precision of 0.845 on a test set of 918 images. In addition, regression models were employed to compare the predicted fish counts, based on the RetinaNet classification of fish in the individual image frames, with the actual catch data obtained from 20 trawl stations.

In their study, Xiang et al. (2020) [7] developed a deep learning model for estimating fish abundance from remotely sensed images. They used a CNN model to extract features from the images and achieved an accuracy of 94%. The extracted features were then fed into a support vector regression (SVR) model to estimate fish abundance. Similarly, Wei et al. (2022) [8] proposed a deep learning-based system for estimating fish abundance from acoustic backscatter data using a CNN model and a random forest model with an accuracy of 90%. Jiang et al. (2021) [9] also presented a deep learning-based system for estimating fish abundance, but using multispectral satellite imagery. They employed a CNN model for feature extraction and a random forest model for estimating fish abundance, achieving an accuracy of 92%.

The studies demonstrate the effectiveness of deep learning models for identifying the location of fish abundance in the ocean. The use of various types of data, such as satellite imagery, oceanographic data, and acoustic backscatter data, can improve the accuracy of these models. These systems can have important applications in the fishing industry, helping fishermen to optimize their fishing efforts and improve their catch.

Literatures are also explored to identify features used for image classification using Mask R-CNN:

In their study, He et al. [10] explained the features utilized for image classification via the Mask R-CNN model. Their novel technique, called Mask R-CNN, can efficiently detect objects in an image and create a segmentation mask for each object it identifies. By adding a branch for predicting an object mask in parallel with the current branch for bounding box identification, Mask R-CNN improves on the Faster R-CNN model with only a slight increase in computation time. Additionally, Mask R-CNN can be easily adapted to different applications, such as estimating human poses, and performs exceptionally well in tasks such as instance segmentation, bounding-box object detection, and person keypoint detection on the COCO dataset.

Sun et al. [11] proposed a new method for RoI pooling in Mask R-CNN, which enhances the model's performance. They introduced Global Mask R-CNN (GM R-CNN), a technique that uses Precise RoI Pooling and Global Mask Head to improve the performance of ship instance segmentation. Tests on the MS COCO dataset and MariShipInsSeg dataset show that GM R-CNN outperforms Mask R-CNN, achieving a 38.7% mask AP on MS COCO test-dev and a 48.6% mask AP on MariShipInsSeg testing sets, gains of 1.6% and 1.9% compared to Mask R-CNN, respectively, without any additional adjustments.

Lin et al.'s [12] "Focal Loss for Dense Object Detection" introduced the Focal Loss function, which is used in Mask R-CNN for object classification. They identified the high foreground-background class imbalance encountered during training of dense detectors as the main cause of poor performance. Their innovative Focal Loss method adjusts the conventional cross-entropy loss so that it downplays the loss attributed to examples with clear classifications, correcting the class imbalance. This approach helps the detector concentrate on a small number of hard examples and avoid being overloaded with easy negatives during training. They used RetinaNet, a simple dense detector, to evaluate the effectiveness of their loss.

Carion et al.'s [13] "End-to-End Object Detection with Transformers" proposed a novel approach to detecting objects as a direct set prediction issue. Their DETR framework simplifies the detection pipeline and eliminates the need for several manually created elements such as a non-maximum suppression mechanism or anchor generation, which explicitly convey prior knowledge about the task. The key components of DETR are a transformer encoder-decoder architecture and a set-based global loss that forces unique predictions through bipartite matching.

Ding et al. [14] proposed a RoI Transformer to address these issues. RoI Transformer applies spatial transformations to RoIs and learns the transformation parameters while being supervised by oriented bounding box (OBB) annotations. For oriented object detection, RoI Transformer is lightweight and easy to incorporate into detectors. They tested the effectiveness of RoI Transformer on two challenging aerial datasets.

Divya et al. [16] in their research had done a comparative study on various deep learning models for identifying fish copiousness. This was the pilot study performed to formulate a model, to recognize the best performing features and algorithms suitable for satellite images.

3 Methodology

Mask R-CNN is a powerful deep learning model that can be used for object detection and image segmentation tasks, including the detection of fish in underwater images. The model architecture is given in Fig. 1. To detect fish abundance using Mask R-CNN, the following steps can be taken:

1. Data Collection: Collect underwater images of fish in different locations and under different conditions. These images should be labeled with the presence/absence of fish in each image.

2. Data Preprocessing: Preprocess the images by resizing them to a common size, normalizing the pixel values, and augmenting the data with techniques such as flipping, rotating, and changing the brightness.
3. Training: Train a Mask R-CNN model on the preprocessed data using a loss function that combines both object detection and segmentation objectives. The model is trained on a GPU to speed up the process.
4. Evaluation: Evaluate the performance of the model on a test set of images by computing metrics such as precision, recall, and F1-score. These metrics can be used to determine the accuracy of the model in detecting fish density.
5. Deployment: Deploy the trained model on new images to detect fish copiousness in real-time. The model can be used to identify the number of fish in an image and estimate the abundance of fish in a particular area.

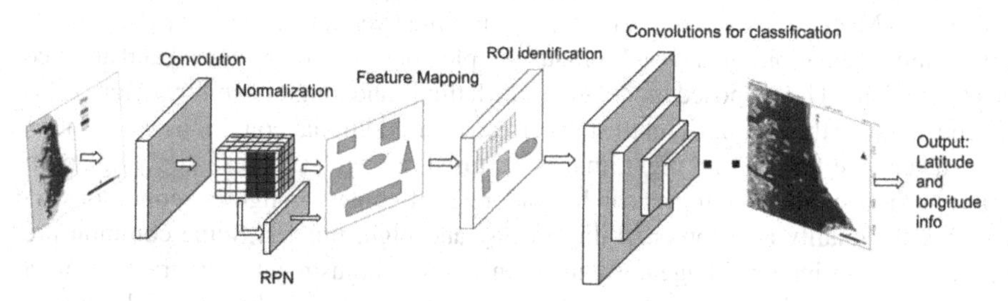

Fig. 1. Proposed Deep learning model for Fish copiousness detection

Mask R-CNN is primarily designed for object detection and instance segmentation, which can be further used for image classification. The algorithm for using Mask R-CNN for image classification is given below:

1. Backbone network (CNN):

   Input: Image I with dimensions (W, H, C), where W is the width, H is the height, and C is the number of channels.

   Output: Feature maps F with dimensions (W', H', D), where W' and H' are the spatial dimensions after downsampling and D is the number of feature channels.

   Mathematical formulation: $F = CNN(I)$
2. Region Proposal Network (RPN):

   Input: Feature maps F with dimensions (W', H', D).

   Output: Region proposals P with bounding box coordinates (x, y, w, h) and objectness scores.

   Mathematical formulation: $P = RPN(F)$
3. ROI Align:

   Input: Feature maps F with dimensions (W', H', D) and region proposals P.

   Output: Aligned feature maps R with fixed spatial dimensions for each ROI.

   Mathematical formulation: $R = ROI\_Align(F, P)$
4. Region Classification:

   Input: Aligned feature maps R with dimensions (W'', H'', D').

   Output: Class probabilities C for each region proposal.

   Mathematical formulation: $C = FC(R)$

5. Mask Prediction:
    Input: Aligned feature maps R with dimensions (W'', H'', D').
    Output: Binary masks M for each region proposal.
    Mathematical formulation: M = Convolutional_Network(R)
    The formula for image classification using Mask R-CNN can be summarized as follows:
    Input -> Backbone network -> RPN -> ROI Align -> Region Classification.

In the above equations, "CNN" represents the convolutional neural network used as the backbone, "RPN" denotes the Region Proposal Network, "ROI_Align" is the operation to align the feature maps with the region proposals, "FC" stands for the fully connected layer for classification, and "Convolutional_Network" represents the convolutional layers for mask prediction.

These equations provide a high-level understanding of the mathematical operations involved in Mask R-CNN. They represent a simplified version of the algorithm, and the actual implementation can have additional complexities, hyperparameters, and architectural variations. The proposed model gives the latitude and longitude information on fish copiousness for the region of interest. There are few additional conditions which needs to be taken care for day and night images captured through satellites. Applying appropriate pre-processing techniques can help address differences in lighting conditions and enhance the quality and comparability of day and night images. Some common pre-processing steps include image normalization, contrast adjustment, histogram equalization, and adaptive filtering. To leverage the advantages of both day and night images, fusion techniques can be employed to combine information from different lighting conditions. Fusion methods, such as multi-sensor fusion or data fusion algorithms, can integrate the complementary information available in day and night images to enhance the overall monitoring capabilities. Also, the image processing algorithms may need to be adjusted or customized to accommodate the variations between day and night images. This could involve modifying parameters, thresholds, or assumptions based on the specific characteristics of each type of image.

The accuracy of the model may depend on various factors, such as the quality of the images, the complexity of the underwater environment, and the diversity of the fish species. Therefore, it is important to fine-tune the model and optimize the hyperparameters to achieve the best possible results. Fish copiousness estimation using satellite images is a relatively new field of research, and there are not many publicly available datasets that specifically focus on this task. However, some research has been done on the topic, and some datasets that could be used for this task are:

1. Global Fishing Watch: This dataset provides satellite imagery and data on fishing activities worldwide. It includes vessel tracks, fishing effort, and other data that could be used to estimate fish abundance.
2. Ocean Health Index: This dataset provides a range of ocean-related data, including satellite imagery and information on fish populations. The data covers over 200 coastal countries and territories worldwide.
3. NOAA Coral Reef Watch: This dataset provides satellite imagery and data on coral reefs worldwide. It includes information on fish populations and can be used to estimate fish abundance in coral reef areas.

4. AquaMaps: This dataset provides information on the distribution of marine species worldwide, including fish. It includes satellite imagery and other data that could be used to estimate fish abundance in different areas.

While these datasets may not provide direct information on fish abundance, local fishing records and environmental data, are combined with satellite imagery to improve the accuracy of fish abundance estimates. Figure 2. Shows the transformed satellite image after applying mask-R CNN.

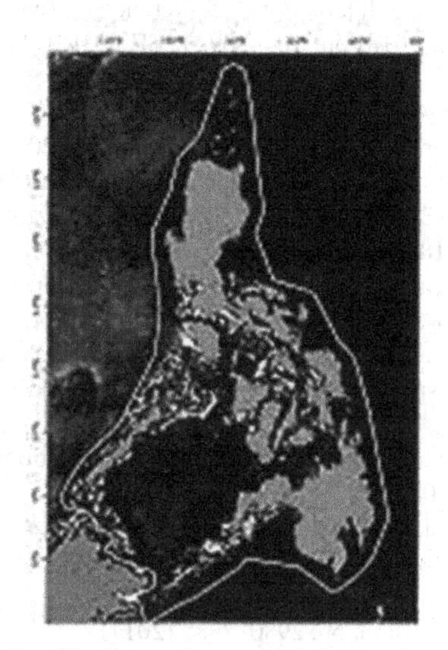

Fig. 2. Sample Satellite image from which fish copiousness is detected

4 Conclusion

Deep learning models like Mask R-CNN have shown promising results in detecting and segmenting fish in underwater images, which can help in estimating fish abundance and monitoring aquatic ecosystems. The use of such models can greatly reduce the time and effort required for manual fish detection and can also provide more accurate and consistent results. However, the performance of these models may vary depending on the quality and variability of the input images, the choice of hyperparameters, and the training dataset. Therefore, further research is needed to optimize the performance of these models and to explore their potential in other applications related to fish and aquatic resources management.

References

1. Merz, J.E., et al.: Comparison of three sampling methods for small-bodied fish in lentic nearshore and open water habitats. Environ. Monit. Assess. **193**(5), 255 (2021). https://doi.org/10.1007/s10661-021-09027-9

2. Saunders, W.C., et al.: Accurate estimation of salmonid abundance in small streams using nighttime removal electrofishing: an evaluation using marked fish. N. Am. J. Fish Manag. **31**(2), 403–415 (2011). https://doi.org/10.1080/02755947.2011.578526
3. Abangan, A.S., et al.: Artificial intelligence for fish behavior recognition may unlock fishing gear selectivity. Front. Mar. Sci. **10** (2023). https://doi.org/10.3389/fmars.2023.1010761
4. Hridayami, P., et al.: Fish species recognition using VGG16 deep convolutional neural network. J. Comput. Sci. Eng. **13**(3), 124–130 (2019). https://doi.org/10.5626/JCSE.2019.13.3.124
5. Tamou, A.B., et al.: Transfer learning with deep convolutional neural network for underwater live fish recognition. In: IEEE International Conference on Image Processing, Applications and Systems (IPAS), vol. 2018, pp. 204–209. IEEE (2018). https://doi.org/10.1109/IPAS.2018.8708871
6. Shen, Z., Savvides, M.: Meal v2: Boosting vanilla resnet-50 to 80%+ top-1 accuracy on imagenet without tricks, arXiv Preprint ArXiv:2009.08453 (2020)
7. Xiang, T.Z., et al.: Mini-unmanned aerial vehicle-based remote sensing: techniques, applications, and prospects. IEEE Geosci. Remote Sens. Mag. **7**(3), 29–63 (2019). https://doi.org/10.1109/MGRS.2019.2918840
8. Wei, Y., et al.: Monitoring fish using imaging sonar: capacity, challenges and future perspective. Fish Fish. **23**(6), 1347–1370 (2022). https://doi.org/10.1111/faf.12693
9. Jiang, Y., et al.: High-resolution mangrove forests classification with machine learning using worldview and uav hyperspectral data. Remote Sens. **13**(8), 1529 (2021). https://doi.org/10.3390/rs13081529
10. He, K., et al.: Mask r-cnn. In: Proceedings of the IEEE International Conference on Computer Vision, pp. 2961–2969 (2017)
11. Sun, Y., et al.: Global Mask R-CNN for marine ship instance segmentation. Neurocomputing **480**, 257–270 (2022). https://doi.org/10.1016/j.neucom.2022.01.017
12. Lin, T.Y., et al.: Focal loss for dense object detection. In: Proceedings of the IEEE International Conference on Computer Vision, pp. 2980–2988 (2017)
13. Carion, N., et al.: End-to-end object detection with transformers. In: Computer Vision–ECCV, Proceedings of the Part I 16: 16th European Conference, 23–28 August, 2020, vol. 2020, pp. 213-229. Springer, Glasgow, UK (2020)
14. Ding, J., et al.: Learning roi transformer for oriented object detection in aerial images. In: Proceedings of the IEEE/CVF Conference on Computer Vision and Pattern Recognition, pp. 2844–2853 (2019). https://doi.org/10.1109/CVPR.2019.00296
15. Allken, V., et al.: A deep learning-based method to identify and count pelagic and mesopelagic fishes from trawl camera images. ICES J. Mar. Sci. **78**(10), 3780–3792 (2021). https://doi.org/10.1093/icesjms/fsab227
16. Divya, M.O., et al.: Artificial intelligent fish abundance detector model for preserving environmental stability amid aquatic sustenance and fishermen. J. Surv. Fish. Sci. 776–784 (2023)

A Study on Smart Contract Security Vulnerabilities

Vaibhav Bhajanka(✉) and Nitisha Pradhan

Department of Computer Science and Engineering, Sikkim Manipal Institute of Technology, Sikkim Manipal University, Majitar, Sikkim, India
vaibhavbhajanka@gmail.com

Abstract. Blockchain has enabled individuals to communicate and make transactions without a central authority in a secure manner. It leverages concepts such as decentralization, immutability, and consensus to achieve the same. A blockchain is a chain of blocks linked through hashes. It is a distributed ledger where each node on the network has a copy of the blockchain. The consensus mechanism allows nodes to verify the validity of a block and decide on the next block for the chain. Smart Contracts have become the most emerging blockchain application. It is a small piece of executable code deployed on the blockchain. A contract gets triggered after a particular event/condition has been fulfilled. This trigger is generated either due to an invocation from a transaction or an explicit call. Since these contracts involve crypto transactions, they are susceptible to various kinds of attacks and exposed to a multitude of security vulnerabilities. This research paper focuses on the study of different categories of vulnerabilities in the blockchain such as Denial of Service with Failed Call, Randomness using 'Block Hash', Mishandled Exception, Immutable Bugs/mistakes, Transaction Ordering Dependency, etc. The paper then discusses various preventive measures and detection tools for the mentioned vulnerabilities.

Keywords: Blockchain · consensus · smart contracts · Denial of Service · Transaction Ordering Dependency

1 Introduction

Blockchain enables peer to peer transfer of digital assets [32] without any intermediaries. The term blockchain is straightforward, meaning a chain of blocks. The first block in the chain is generally called the genesis block/state [6]. Each block has its block hash, which is randomly calculated using a hashing algorithm [32]. Each subsequent block is linked to the previous block using the hash of the previous block. The components of a block include - block number, block hash, previous block hash, nonce (number generated only once to vary the hash value), transaction data (using Merkle tree [5, 31] root), and timestamp [32].

The characteristics of blockchain include decentralization [18], immutability [6], transparency, anonymity, and cryptographical security [32]. Blockchain is a distributed ledger where numerous computers (distributed throughout the globe) are connected on

P. Das et al. (Eds.): AMRIT 2023, CCIS 1953, pp. 105–115, 2024.
https://doi.org/10.1007/978-3-031-47224-4_10

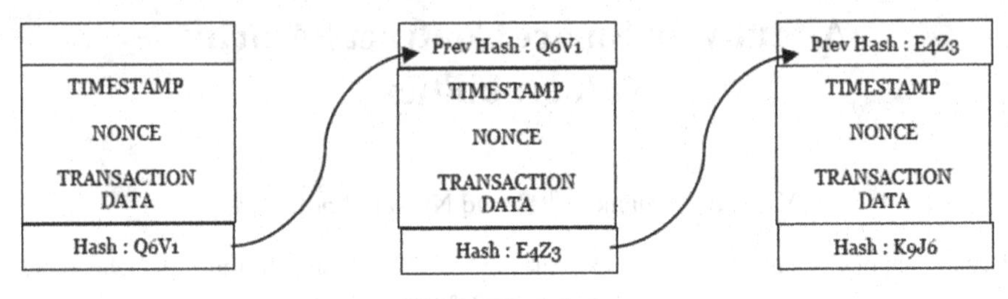

Fig. 1. Blockchain

the blockchain network. Therefore, there is no single central authority that operates the blockchain. Once a block is added to the chain, all verified transactions [32] in that block become irreversible and cannot be undone. All nodes on the network can view transactions made between peers. But the public addresses of these nodes are hashes, therefore no personal data is exposed. All transactions are hashed and stored as a Merkle tree, and blocks are also identified using hashes. Cryptography is implemented for a secure framework.

Centralized organizations consist of entities that can be held accountable in case of any conflict. But the same is not true for decentralized frameworks. To realize trust and security in blockchain, we require a mechanism for validation, verification, and some sense of agreement between peers to decide the to-be added transactions and the next valid block [32].

A consensus(agreement) mechanism [32] is put in place to address the above- mentioned issues. Some consensus mechanisms are Proof of Work [32], Proof of Stake [6], Proof of Burn [22], Proof of Capacity [13], etc. Bitcoin [32] uses Proof of Work. After a transaction has been successfully executed and verified, it is added by miners to their block.

Miners are nodes that compete to add a block to the blockchain. The decision of which block will be added to the chain is made by the consensus mechanism. The first miner to add the next block to the chain is rewarded. A miner uses computational power to solve a mathematical problem to prove to other nodes that it has spent computational power and receives the right to add their block on the blockchain.

Out of the diverse applications of Blockchain, Smart contracts [10] comprise the most powerful applications. They perform the autonomous execution of agreed terms between two untrusted parties [21]. The terms of the agreement between buyer and seller are coded into a smart contract and operated on a dedicated virtual machine [34] embedded in a blockchain. The contract code is executed, as and when the condition of the terms is fulfilled. Ethereum [6] is the most famous and widely used platform for implementing Smart Contracts. Ethereum Smart Contracts consist of - a contact address, executable code, a state consisting of private storage, and a balance in ether [33]. Contract address is used to uniquely identify Smart Contracts. Executable code is the program written by the developer. The private storage stores variables/data for invoking other Smart Contracts. The balance decides what/how many instructions can be executed since each instruction in Ethereum Smart Contract is associated with a gas fee [43]. If the caller function does not provide enough ether, callee Smart Contract will fail to execute [28].

Smart Contracts can be invoked either internally or externally [6]. The unique contact address is used for invocation along with certain parameters such as gas fees and invocation data. Ethereum supports stateful [27] Smart Contracts which means it allows value to stay on the blockchain and be utilized for multiple invocations. For the same reason, they have a wide and versatile range of applications.

Smart Contracts are still being studied and explored since they have such a wide application range. Sometimes Smart Contracts do not behave as they are intended to because there is a semantic gap [27] between the assumption of contract writers regarding the underlying structure and execution of semantics and the actual smart contract's semantic system.

Ethereum Smart Contracts are exposed to various vulnerabilities [25] because of certain features of the Solidity Programming Language, Ethereum Virtual Machine, and design of the Ethereum blockchain. This paper discusses such vulnerabilities and how to tackle/detect them.

2 Motivation

Many vulnerabilities exist in Smart Contracts based on various factors. The primary motivation for this study is to aid students, research enthusiasts, and developers attempting to get a brief idea of the most common and unfavorable vulnerabilities in Smart Contracts. This survey has been planned to be put together after substantial researches, as the existing surveys do not provide detail in such a concise manner. This survey focuses on the security vulnerabilities that may cater to dangerous attacks and can cause harm soon. It briefly discusses the generalized detection and preventive measures to tackle the vulnerabilities.

3 Background

The term blockchain was coined in 2008 after Satoshi Nakamoto introduced Bitcoin to the public. According to the Bitcoin whitepaper penned by Nakamoto, Bitcoin enables peer-to-peer transfer of digital assets without any intermediaries. He also mentioned a "chain of blocks" referring to the blockchain [32]. Blockchain and its many applications have come to light recently, but many of its underlying technologies were being researched upon since 1980s.

Ralph Merkle described his idea of Merkle trees in 1979 and how they could be used in public key distribution and digital signatures [31]. Later, David Chaum(widely recognized as inventor of first digital cash) came up with a concept of a vault system for establishing, maintaining and trusting mutually suspicious groups in 1982 [8]. In 1991, Stuart Haber and W. Scott Stornetta discussed a cryptographically secure chain of blocks and timestamping of digital documents in their article. They later integrated their concept with Merkle Trees [5]. "Smart Contracts" was introduced by Nick Szabo in 1997 [38].

Around the same time other activities took place which propelled the entire Blockchain concept, Napster [7] introduced in 1999. It was used for P2P music sharing. Later, Adam Back used Proof of Work (PoW, consensus mechanism) in hashcash algorithm [4] (provides DoS counter measures to limit email spamming). Hal Finney in 2004, introduced reusable PoW [17]. Apart from PoW, other consensus mechanisms can be used to run a crypto network. Proof of Stake (PoS) allows nodes to stack their cryptocurrency. Nodes with the highest stake validate new blocks. In case of any misdeed by the validator, their stacked crypto is taken away. Proof of Burn allows miners to burn coins (eliminate specific coins from circulation). It prevents inflation and validates new transactions. Other consensus mechanisms include Proof of Capacity, Proof Activity, etc. [13].

Smart Contracts came much later into application when Vitalik Buterin (Founder of Ethereum, initial contributor to Bitcoin codebase) got frustrated with Bitcoin's limitations and launched Ethereum. It allows blocks to record contracts, not only currency. Smart Contracts have enabled developers to create decentralized apps (DApps). The transfer of values between untrusted parties is implemented autonomously without a centralized authority on the condition that the terms of the contract are fulfilled [6]. Smart Contracts save both time and money. They have a wide range of applications as well.

An action triggered by an externally owned account (account managed by a human, not a contract) in Ethereum is called a transaction. Smart contracts can be invoked using transactions. Each transaction is associated with a certain gas fee required for its execution [6]. The gas limit is the amount of gas the user is willing to spend for the transaction. If the gas limit is less than the required gas fee, the transaction will fail with an exception. Gas Price is the price of each gas unit [43].

Miners are rewarded with a transaction fee when they include a transaction in a particular block [6, 32].

$$\text{Transaction Fee} = \text{Amount of gas burned} \times \text{Gas Price} \tag{1}$$

Smart contracts deal with a huge amount of cryptocurrency. Therefore, attackers might want to find loopholes in the smart contract infrastructure. This paper talks about 23 vulnerabilities categorized under three root causes - Solidity Programming Language, features of Ethereum Virtual Machine, and design features of Ethereum Blockchain [25].

4 Related Work

Systematic Review of Security Vulnerabilities in Ethereum Blockchain Smart Contract by S. Singh Kushwaha et al. [25], presents a brief about 23 vulnerabilities of Ethereum-based Smart Contracts categorized under 3 main root causes and 17 sub root causes. It also mentions various attacks that have exploited these vulnerabilities (such as the famous DAO attack where hackers stole 60 million US dollars), multiple detection tools and preventive measures.

Making Smart Contracts Smarter, Proceedings of the 2016 ACM SIGSAC Conference on Computer and Communications Security by L. Luu et al. [27], presents an approach to make Smart Contracts more secured. It discusses various vulnerabilities and missing abstractions in Smart Contracts semantic that lead developers to a false sense of security. The paper proposes refinement in current Ethereum protocols and an upgrade for all clients on the network. Also, it presents a Detection Tool called Oyente for the detection of Smart Contract bugs.

5 Overview of Vulnerabilities

A large number of vulnerabilities occur in the Smart Contracts. This survey focuses on elaborated study of various vulnerabilities while categorizing them into three main classes based on the cause of vulnerability.

- Solidity Programming Language
- Ethereum Virtual Machine
- Design Features of Ethereum Blockchain

5.1 Solidity Programming Language

Solidity [18] is the most prevalently used programming language to code Smart Contracts. An attacker may manipulate the features, variables, syntax, semantics, etc. to leverage a solidity bug/design fault into an attack [25, 27].

Denial of Service with Block Gas Limit. [43] Each block in Ethereum blockchain has a gas limit (upper bound beyond which no gas can be spent) to prevent attackers from creating an infinite transaction loop. If a particular transaction exceeds this limit, Denial of Service takes place, and the transaction fails with an exception. One use case is unknown array size which grows over time for storing transactions.

Denial of Service with Failed Call. [37] This vulnerability occurs when contracts fail accidentally due to programming errors or an attack by an adversary. For instance, when an attacker combines multiple calls in a single transaction and runs a loop, which prevents others from calling that Smart Contract.

Randomness using 'Block Hash'. [3, 30] Block Hash can be used for randomizing operations when needed. Block Hash has high entropy as it is calculated using a hashing algorithm. However, Block Hash is public to all nodes on the network (miners as well) so if a dishonest miner is aware of the same, they can manipulate the Block Hash for personal gains.

'tx.origin' Usage. [30] 'tx.origin' is a global variable in solidity programming language which returns the address of the external account which initiated the transaction and not the immediate sender. Contracts using this variable are vulnerable to phishing (people can imitate the transaction initiator) attacks.

Integer Overflow/ Underflow. [11, 19] Overflow occurs when a calculation value exceeds the upper range of variable size. Similarly, underflow occurs when this value declines below the lower range of variable size. When a particular data does not fit in the size range of Integer data type, it is termed Integer overflow/ underflow. Consequently, the given data cannot be represented using Integer data type.

Re-Entrancy. [26, 35, 36] It is also called recursive call attack. An adversarial contract calls into the callee contract before its first invocation finishes (re-entrance into the calling contract). A malicious contract can exploit this vulnerability to call another contract to withdraw infinitely without affecting its own balance.

Mishandled Exception. [27] A contract may throw various kinds of exceptions. If a callee contract raises an error during execution and the same is not reported back to the caller contract, the caller contract may not behave as intended. Such situations become a threat.

Gasless 'Send'. [2, 43] When the send function is used to transfer ether from one contract to another, the fallback function (to ensure the program does not throw an error) is invoked. Usually, the gas limit for the fallback function is 2300 units if the amount sent is non-zero. If the target contract has an expensive fallback function, an out-of-gas exception is returned. If such an exception is not handled, the callee contract can keep the non-transferred amount(ether) wrongfully.

Gas Costly Pattern. [2, 41] Programs/Codes in Smart Contracts which are not optimized [9] properly may contain unnecessary and irrelevant lines of code. The code may be following an expensive pattern which drives up the execution cost.

Call to the Unknown. [3] When a particular contract calls a function in another contract without providing the correct function signature or enough data for the callee contract function execution, the fallback function is called. There are primitives (send() function, Delegate Call) in the Solidity language which may also have the side effect of invoking the fallback function. Attackers may embed malicious code into the fallback function.

The DAO (Decentralized Autonomous Organization) attack was accomplished by exploiting the same vulnerability [29].

Hash Collision with Multiple Variable Length Arguments. [25] Application Binary Interface [12] (ABI) is used in the Ethereum ecosystem to standardize the interaction for Smart Contracts, both from outside the blockchain and from contract to contract. It provides the specification for encoding and decoding data.

abi.encodePacked(...) allows a non-standard encoding mode of data in raw bytes. This function drops conventions specified by ABI. It encodes parameters in the same order irrespective of their position in the parameter list. For example, both [ab,c] and [a,bc] will be packed as [abc]. Attackers can change the position of elements in signature verification and unethically gain access.

Insufficient Gas Griefing. [42] It is implemented in attacks that prevent transactions from being performed as intended. Contracts may use data from a caller contract to make a sub-call to another contract. This vulnerability can be exploited if a sub-call [28] fails, and the entire transaction is reverted. An attacker can deliberately fail transactions [16] by providing just enough gas for transaction execution and not for the sub-call.

Unprotected Ether Withdrawal. [25] When access control is missing or inadequate, attackers can exploit this vulnerability to withdraw ether unethically. For example, Constructors are special functions used to perform critical operations during the initialization of contracts. During development, if the contract name gets changed, the constructor is exposed to malicious users as it starts behaving like any other callable function.

Floating Pragma. [20] An Ethereum Smart Contract compiled using an outdated compiler can lead to bugs and errors in code. The compiler version should be checked in the original code before the deployment of the contract. Pragma version should be locked before deployment.

Function's Default Visibility. [3, 42] Solidity programming language has four function visibility specifiers for smart contract functions - external, internal, public, and private. By default, all functions have public visibility that allows users to call this function externally. Often developers tend to forget that functions are public by default and ignore changes that should be made to functions for privileged access.

DELEGATE CALL. [24] ABI of the target function (in another contract) is required to call the target function. If ABI is not available, we can use a special variant of message call known as DELEGATE CALL. A delegate-call from contract A delegates any other contract (say contract B) to modify the storage of the calling contract (Contract A). Such context-preserving nature of DELEGATE-CALL can be harmful, and lead to vulnerabilities and unexpected code execution.

Unprotected 'Self-Destruct()'. [24] Self-destruct opcode removes all code from a contract and sends the balance ether to a predefined address. If this address points to another contract, no functions (in the called contract) get called, not even the fallback function. If a malicious user gets access to such a contract, they can use selfdestruct() function to send ether to any target address.

5.2 Ethereum Virtual Machine

Ethereum Virtual Machine (EVM) helps developers deploy the Smart Contracts on the Ethereum Network. Some features of the EVM make it possible for attackers to extend a vulnerability into an attack.

Immutable Bugs/Mistakes. [3, 42] Immutability stands out to be an important feature of Ethereum Blockchain that does not allow developers to change data/transaction details in a Smart Contract once it has been deployed on the EVM. If bugs are present in the code, they cannot be debugged and may become an exploitable vulnerability.

Ether Lost in Transfer. [14] Ether transfer from one contract to the other is carried out using the recipient contract address. If the ether is transferred to an orphan address (that does not belong to any user or contract), the ether is lost forever as it cannot be recovered. There is no way to check if an address is an orphan or not. The recipient address should be checked manually to avoid loss.

5.3 Design Features of Ethereum Blockchain

Ethereum Blockchain is designed such that nodes can make transactions using the Ethereum currency, ether. Ethereum is programmable, meaning nodes can deploy Smart Contracts and build decentralized applications. Certain features of Ethereum Blockchain such as block components and semantics can be altered by adversaries for unethical purposes.

Timestamp Dependency. [1, 42] Block timestamp is the time at which a particular block is generated. Miners get a window during which they collect all transactions and add them to the block. Next, the block is added to the chain. If a dishonest miner collects all transactions, they can manipulate (delay/prepone) the timestamp, provided the block timestamp is set as a triggering condition for a particular transaction.

Lack of Transactional Privacy. [1, 23] Ethereum Blockchain is anonymous yet transparent. Nodes are unaware of the actual identity of other nodes but any node on the network can view transactions made from one node to the other. Transaction details are available publicly. These details can be used unethically by attackers.

Transaction Ordering Dependency. [15] The order of transaction execution is an important aspect. When two or more dependent transactions call the same Smart Contract, they are to be executed in proper order. Failure to maintain the order may lead to unexpected behavior of the smart contract. An unethical miner has an accessibility to manipulate the order allowing attacks.

Untrustworthy Data Feeds. [44] Many smart contracts tend to rely on external information/data to perform certain functions. If the external agent is corrupt, they might intentionally send wrong information to fail the Smart Contract leading to a disastrous situation.

6 Discussion

The above-mentioned vulnerabilities raise serious security concerns as Smart Contracts involve transactions and are associated with digital currency, exchange of values and other applications. Users would not want to get compromised by any means. Detection and preventive measures need to be put in place to keep a check on these vulnerabilities for a safe and secure environment.

Various detection tools are available in the market that can identify the major vulnerabilities. SmartCheck [39], Oyente [27], Remix [15], Securify [40] are some of these tools. Oyente is a symbolic execution tool that helps to find security bugs in Smart Contracts. It flags about half of the existing Ethereum Smart Contracts as vulnerable. It has also been able to detect the famous DAO bug. It takes input in the form of EVM byte code. Oyente also helps in writing better contracts for developers and avoid calling risky contracts.

Developers should be cautious while writing Smart Contracts and adhere to proper semantics [27] and design to ensure there is no leakage of confidential information. Functions inside contracts must be supplied with required parameters and fees for proper execution. Exception Handling is an important factor that needs to be taken care of.

Ethereum is still expanding, and more security flaws may come up. Therefore, it is essential that existing vulnerabilities and faulty design are understood and resolved for scaling.

7 Conclusion

The survey provides a brief overview of the major vulnerabilities in Smart Contracts that are prone to security attacks. These vulnerabilities are classified under three root causes - Solidity Programming Language, Features of Ethereum Virtual Machine, and Design features of Ethereum Blockchain. It also discusses the fundamentals of blockchain and the consequent emergence of Smart Contracts. It also provided general detection and analysis methods with certain preventive measures to curtail vulnerabilities and contain attacks. Developers and users of Smart Contracts must be aware of these risks as digital assets and currency are involved.

References

1. Alharby, M., Van Moorsel, A.: Blockchain-based smart contracts: a systematic mapping study. arXiv preprint arXiv:1710.06372 (2017). https://doi.org/10.48550/arXiv.1710.06372
2. Ashraf, I., Ma, X., Jiang, B., Chan, W.K.: GasFuzzer: fuzzing ethereum smart contract binaries to expose gas-oriented exception security vulnerabilities. IEEE Access **8**, 99552–99564 (2020). https://doi.org/10.1109/ACCESS.2020.2995183
3. Atzei, N., Bartoletti, M., Cimoli, T.: A survey of attacks on ethereum smart contracts (sok). In: International Conference on Principles of Security and Trust, pp. 164–186. Springer, Berlin, Heidelberg (2017). https://doi.org/10.1007/978-3-662-54455-6_8
4. Back, A.: Hashcash – a denial of service counter-measure. https://www.hashcash.org/papers/hashcash.pdf
5. Bayer, D., Haber, S., Stornetta, W.S.: Improving the efficiency and reliability of digital time-stamping. In: Sequences II, pp. 329–334. Springer, New York, NY (1993). https://doi.org/10.1007/978-1-4613-9323-8_24
6. Buterin, V.: Ethereum white paper: a next-generation Smart Contract & decentralized application platform. https://ethereum.org/en/whitepaper/
7. Carlsson, B., Gustavsson, R.: The rise and fall of napster - an evolutionary approach. In: Liu, J., Yuen, P.C., Li, Ch., Ng, J., Ishida, T. (eds.) Active Media Technology. AMT 2001. LNCS, vol. 2252. Springer, Berlin, Heidelberg (2001). https://doi.org/10.1007/3-540-45336-9_40
8. Chaum, D.L.: Computer Systems Established, Maintained and Trusted by Mutually Suspicious Groups. University of California, Electronics Research Laboratory (1979)
9. Chen, T., Li, X., Luo, X., Zhang, X.: Under-optimized smart contracts devour your money. In: 2017 IEEE 24th International Conference on Software Analysis, Evolution and Reengineering (SANER), pp. 442–446. IEEE (2017). https://doi.org/10.1109/SANER.2017.7884650
10. Clack, C.D., Bakshi, V. A., Braine, L.: Smart contract templates: foundations, design landscape and research directions. arXiv preprint arXiv:1608.00771 (2016). https://doi.org/10.48550/arXiv.1608.00771
11. Common Vulnerabilities and Exposures. CVE-2018-10299 (2018). https://nvd.nist.gov/vuln/detail/CVE-2018-10299
12. Contract ABI Specification. https://docs.soliditylang.org/en/v0.8.11/abi-spec.html
13. Debus, J.: Consensus methods in blockchain systems. Frankfurt School of Finance & Management, Blockchain Center, Technical Report (2017)
14. Demir, M., Alalfi, M., Turetken, O., Ferworn, A.: Security smells in smart contracts. In: 2019 IEEE 19th International Conference on Software Quality, Reliability and Security Companion (QRS-C), pp. 442–449. IEEE (2019). https://doi.org/10.1109/QRS-C.2019.00086

15. Di Angelo, M., Salzer, G.: A survey of tools for analyzing ethereum smart contracts. In: 2019 IEEE International Conference on Decentralized Applications and Infrastructures (DAPPCON), pp. 69–78. IEEE (2019). https://doi.org/10.1109/DAPPCON.2019.00018
16. Eskandari, S., Moosavi, S., Clark, J.: SoK: transparent dishonesty: front-running attacks on Blockchain. In: 3rd Workshop on Trusted Smart Contracts (WTSC) (2019). https://doi.org/10.1007/978-3-030-43725-1_13
17. Finney, H.: Rpow-reusable proofs of work (2004). https://nakamotoinstitute.org/finney/rpow/index.html
18. Frantz, C.K., Nowostawski, M.: From institutions to code: towards automated generation of smart contracts. In: 2016 IEEE 1st International Workshops on Foundations and Applications of Self* Systems (FAS* W), pp. 210–215. IEEE (2016). https://doi.org/10.1109/FAS-W.2016.53
19. Gao, J., Liu, H., Liu, C., Li, Q., Guan, Z., Chen, Z.: Easyflow: keep ethereum away from overflow. In: 2019 IEEE/ACM 41st International Conference on Software Engineering: Companion Proceedings (ICSE-Companion), pp. 23–26. IEEE (2019). https://doi.org/10.1109/ICSE-Companion.2019.00029
20. Gupta, B.C., Shukla, S.K.: A study of inequality in the ethereum smart contract ecosystem. In: 2019 Sixth International Conference on Internet of Things: Systems, Management and Security (IOTSMS), pp. 441–449. IEEE (2019). https://doi.org/10.1109/IOTSMS48152.2019.8939257
21. Huang, Y., Bian, Y., Li, R., Zhao, J.L., Shi, P.: Smart contract security: a software lifecycle perspective. IEEE Access **7**, 150184–150202 (2019). https://doi.org/10.1109/ACCESS.2019.2946988
22. Karantias, K., Kiayias, A., Zindros, D.: Proof-of-burn. In: Bonneau, J., Heninger, N. (eds.) Financial Cryptography and Data Security. FC 2020. LNCS, vol. 12059. Springer, Cham (2020). https://doi.org/10.1007/978-3-030-51280-4_28
23. Kosba, A., Miller, A., Shi, E., Wen, Z., Papamanthou, C.: Hawk: the blockchain model of cryptography and privacy-preserving smart contracts. In: 2016 IEEE Symposium on Security and priVacy (SP), pp. 839–858. IEEE (2016). https://doi.org/10.1109/SP.2016.55
24. Krupp, J., Rossow, C.: {teEther}: Gnawing at ethereum to automatically exploit smart contracts. In: 27th USENIX Security Symposium (USENIX Security 18), pp. 1317–1333 (2018)
25. Kushwaha, S.S., Joshi, S., Singh, D., Kaur, M., Lee, H.N.: Systematic review of security vulnerabilities in ethereum blockchain smart contract. IEEE Access 6605–6621 (2022). https://doi.org/10.1109/ACCESS.2021.3140091
26. Liu, C., et al.: Reguard: finding reentrancy bugs in smart contracts. In: 2018 IEEE/ACM 40th International Conference on Software Engineering: Companion (ICSE-Companion), pp. 65–68. IEEE (2018). https://doi.org/10.1145/3183440.3183495
27. Luu, L., Chu, D.H., Olickel, H., Saxena, P., Hobor, A.: Making smart contracts smarter. In: Proceedings of the 2016 ACM SIGSAC Conference on Computer and Communications Security, pp. 254–269 (2016). https://doi.org/10.1145/2976749.2978309
28. Marchesi, L., Marchesi, M., Destefanis, G., Barabino, G., Tigano, D.: Design patterns for gas optimization in ethereum. In: 2020 IEEE International Workshop on Blockchain Oriented Software Engineering (IWBOSE), pp. 9–15. IEEE (2020). https://doi.org/10.1109/IWBOSE50093.2020.9050163
29. Mehar, M.I., et al.: Understanding a revolutionary and flawed grand experiment in blockchain: the DAO attack. J. Cases Inform. Technol. **21**(1), 19–32 (2019). https://doi.org/10.4018/JCIT.2019010102
30. Mense, A., Flatscher, M.: Security vulnerabilities in ethereum smart contracts. In: Proceedings of the 20th International Conference on Information Integration and Web-Based Applications and Services, pp. 375–380 (2018). https://doi.org/10.1145/3282373.3282419

31. Merkle, R.C.: Secrecy, Authentication, and Public Key Systems. Stanford university (1979)
32. Nakamoto, S.: Bitcoin: a peer-to-peer electronic cash system [White Paper] (2008). https://bitcoin.org/bitcoin.pdf
33. Oliva, G.A., Hassan, A.E., Jiang, Z.M.J.: An exploratory study of smart contracts in the Ethereum blockchain platform. Empir. Softw. Eng. **25**(3), 1864–1904 (2020). https://doi.org/10.1007/s10664-019-09796-5
34. Pinna, A., Ibba, S., Baralla, G., Tonelli, R., Marchesi, M.: A massive analysis of ethereum smart contracts empirical study and code metrics. IEEE Access **7**, 78194–78213 (2019). https://doi.org/10.1109/ACCESS.2019.2921936
35. Rodler, M., Li, W., Karame, G. O., Davi, L.: Sereum: Protecting existing smart contracts against re-entrancy attacks. arXiv preprint arXiv:1812.05934 (2018). https://doi.org/10.48550/arXiv.1812.05934
36. Samreen, N.F., Alalfi, M.H.: Reentrancy vulnerability identification in ethereum smart contracts. In: 2020 IEEE International Workshop on Blockchain Oriented Software Engineering (IWBOSE), pp. 22–29. IEEE (2020). https://doi.org/10.1109/IWBOSE50093.2020.9050260
37. Solmaz, O.: The Anatomy of a Block Stuffing Attack (2018). https://solmaz.io/2018/10/18/anatomy-block-stuffing/
38. Szabo, N.: Formalizing and securing relationships on public networks. First Monday (1997). https://doi.org/10.5210/fm.v2i9.548
39. Tikhomirov, S., Voskresenskaya, E., Ivanitskiy, I., Takhaviev, R., Marchenko, E., Alexandrov, Y.: Smartcheck: static analysis of ethereum smart contracts. In: Proceedings of the 1st International Workshop on Emerging Trends in Software Engineering for Blockchain, pp. 9–16 (2018). https://doi.org/10.1145/3194113.3194115
40. Tsankov, P., Dan, A., Drachsler-Cohen, D., Gervais, A., Buenzli, F., Vechev, M.: Securify: practical security analysis of smart contracts. In: Proceedings of the 2018 ACM SIGSAC Conference on Computer and Communications Security, pp. 67–82, (2018). https://doi.org/10.1145/3243734.3243780
41. Wang, X., Wu, H., Sun, W., Zhao, Y.: Towards generating cost-effective test-suite for Ethereum smart contract. In: 2019 IEEE 26th International Conference on Software Analysis, Evolution and Reengineering (SANER), pp. 549–553. IEEE (2019). https://doi.org/10.1109/SANER.2019.8668020
42. Wöhrer, M., Zdun, U.: Design patterns for smart contracts in the ethereum ecosystem. In: 2018 IEEE International Conference on Internet of Things (iThings) and IEEE Green Computing and Communications (GreenCom) and IEEE Cyber, Physical and Social Computing (CPSCom) and IEEE Smart Data (SmartData), pp. 1513–1520. IEEE (2018). https://doi.org/10.1109/Cybermatics_2018.2018.00255
43. Yang, R., Murray, T., Rimba, P., Parampalli, U.: Empirically analyzing ethereum's gas mechanism. In: 2019 IEEE European Symposium on Security and Privacy Workshops (EuroS&PW), pp. 310–319. IEEE (2019). https://doi.org/10.1109/EuroSPW.2019.00041
44. Zhang, F., Cecchetti, E., Croman, K., Juels, A., Shi, E.: Town crier: an authenticated data feed for smart contracts. In: Proceedings of the 2016 aCM sIGSAC Conference on Computer and Communications Security, pp. 270–282 (2016). https://doi.org/10.1145/2976749.2978326

A GUI-Based Study of Weather Prediction Using Machine Learning Algorithms

Debdeep Mukherjee, Rupak Parui, and Lopamudra Dey(✉)

Department of Computer Science and Engineering, Heritage Institute of Technology, Kolkata, India

{debdeep.mukherjee.cse22,rupak.parui.cse22}@heritageit.edu.in, lopamudra.dey@heritageit.edu

Abstract. The current climate is changing dramatically, making it more difficult and less accurate to anticipate the weather using traditional approaches. Improved and trustworthy weather forecast techniques are needed to get beyond these obstacles. Machine learning has tremendous influence in the modern-day world. Deep Learning and machine learning-based forecast systems can predict general weather patterns as well as numerical weather prediction models while using only a fraction of the computing power the models require. These models are developed in order to forecast weather variables such as solar radiation, temperature, and wind speed one to 24 h in advance. We have used supervised machine learning algorithms to predict the weather condition depending upon some parameters like temperature, humidity, precipitation, and wind speed. We have pre-processed the datasets and then after cross-validation, selected the most appropriate algorithm to predict the weather. We also developed a user interface to display the predicted results. The codes and website are available at https://github.com/98rupak/weather_project.

Keywords: Weather Prediction · Machine Learning · Class-imbalance · Accuracy · SVM · KNN

1 Introduction

The remarkable improvement in the quality of weather forecasts is one of the great successes of environmental science in the 20th century, which continues at a sustained pace at the beginning of the 21st century. This is due to the progress of numerical prediction systems and the increasing number and variety of observations of the state of the atmosphere and related media, including observations from Earth observation satellites. The rapid development of supercomputers has been one of the keys to this success, which has also required significant scientific work. Each country in the world has a National Meteorological Service (NMS), whose mission is to make regular observations of the atmosphere and to issue forecasts for government, industry, and the public. But only the most advanced countries have Numerical Weather Prediction (NWP) centres, whose products are also distributed to other countries, in exchange for their observations. There are mainly three types of weather forecasts that are short-range forecasts, medium-range,

P. Das et al. (Eds.): AMRIT 2023, CCIS 1953, pp. 116–126, 2024.
https://doi.org/10.1007/978-3-031-47224-4_11

and extended-range forecasts. The important process of the weather is built into the prediction model. The model starts with the present weather and tries to simulate how the weather will develop in the future.

In this topic there are several works done by various researchers. The relation between temperature and latitude, which are considered as prime important factors for weather forecasting have been discussed in [1]. Prediction of heavy rain damage with the help of big data and other machine learning algorithms which maximized the performance of the model [2]. Machine learning models can predict weather features accurately enough to compete with traditional models, utilizing the historical data from surrounding areas to predict weather of a particular area [3]. Date, Minimum Temperature, Humidity and Wind Direction can be considered as predictors for the rainfall and also the correlation coefficient can give a chance of rainfall if the value of the correlation coefficient is close to 1 [4]. Intelligent models can be created using machine learning technologies that are much simpler than conventional physical models. They use less resources and can be run on almost any machine, including mobile devices [5]. In [6] Weyn et al. proposed a seasonal forecasting utilizing a revolutionary method of weather prediction. The authors created a machine learning weather prediction system dubbed Deep Learning Weather Prediction using a convolution neural network (DLWP). In contrast to conventional numerical weather prediction models, which produce mathematical representations of physical laws, the model is trained on historical weather data. For the entire world, DLWP makes predictions for the next 2–6 weeks. A portable, low-cost weather forecasting system is developed using random forest categorization in [7]. In [8], to create models that enable precise weather forecasts, this research attempted to use historical weather observation data and machine learning (ML) approaches. The results of applying various model settings were compared and examined in terms of training expense and prediction accuracy.

In this paper, four machine learning algorithms, SVM, KNN, Random forest and Naive Bayes classifier have been considered which have been run in three processed datasets. Then the most suitable dataset has been chosen for developing an user interface which predicts the weather when the parameters of the weather are fed to it. We have build a user interface using python and we have seen that the model works accurately on real-life datasets and it predicts the weather of the given location of the earth when the weather parameters are entered as input.

2 Proposed Methodology

2.1 Datasets

2.1.1 Finding Datasets

The first step in the work is to find the dataset to work on. Since all the algorithms are a part of supervised form of machine learning, hence a dataset is required for training and testing purpose. Therefore, various sites of the internet are surfed like kaggle, github, etc. and various datasets are collected.

2.1.2 Comparing Datasets

The various datasets that we have got might not be according to our need. They might be containing stray columns and the dataset might be either insufficient, may contain various excess, unnecessary columns or in rare cases may be meeting the required conditions appropriately. So, we need to compare the dataset obtained from the various domains.

2.1.3 Choosing Appropriate Dataset

After comparing the chunks of datasets from the various domains on the internet, we need to filter out the unnecessary datasets, and select the appropriate datasets that we will be working on. These datasets contain higher number and variety of data for training and testing the model. In the below figure (Table 1) the chosen datasets have been shown in red and bold.

Table 1. The datasets considered in this study with links and comments.

Dataset Name	Link	Comments
weather	https://www.kaggle.com/ananthr1/weather-prediction	contains string values and Rain Tomorrow column is biased towards 0
training_data_with_weather_info_week1	https://www.kaggle.com/davidbnn92/weather-data/data	contains a lot of null values, some columns have string values
forecast_data	https://www.kaggle.com/chitwanmanchanda/weather-api-data	incomplete data and contains unimpactful coloumns
Weather Data in India from 1901 to 2017	https://www.kaggle.com/mahendran1/weather-data-in-india-from-1901-to-2017	less no. of data, only month wise avg temp given
Temp_and_rain	https://www.kaggle.com/yakinrubaiat/bangladesh-weather-dataset	only month wise avg temp and total rain is given for every year
weatherHistory	https://www.kaggle.com/muthuj7/weather-dataset	contains string values and one column(Loud Cover)with all values as 0
austin_weather	https://www.kaggle.com/grubenm/austin-weather	Lots of null values and Precipitation column contain string as well as integer in different rows
seattle-weather	https://www.kaggle.com/ananthr1/weather-prediction	contains string values in weather column

2.1.4 Data Preprocessing

We have performed the following pre-processing steps on the dataset before applying machine learning algorithms.

a. String data: We mapped the string data to corresponding numerical data by assigning integer values to all the string values.
b. Missing values: There were many missing values in the dataset. We filled those missing values with the average values of the corresponding column data.
c. NaN values: Some integer columns contained some string values that were shown as errors. We changed those values by replacing those values with the average value of the column.
d. Outliers: Some values were there in some parameter which were not feasible for that particular column. So those rows were deleted.
e. Noisy data: There were some rows where the all the values of the parameters were same but they belonged to two different classes.

f. Unbalanced classes: There were classes that were biased towards one class and hence there was error in the accuracy of the various algorithm in that dataset. It is the problem in machine learning where the total number of a class of data (positive) is far less than the total number of another class of data (negative). This problem is extremely common in practice and can be observed in various disciplines including fraud detection, anomaly detection, medical diagnosis, oil spillage detection, facial recognition, etc.

To chalk out the imbalanced class problem we have used Re-sampling method. This technique is used to up sample or down sample the minority or majority class. When we are using an imbalanced dataset, we can oversample the minority class using replacement. This technique is called oversampling. Similarly, we can randomly delete rows from the majority class to match them with the minority class which is called under sampling. After sampling the data we can get a balanced dataset for both majority and minority classes. Therefore, when both classes have a similar number of records present in the dataset, we can assume that the classifier will give equal importance to both classes.

g. Pre-processing: Now the dataset selected above needs to be pre-processed. There were various columns for the class label. So, the classes have been mapped to a common class and the number of classes are reduced by re-mapping those mapped classes to some common group classes. After the various preprocessing steps, the modified datasets are ready to be incorporated within the algorithm. The various modifications done for each of the datasets have been mentioned in the middle column (Fig. 3) (Fig. 1).

ACTUAL DATA SETS	MODIFICATIONS	MODIFIED DATA SET
seattle-weather	weather types like sun, drizzle, rain have been mapped to integers like 0,1,2	seattle-weather
weatherHistory	Columns that contain string values have been mapped to integer values	weather_data
weather	strings values have been mapped to integers, RainTomorrow column is made unbiased	predict

Fig. 1. It shows the modified datasets we have used after preprocessing to further analysis.

2.2 Select Algorithm

After modifying the datasets, now we need to select an algorithm where we will divide the dataset into training and testing data, train the algorithm with the training data and then test the algorithm by feeding the testing data. Thereby, we will compare the results given by the model with the actual data of the dataset and hence we will be able to calculate the efficiency of the model.

In the work there are four algorithms for each of the datasets:

1. Knn
2. Random Forest.
3. Support Vector Machine.
4. Naïve-Bayes Classifier.

A flowchart of the above proposed methodology is shown in Fig. 2.

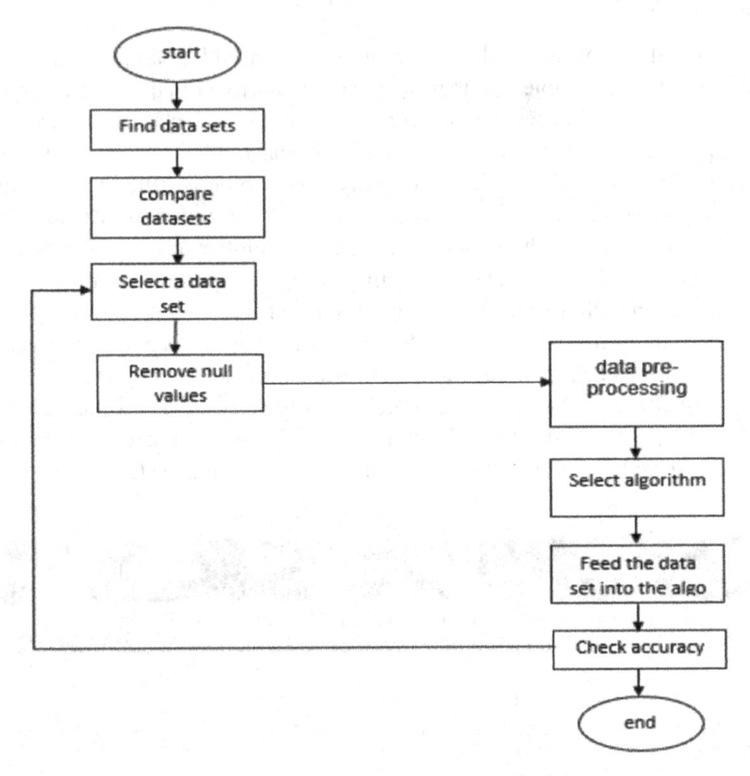

Fig. 2. Flowchart of the proposed methodology

3 Results and Analysis

3.1 Results on Various Datasets

On training each of the selected dataset with all the selected algorithms, an accuracy chart is created for each of the dataset and the various observations are made for each of the dataset for the behavior of various algorithms. The accuracy chart for each dataset is provided below.

In the weather.csv dataset (Fig. 3), we observe that Random forest algorithm gives the highest accuracy and the lowest accuracy is given by the Naïve-Bayes classifier. KNN and SVM almost giving the same accuracy in this dataset.

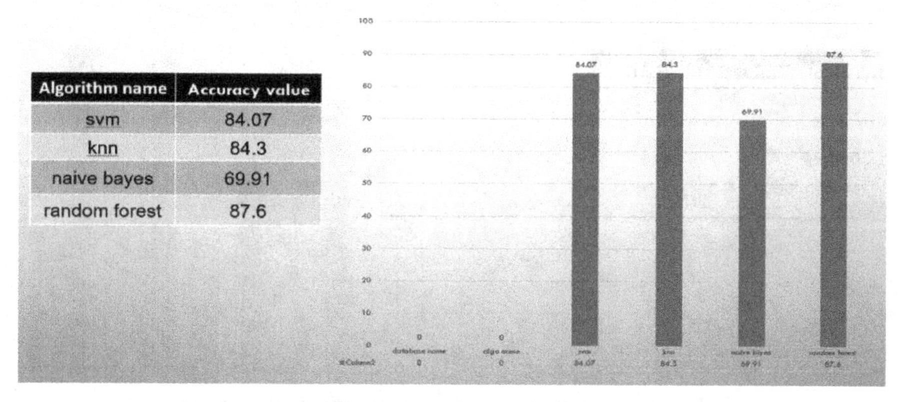

Algorithm name	Accuracy value
svm	84.07
knn	84.3
naive bayes	69.91
random forest	87.6

Fig. 3. Accuracy Comparison on weather_data.csv dataset

Database name : seattle-weather.csv

Algorithm name	Accuracy value
svm	85.19
knn	77.9
naïve bayes	86.56
random forest	83.14

Fig. 4. Accuracy Comparison on seattle-weather.csv dataset

Now there is a difference in the prediction accuracy for the seattle-weather.csv dataset (Fig. 4). Here we are observing a sudden rise in accuracy of the Naïve-Bayes classifier algorithm. The minimum accuracy is given by KNN algorithm.

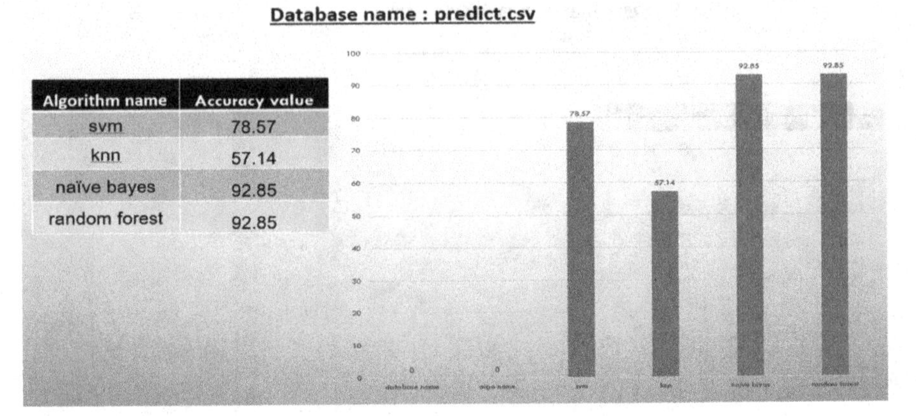

Algorithm name	Accuracy value
svm	78.57
knn	57.14
naïve bayes	92.85
random forest	92.85

Fig. 5. Accuracy Comparison on predict.csv dataset

In this predict.csv (contains 2 classes) dataset interestingly Naïve-Bayes Classifier and Random Forest Classifier gives the same and highest accuracy whereas KNN gives the lowest accuracy of all of them almost close to 50% (Fig. 5). This dataset was a binary classification dataset and in this dataset there were only 44 rows of data hence we see that Naïve-Bayes in giving such a high accuracy and also Naïve Bayes works well on binary dataset.

The other datasets were multi class datasets where the binary codes of the aforesaid algorithms have been implemented. Random forest is showing consistent good accuracy in all the datasets. KNN is giving the lowest accuracy among all the algorithms in seattle-weather.csv and predict.csv datasets which may be taken as an inference that KNN might not work as an efficient algorithm in this weather prediction purpose. Naïve-Bayes classifier is showing fluctuating accuracies in the datasets. SVM, among all the algorithms is showing almost same accuracies in all the datasets may it be multiclass or binary.

3.2 Multiclass Datasets Analysis

The three datasets viz. Seattle-weather.csv, predict.csv and weather-data.csv contains multiple classes and therefore their results were studied using various classes. Now the predict.csv dataset originally contained binary class so it was unchanged. The seattle-weather.csv dataset was separated into 2 class and 5 class datasets. The weather-data.csv dataset was separated into 2 class, 5 class and 11 class datasets and their accuracy was studied by feeding into the algorithms. Here we considered KNN and Random Forest for this purpose.

On feeding the algorithms with this datasets we obtained the following accuracies (Fig. 6). It can be seen that accuracies are decreasing with increase of number of classes. However, Random forest is giving better accuracies compared to other algorithms.

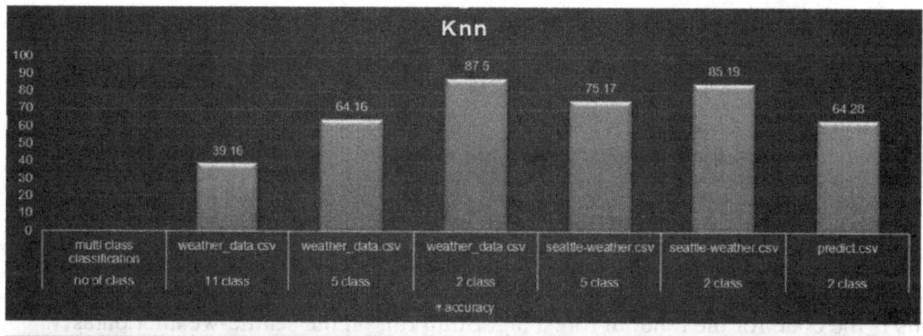

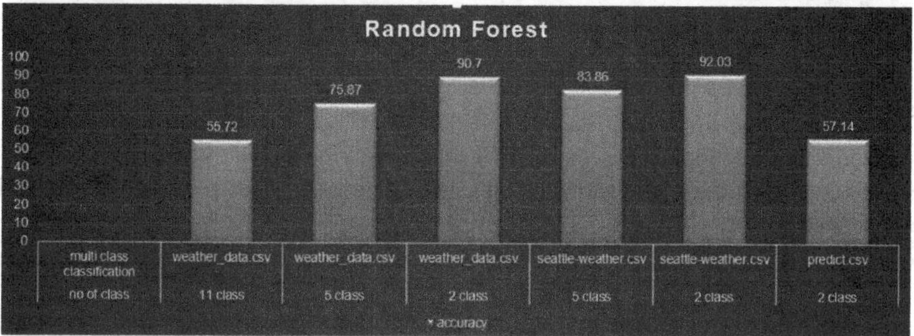

Fig. 6. Accuracy Comparison considering different classes on various datasets

4 Developing a GUI

A user interface was developed using front end technology and the full stack environment was done in a software called Geany. The algorithm chosen for the UI was Random forest algorithm as it is giving better accuracies comparing to other algorithms. In the backend we have used seattle-weather.csv file with 5 classes as training dataset. The other python files that were incorporated at the backend in Geany are DbPop.py and mycsvlibrary.py.

DbPop.Py

This file mainly imports two functions from mycsvlibrary.py. The index function returns the code of the home page that is written in the file edit_task.tpl. The login function returns input to the main page. Page transfer is done by the 'post' command. The do_login()

takes values of different text boxes as input. Bottle that is used in this program provides the localhost environment.

Mycsvlibrary.Py

This file holds the definition of both the functions called by DbPop.py. The Popucsv function holds the definition of file writing. The start_learning function returns the accuracy of random forest algorithm into the DbPop.py file from where it was called.

Randomforest.Py

This is the code for the random forest algorithm run on the seattle-weather dataset.

The other tpl files contains the design of the elements of the UI.

Output of the UI

We have used real-life parameters as test dataset in our website and the weather is predicted using Random forest classifier. A snapshot of the GUI is shown in Fig. 7. User needs to enter the parameters and click on submit button (a). Then predicted output will be generated (b).

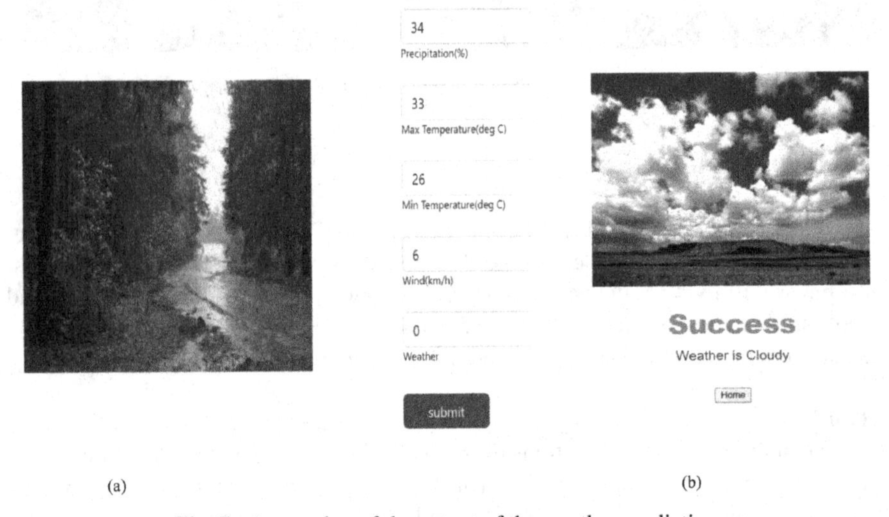

Fig. 7. A snapshot of the output of the weather prediction

This is the output of the weather prediction is based on real life parameters of the location: Jalpaiguri, West Bengal. Finally, we have compared the predicted results with the weather available in Google dated 12th March, 2022on those weather parameters (Fig. 8). It can be noted that our result (cloudy) matches with Google result.

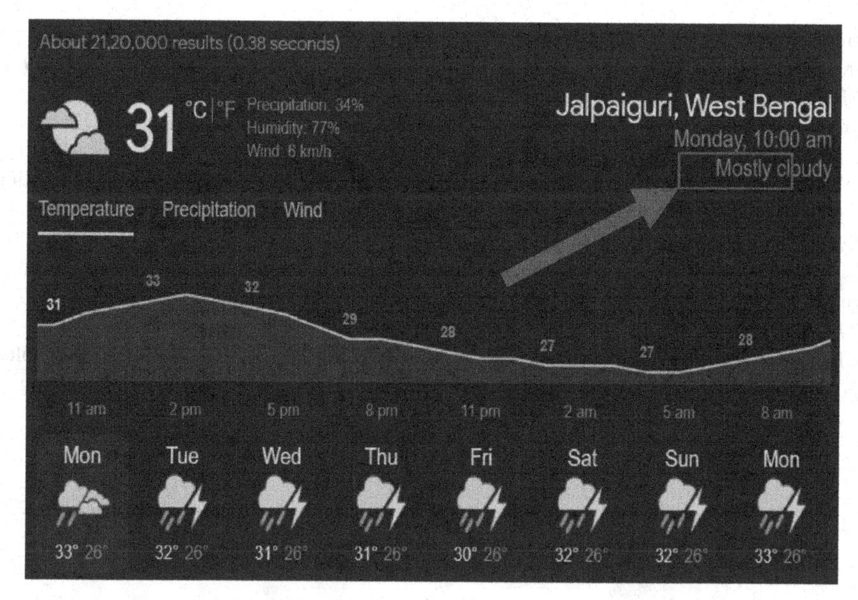

Fig. 8. Actual weather available in Google (12th March, 2022)

5 Conclusion

Weather prediction using Machine learning involved dealing with numerous datasets and training those datasets with various Machine Learning algorithms. The datasets were made suitable to be used for training the algorithms. From the ancient times people have been trying to predict the weather and in the present generation there are devices and gadgets with the help of which we could predict the weather appropriately. In spite of all these, there remains scope of improvement in this field though the accuracy and usefulness of weather prediction have spread across the economy and it is helping people in fighting with the natural calamities like thunderstorm, floods, drought, cloud bursts etc. We have also observed that when the number of classes in the labelled dataset increases, the accuracy decreases and vice-versa. Hence there is an inverse relationship between the number of classes and the accuracy of the algorithm.

References

1. Agarwal, A K., Shrimali, M., Saxena, S., Sirohi, A., Jain, A.: Forecasting using machine learning: Int. J. Recent Technol. Eng. (IJRTE) ISSN: 2277–3878, Vol-**7** Issue-6C, (2019)
2. Choi, C., Kim, J., Kim, J., Kim, D., Bae, Y., & Kim, H S .: development of heavy rain damage prediction model using machine learning based on big data. In: hindawi advances in meteorology Vol **2018**, Article ID 5024930, 11, p https://doi.org/10.1155/2018/5024930 (2018)
3. Jakaria, A.H.M., Hossain, M M., & Rahman, M A.: smart weather forecasting using machine learning: a case study in tennessee. In: Proceedings of ACM Mid-Southeast Conference (Mid-Southeast'18). ACM, NY, USA, 4 p. (2018) https://doi.org/10.1145/nnnnnnn.nnnnnnn

4. Bushara, N O., Abraham, A.: Weather forecasting in sudan using machine learning schemes. In: Journal of Network and Innovative Computing ISSN 2160–2174 Vol **2**, pp. 309–317 © MIR Labs, www.mirlabs.net/jnic/index.html (2014)
5. Tandon, S., Singh, P K., Patel, A.: Weather prediction using machine learning.In: Electronic copy available at: https://ssrn.com/abstract=3836085
6. Weyn, J. A., Durran, D. R., Caruana, R., & Cresswell-Clay, N.: Sub-seasonal forecasting with a large ensemble of deep-learning weather prediction models.In: .J. Adv. Model. Earth Syst., 13(**7**), e2021MS002502 (2021)
7. Singh, N., Chaturvedi, S., & Akhter, S.: Weather forecasting using machine learning algorithm. In 2019 International Conference on Signal Processing and Communication (ICSC), pp. 171–174. IEEE (2019)
8. Chen, I-C., Hu, S-C.: Realizing specific weather forecast through machine learning enabled prediction model. In: The 2019 3rd High Performance Computing and Cluster Technologies Conference, pp.71–74 (2019)

A Systematic Review on Latest Approaches of Automated Sleep Staging System Using Machine Intelligence Techniques

Suren Kumar Sahu[1(✉)], Santosh Kumar Satapathy[2(✉)], and Sudhir Kumar Mohapatra[1]

[1] Faculty of Emerging Technologies, Sri Sri University, Cuttack 754006, Odisha, India
suren.sahu2602@gmail.com

[2] Information and Communication Technology, Pandit Deendayal Energy University, Gandhinagar 382007, Gujarat, India
Santosh.Satapathy@sot.pdpu.ac.in

Abstract. Background and Objective: Sleep staging plays a vital role in sleep research because sometimes sleep recording errors may cause severe problems like misinterpretations of the changes in characteristics of the sleep stages, medication errors, and, finally, errors in the diagnosis process. Because of these errors in recordings and analysis, the sleep behavior and automated sleep staging system are adopted by different researchers with different methodologies. This study identifies specific challenges with the existing studies and highlights certain points that support the improvement of automated sleep staging-based polysomnography signals.

Methods: This work provides a comprehensive review of an automated sleep staging system, which was contributed by the different researchers in the recent research developments using Electroencephalogram, Electrocardiogram, Electromyogram, and combinations of these signals.

Results: Our review in this research area shows that single-model and multi-modal signals are used for sleep staging, and also we have observed some great points from the existing methodologies: (1) It has been noticed that 30-s length of the epoch of EEG signals may not be sufficient to extract enough information for discriminating the sleep patterns but in the other hand that a 10-s and 15-s length epoch is well suitable for sleep staging, (2) due to similar characteristics on N1 and REM sleep stages, most of the traditional classification models misclassified N1 sleep stages as REM stage, which alternatively degrades the sleep stagging performance, (3) consideration of heterogeneous form signal fusions can give the improvement results on sleep staging, and (4) applying deeplearning based models, combinedly to significant PSG signals can lead to more robust automatic sleep staging results.

Conclusions: The review mentioned above points simultaneously improves automated sleep staging by polysomnography signals. These points can help to focus our research work from the traditional feature extraction method to systematic improvements such as automatic feature recognition without explicit features, a proper characterization of the sleep stage's behavior, safety, and reduced cost.

P. Das et al. (Eds.): AMRIT 2023, CCIS 1953, pp. 127–136, 2024.
https://doi.org/10.1007/978-3-031-47224-4_12

Keywords: Sleep Staging · EEG · Feature Extraction · Machine Learning · Deep Learning

1 Introduction

The Significance of sleep in a person's life cycle cannot be overstated. It allows the body to rest and rejuvenate, strengthening neural connections and improving learning and memory [1]. However, with the fast-paced lifestyle and increased stress levels in modern society, sleep disorders have become a growing problem. Studies indicate that up to 24% of adults experience sleep issues regularly. Sleep is a significant part of a person's life, taking up nearly a third of it. Sleep directly and indirectly impacts physiological performance, making a good night's sleep essential. Unfortunately, sleep disturbances are becoming more common due to the stressful nature of modern life and the increased use of technology [2]. There are two types of sleep, non-rapid eye movement (NREM) and rapid eye movement (REM). REM sleep occurs in 90-min cycles and lasts from 5 to 30 min. It is characterized by high neural activity [3]. NREM sleep, on the other hand, reduces metabolic rate, blood pressure, heart rate, and sympathetic activity while increasing parasympathetic activity [4]. To assess sleep stages, standardized organizations have established sleepscoring criteria. Polysomnography (PSG) is the most reliable method for determining sleep stages. PSG involves simultaneously measuring several parameters such as electroencephalograms (EEG), electrocardiograms (ECG), electrooculograms (EOG), electromyograms (EMG), blood oxygenation, airflow, and respiratory effort. PSG data is typically collected in a sleep laboratory and processed into 30-s epochs, classified as alertness, REM sleep, or one of three NREM sleep stages (N1, N2, N3) [5, 6]. Each of the five stages is defined as follows in the American Academy of Sleep Medicine (AASM) manual [7]:

Stage W: distinguished by an alpha rhythm in the EEG signal, typically above the occipital area, as well as other activities such as rapid eye movements with normal or increased chin muscle activity, reading eye motions, or eye blinking.

Stage N1: distinguished by slowly moving eyes and a high probability of awakenings or arousals. Alpha bands must not surpass half of the broad spectral band, and the amplitude of the EEG signal should be in the frequency range of 2 to 7 Hz and remain less than 200 mV. N1 usually accounts for 5% of total sleep time.

Stage N2: Arousals and awakening stage are less seen in N2 than in N1, and slowly moving eyes begin fading. N2 is awarded if there are one or more K-complexes or spindle trains in the last 50% of the previous or early 50% of the current epoch. N2 typically accounts for half of the sleep cycle in a single night.

Stage N3: It accounts for predominant slow waves and delta in the EEG signal and periodic arousals or awakenings. K-complex and Spindles may arise, and N3 is assigned if slow waves account for more than 20% of the period. N3 usually accounts for 20% of overall sleep. It depicts a deep sleep mode.

Stage REM: The dreaming stage is distinguished by the movement of the fast eye and active brain waves of N3. In REM, arousals and awakenings are common. The EEG exhibits mixed frequency and low voltage. REM is scored as such until there is a transition to stage W or N3, or a K-complex without arousal or a spindle occurs in the first 50% of the period with no REMs. REM sleep accounts for one-fourth of total sleep time.

2 Clinical Importance

Across the world, most sleep centers obtain manual sleep staging practices in the early years. This has a good solution during that period. Still, several drawbacks are observed in the sleep staging process: 1) This approach consumes much time for manually interpreting the sleep stages' behavior throughout the night 2) There are also huge variations caught between experts in sleep staging annotations 3) Laboratory setup is also too expensive [8]. These limitations forced the existence of an automated sleep staging process in the field of sleep research. It helps to reduce the cost and also provides accurate analysis of sleep patterns and less manual interpretation. Common standards are followed, and no variations in the sleep scoring [9]. This automated sleep scoring process also opens a new way for early detection of sleep-related diseases and quantifying sleep patterns. This automatic sleep-scoring process is vital in diagnosing different sleep-related disorders.

3 Visual Scoring Procedures

The inadequate understanding of the underlying physiological mechanisms that shape the physiological signs acquired during sleep is one of the hurdles in sleep stage analysis. Sleep patterns are known to change with age, but the source of these changes is unknown. The absence of apparent causality between physiological processes and observable signs complicates data interpretation. Furthermore, new and refined correlations are being uncovered in this dynamic area of research. While human experts can still make diagnoses based on medical data, computational tools can help by detecting minor discrepancies, speeding up analysis, and lowering expenses. These models can deliver a range of outputs, from event labeling to feature extraction and even diagnosis suggestions [10].

This study emphasizes a range of physiological signals that contain information on sleep stages and their classification and can be helpful for diagnosis, monitoring treatment, and evaluating drug effectiveness. However, before these advantages can be realized, the signals must be measured and the information extracted. Because there are many digital signal processing systems, there is no standard method for obtaining information, and it needs to be clarified which signals contain enough information for diagnosis [11]. To address this issue, the study analyses the information extraction techniques for various physiological signals to provide insight into the information contained in the data. A preliminary study of the literature found that the most widely used signals in automated sleep stage grading systems were EEG, ECG, and EOG data. The following sections summarize the review's findings for sleep grading systems that use these signals [12]. A physiological marker can only indicate one stage of sleep. Measurement of many signals yields duplicates information and potentially more uncorrelated information. As a result, multiple studies on automatic sleep stage evaluation using various movements have been conducted [13].

Electroencephalogram (EEG)

The EEG captures the electrical activity of the brain, and the patterns of activity during different stages of sleep are diverse enough that they have been utilized to establish numerous sleep stage classification systems. Time-domain (TD) features, spectral features (Wave bands), time-frequency (TF) features, and non-linear (NL) features are all ways to extract sleep-related information from EEG data [14]. Several classification approaches, including ensemble classification methods like Random Forest and Bootstrap Aggregating, K-means, and SVM, have been employed in sleep classification research to aid medical practitioners.

4 Automatic Sleep Scoring by Artificial Intelligence

4.1 Shallow Learning - Expert Knowledge

A summary of the research and review findings for the selected studies that employed Electroencephalogram (EEG) signals for automatic sleep stage classification.

Sharma et al. [9] proposed an automatic sleep scoring system based on feature-based approaches, achieving high accuracy and performance compared to traditional methods.

Seifpour et al. [10] developed a novel sleep-scoring approach based on sparse representation and dictionary learning, showing promising results for accurate and efficient sleep stage classification. Hassan and Bhuiyan (2017) [9] proposed an automated sleep scoring method using a hybrid feature selection approach and a classifier ensemble technique, achieving high accuracy and outperforming traditional methods.

The contribution of automatic sleep stage scoring using EEG signals so far includes improving the accuracy and reliability of sleep stage classification, enabling the diagnosis and treatment of sleep disorders, aiding in the study of the relationship between brain activity during sleep and health outcomes, facilitating the development of personalized sleep interventions, and providing a non-invasive and cost-effective tool for sleep analysis. Table 1 presents the details of the work carried out by the different researchers in between 2022 to 2016.

4.2 Deep Learning - Knowledge from Raw Data

Ranqi Zhao et al. [18] proposed a dual-modal and multiscale deep neural network to classify sleep stages. They used EEG and ECG signals to have three binary classifications: sleep and wake, light and deep sleep, and REM and NREM. They are divided into four-class sort wake, light sleep, deep sleep, and REM. Using the MIT-BIH database, they trained and implemented their algorithm to obtain accuracies of 88.80%, 97.97%, and 98.84%, respectively, for binary classification and 80.40% for four-class types. The MIT-BIH dataset, used for training and testing the proposed model, contains 18-night periods of sleep of 16 healthy adults, ranging from age 32 to 56. The proposed model uses nine convolution layers to classify EEG signals and 13 convolutional layers to classify ECG signals.

Yamei Li et al. [19] worked on a semi-supervised pediatric sleep staging classification based on a single-channel EEG signal. BiSALnet adopts an SPD manifold structure to

Table 1. Summary of existing sleep staging studies based on ML techniques with EEG signals

Existing Studies	Datasets	Feature extraction and selection method	Classification method	Accuracy Results
Ref [9]	Subjects: 20 Domain: healthy Age Group: 25 to 35 years	Linear Evaluation Technique	Support Vector Machine	88%
Ref [10]	EDF dataset (The benchmark Sleep)	Signal-sampleentropy and Timefrequency wavelet filter	Support Vector Machine for a two-class problem	98%
Ref [11]	EDF dataset (The benchmark Sleep)	Time domain feature (Local Extrema)	Multi class Support Vector Machine	98%
Ref [12]	Subjects: 23 Domain: healthy adults Group: Male Age Range: 23 to 45	Synchronization Likelihood and Relative Entropy	SVM and KNN	93%
Ref [13]	Sleep-EDF database	Nested 5 classes of sleep stage classification	RF	95%
Ref [14]	Sleep-EDF database	Coupling techniques based on frequency features	Naive Bayes	94%
Ref [15]	Sleep-EDF database	Wavelet transform Techniques	Boosting	92%
Ref [16]	Sleep-EDF database	Discrete Fourier Transform aka DFT	RF	90%
Ref [17]	Sleep-EDF dataset	Ensemble Empirical Mode Decomposition	RUSBoost	98%

encapsulate desired feature distribution properties in the proposed architecture. BiSALnet is tested upon two databases, the NPH, and the Sleep-EDF databases. The NPH database comprises 15 patient recordings, with classification done every 30s among five classes: Wake, N1, N2, N3, and REM. The EEG signal recorded was from O1-M2. The Sleep-EDF database consisted of 8 patients' 30s EEG epochs sampled at 100 Hz. The EEG signal used in Sleep-EDF is Fpz-Cz. BiSALnet includes a BiStream Adversarial Learning Framework constituting a Teacher and a Student. The proposed model achieves an accuracy of 80.00% when tested on the NPH dataset, while sensitivity and F1-Scores are 76.00% each. The accuracy improves when tested on the Sleep-EDF dataset, reaching 91.00%, while sensitivity and F1-Score are 75.00% and 77.00%, respectively.

Chuan Shao Zhang et al. [20] propose a model, RMH-net, for sleep staging analysis. The RMH model consists of 3 parts: backbone, Manifold Learning Block, and Hyperbolic Block. The Backbone focuses on latent discriminative features since it has more receptive fields. The Hyperbolic parts perform well in detecting the local information spots, and Manifold Learning Block working in cohesion with Hyperbolic Block helps in sleep feature extraction. The database used is the Sleep-EDF database which includes 197 PSG recordings. The EEG signal is divided into two parts based on electrode location, i.e., Fpz-Cz and Pz-Oz. The Sleep Cassettes database consists of 78 healthy Caucasians ranging from age 25 to 101. Of these extensive subjects, eight are chosen at random to experiment. RMH-net achieves an accuracy of 89.00% when tested on the Sleep-EDF database and a Kappa value of 78.00%

Hai Wang et al. [21] propose an automatic sleep staging analysis model using transfer learning and network fusion. The model, Seq-Deepsleepnet, consists of 3 networks: Seqsleepnet subnetwork, deepsleepnet subnet, and LightBGM classifier. The seqsleepnet subnetwork consists of a sequence-to-sequence classifier having a bidirectional RNN. Deepsleepnet consists of CNNs, and bi-directional LSTM and LightBGM merge the output of the other two networks as input to the classifier. The Sleep-EDF expanded dataset is used to evaluate the proposed model. For pre-training MASS dataset is used, which consists of 200 nights of PSG of 97 males ranging from age 19 to 49 and 103 females ranging from age 18 to 38; all recordings have a sampling frequency of 256 Hz. For evaluation Sleep-EDF expanded database consists of 153 Sleep Cassette files of 20 healthy Caucasians ranging from age 25 to 101; all recordings have a sampling rate of 100 Hz. The model on the Sleep-EDF database gives an accuracy of 87.84%.

Ritika Jain et al. [22] proposed a model to classify the sleep stages of unknown subjects using single and multiple-channel EEG signals. The classification was performed for both six-class and two-class sets. The algorithm suggested combines Temporal, Non-linear, Spectral, Time-frequency, and Statistical features of EEG signal and RUSBoost (random under-sampling and boosting technique). The model was trained using three datasets, namely Sleep-EDF (8 subjects; 4 healthy and 4 having sleep disorders), DREAMS (20 subjects; 16 females and four males), and Expanded Sleep-EDF (197 topics; 153 healthy and 44 having sleep disorders). Along with issues from these three datasets, the model was tested using subjects' unseen by the model. The model's performance and accuracy were accurate and reliable for unknown problems. The testing subjects from Sleep EDF obtained 92.6% and 97.9% accuracies for six stages and two stages. In contrast, the topics from the expanded version of the same dataset obtained more efficient accuracies of 96.3% for six steps and 99.85 for two sets.

The review of different sleep staging methods shows that the authors obtained deep learning models and improved accuracy for multi-class classification problems. Figure 1 and Fig. 2 present the accuracy results for different research studies carried out by the researchers in the last six to seven years.

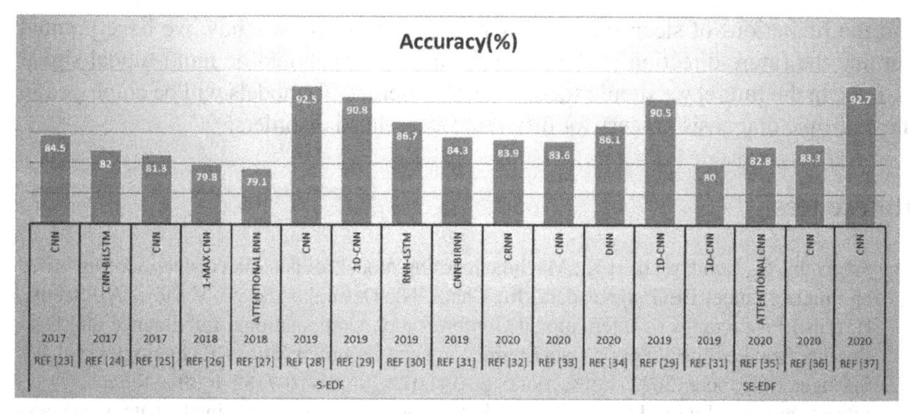

Fig. 1. Overall accuracy performances of the different deep learning classification models using S-EDF and SE-EDF datasets

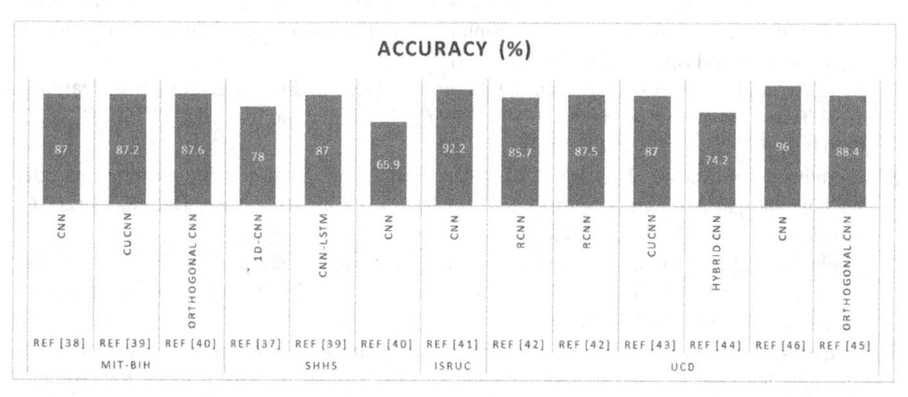

Fig. 2. Overall accuracy performance using different deep learning models with different public datasets

5 Conclusion

In this review, the recently contributed methods in sleep scoring systems based on ML and DL have been sincerely exercised, comparing these studies in different aspects. The main limitation concerning ML algorithms is the required hand-crafted features for inspecting the changes in the behavior of sleep stages over the individual sleep stages. Another challenging job for ML algorithms is, identifying the relevance of the feature subject to the patient's behavior. In ML based sleep-scoring system, the selection of classification algorithms is also of difficult job. Sometimes it may directly impact the classification performance. Due to these limitations, most of the researchers adopted the DL approaches, making it more comfortable and accurate to discriminate the changes in sleep behavior over individual sleep stages. The biggest advantage of DL approach features is that they are automatically recognized without using explicit features. From this review, we have found that presently DL models achieved improved results compared to ML models. This review helps to understand the significant improvements achieved and the limitations of sleep staging. Therefore, in this review study, we have pointed out that the future direction of the sleep scoring system should be multi-modal signal fusions. In the future, we should focus on how different DL models will be employed in the real-time diagnosis process for different sleeprelated disorders.

References

1. Satapathy, S., Kondaveeti, H.K.: Machine learning model for automated sleep scoring based on single-channel EEG signal data. In: Chaki, N., Devarakonda, N., Cortesi, A., Seetha, H. (eds.) Proceedings of International Conference on Computational Intelligence and Data Engineering. Lecture Notes on Data Engineering and Communications Technologies, vol. 99. Springer, Singapore (2022).https://doi.org/10.1007/978-981-16-7182-1_30
2. Smaldone, A., Honig, J.C., Byrne, M.W.: Sleepless in America: inadequate sleep and relationships to health and well-being of our nation's children. Pediatrics **119**, 29–37 (2007)
3. Satapathy, S.K., Loganathan, D., Narayanan, P., Sharathkumar, S.: Convolutional neural network for classification of multiple sleep stages from dual-channel EEG signals. In: 2020 IEEE 4th Conference on Information & Communication Technology (CICT), pp. 1–16 (2020). https://doi.org/10.1109/CICT51604.2020.9312078
4. Phan, H., et al.: Towards more accurate automatic sleep staging via deep transfer learning. IEEE Trans. Biomed. Eng. **68**(6), 1787–1798 (2021). https://doi.org/10.1109/TBME.2020.3020381
5. Surantha, N., Lesmana, T.F., Isa, S.M.: Sleep stage classification using extreme learning machine and particle swarm optimization for healthcare big data. J Big Data **8**, 14 (2021). https://doi.org/10.1186/s40537-020-00406-6
6. Eldele, E., et al.: An attention-based deep learning approach for sleep stage classification with single-channel EEG. IEEE Trans. Neural Syst. Rehabil. Eng. **29**, 809–818 (2021). https://doi.org/10.1109/TNSRE.2021.3076234
7. Satapathy, S.K., Kondaveeti, H.K., Malladi, R.: Automated sleep staging system based on ensemble learning model using single-channel EEG signal. In: Misra, R., Shyamasundar, R.K., Chaturvedi, A., Omer, R. (eds.) Machine Learning and Big Data Analytics (Proceedings of International Conference on Machine Learning and Big Data Analytics (ICMLBDA) 2021). ICMLBDA 2021. LNNS, vol. 256. Springer, Cham (2022). https://doi.org/10.1007/978-3-030-82469-3_17
8. Satapathy, S.K., Shah, K., Shah, S., Shah, B., Panchal, A.: A deep neural model CNN-LSTM network for automated sleep staging based on a single-channel EEG signal. In: Thakur, M., Agnihotri, S., Rajpurohit, B.S., Pant, M., Deep, K., Nagar, A.K. (eds.) Soft Computing for Problem Solving. LNNS, vol. 547. Springer, Singapore (2023). https://doi.org/10.1007/978-981-19-6525-8_6
9. Sharma, M., Goyal, D., Achuth, P., Acharya, U.R.: An accurate sleep stages classification system using a new class of optimally time-frequency localized three-band wavelet filter bank. Comput. Biol. Med. **98**, 58–75 (2018)
10. Seifpour, S., Niknazar, H., Mikaeili, M., Nasrabadi, A.M.: A new automatic sleep staging system based on statistical behavior of local extrema using single channel EEG signal. Expert Syst. Appl. **104**, 277–293 (2018)
11. Chriskos, P., Frantzidis, C.A., Gkivogkli, P.T., Bamidis, P.D., Kourtidou-Papadeli, C.: Achieving accurate automatic sleep staging on manually pre-processed EEG data through synchronization feature extraction and graph metrics. Front. Human Neurosci. **12**, 110 (2018)
12. Memar, P., Faradji, F.: A novel multi-class EEG-based sleep stage classification system. IEEE Trans. Neural Syst. Rehabilit. Eng. **26**(1), 84–95 (2018)

13. Dimitriadis, S.I., Salis, C., Linden, D.: A novel, fast and efficient single-sensor automatic sleep-stage classification based on complementary cross-frequency coupling estimates. Clin. Neurophysiol. **129**(4), 815–828 (2018)
14. Hassan, A.R., Subasi, A.: A decision support system for automated identification of sleep stages from single-channel EEG signals. Know.-Based Syst. **128**, 115–124 (2017)
15. da Silveira, T.L., Kozakevicius, A.J., Rodrigues, C.R.: Single-channel EEG sleep stage classification based on a streamlined set of statistical features in wavelet domain. Med. Biol. Eng. Comput. **55**(2), 343–352 (2017)
16. Hassan, A.R., Bhuiyan, M.I.H.: Automated identification of sleep states from EEG signals by means of ensemble empirical mode decomposition and random under sampling boosting. Comput. Method. Progr. Biomed. **140**, 201–210 (2017)
17. Hassan, A.R., Bhuiyan, M.I.H.: A decision support system for automatic sleep staging from EEG signals using tunable q-factor wavelet transform and spectral features. J. Neurosci. Method. **271**, 107–118 (2016)
18. Zhao, R., Xia, Y., Wang, Q.: Dual-modal and multi-scale deep neural networks for sleep staging using EEG and ECG signals. Biomed. Sig. Process. Control **66**, 102455 (2021). ISSN 1746–8094, https://doi.org/10.1016/j.bspc.2021.102455
19. Li, Y., Peng, C., Zhang, Y., Zhang, Y., Lo, B.: Adversarial learning for semi-supervised pediatric sleep staging with single-EEG channel. Methods **204**, 84–91 (2022). ISSN 1046–2023, https://doi.org/10.1016/j.ymeth.2022.03.013
20. Zhang, C., Liu, S., Han, F., Nie, Z., Lo, B., Zhang, Y.: Hybrid manifold-deep convolutional neural network for sleep staging. Methods **202**, 164–172 (2022). ISSN 1046–2023, https://doi.org/10.1016/j.ymeth.2021.02.014
21. Wang, H., Guo, H., Zhang, K., Gao, L., Zheng, J.: Automatic sleep staging method of EEG signal based on transfer learning and fusion network. Neurocomputing **488**, 183–193 (2022). ISSN 0925–2312, https://doi.org/10.1016/j.neucom.2022.02.049
22. Jain, R., Ganesan, R.A.: Reliable sleep staging of unseen subjects with fusion of multiple EEG features and RUSBoost. Biomed. Sig. Process. Control **70**, 103061 (2021). ISSN 1746–8094, https://doi.org/10.1016/j.bspc.2021.103061
23. Wei, L., Lin, Y., Wang, J., Ma, Y.: Time-frequency convolutional neural network for automatic sleep stage classification based on single-channel EEG. In Proceedings of the 2017 IEEE 29th International Conference on Tools with Artificial Intelligence, Boston, MA, USA, 6–8 November 2017; IEEE: Piscataway, NJ, USA, pp. 88–95 (2017)
24. Supratak, A., Dong, H., Wu, C., Guo, Y.: DeepSleepNet: a model for automatic sleep stage scoring based on raw single-channel EEG. IEEE Trans. Neural Syst. Rehabil. Eng. **25**, 1998–2008 (2017)
25. Vilamala, A., Madsen, K.H., Hansen, L.K.: Deep convolutional neural networks for interpretable analysis of EEG sleep stage scoring. In: Proceedings of the 2017 IEEE 27th International Workshop on Machine Learning for Signal Processing (MLSP), Tokyo, Japan, 25–28 September 2017, pp. 1–6 (2017)
26. Phan, H., Andreotti, F., Cooray, N., Chen, O.Y., de Vos, M.: DNN filter bank improves 1max pooling CNN for single-channel EEG automatic sleep stage classification. In: Proceedings of the 2018 40th Annual International Conference of the IEEE Engineering in Medicine and Biology Society (EMBC), Honolulu, HI,USA, 18–21 July 2018, pp. 453–456 (2018)
27. Phan, H., Andreotti, F., Cooray, N., Chen, O.Y., De Vos, M.: Automatic sleep stage classification using single-channel EEG: learning sequential features with attention-based recurrent neural networks. In: Proceedings of the 2018 40th Annual International Conference of the IEEE Engineering in Medicine and Biology Society (EMBC), Honolulu, HI, USA, 18–21 July (2018)

28. Qureshi, S., Karrila, S., Vanichayobon, S.: GACNN SleepTuneNet: a genetic algorithm designing the convolutional neural network architecture for optimal classification of sleep stages from a single EEG channel. Turk. J. Electr. Eng. Comput. Sci. **27**, 4203–4219 (2019)
29. Yıldırım, Ö., Baloglu, U.B., Acharya, U.R.: A deep learning model for automated sleep stages classification using PSG signals. Int. J. Environ. Res. Public Health **16**, 599 (2019)
30. Michielli, N., Acharya, U.R., Molinari, F.: Cascaded LSTM recurrent neural network for automated sleep stage classification using single-channel EEG signals. Comput. Biol. Med. **106**, 71–81 (2019)
31. Mousavi, S., Afghah, F., Acharya, U.R.: SleepEEGNet: automated sleep stage scoring with sequence to sequence deep learning approach. PLoS ONE **14**, e0216456 (2019)
32. Seo, H., Back, S., Lee, S., Park, D., Kim, T., Lee, K.: Intra- and inter-epoch temporal context network (IITNet) using sub-epoch features for automatic sleep scoring on raw singlechannel EEG. Biomed. Signal Process. Control **61**, 102037 (2020)
33. Zhang, X., et al.: Automated multi-model deep neural network for sleep stage scoring with unfiltered clinical data. Sleep Breath. **24**, 581–590 (2020)
34. Xu, M., Wang, X., Zhangt, X., Bin, G., Jia, Z., Chen, K.: Computation-efficient multi- model deep neural network for sleep stage classification. In: Proceedings of the ASSE 2020: 2020 Asia Service Sciences and Software Engineering Conference, Nagoya, Japan, 13–15 May 2020; Association for Computing Machinery: New York, NY, USA, pp. 1–8 (2020)
35. Zhu, T., Luo, W., Yu, F.: Convolution-and attention-based neural network for automated sleep stage classification. Int. J. Environ. Res. Public Health **17**, 4152 (2020)
36. Jadhav, P., Rajguru, G., Datta, D., Mukhopadhyay, S.: Automatic sleep stage classification using time-frequency images of CWT and transfer learning using convolution neural network. Biocybern. Biomed. Eng. **40**, 494–504 (2020)
37. Fernandez-Blanco, E., Rivero, D., Pazos, A.: Convolutional neural networks for sleep stage scoring on a two-channel EEG signal. Soft. Comput. **24**, 4067–4079 (2019)
38. Sors, A., Bonnet, S., Mirek, S., Vercueil, L., Payen, J.-F.: A convolutional neural network for sleep stage scoring from raw single-channel EEG. Biomed. Signal Process. Control **42**, 107–114 (2018)
39. Zhang, L., Fabbri, D., Upender, R., Kent, D.T.: Automated sleep stage scoring of the Sleep Heart Health Study using deep neural networks. Sleep **42**(11), zsz159 (2019)
40. Li, Q., Li, Q.C., Liu, C., Shashikumar, S.P., Nemati, S., Clifford, G.D.: Deep learning in the cross-time frequency domain for sleep staging from a single-lead electrocardiogram. Physiol. Meas. **39**, 124005 (2018)
41. Cui, Z., Zheng, X., Shao, X., Cui, L.: Automatic sleep stage classification based on convolutional neural network and fine-grained segments. Complexity **2018**, 1–13 (2018)
42. Biswal, S., et al.: SLEEPNET: Automated Sleep Staging System via Deep Learning. arXiv 2017. arXiv:1707.08262
43. Zhang, J., Wu, Y.: Complex-valued unsupervised convolutional neural networks for sleep stage classification. Comput. Methods Programs Biomed. **164**, 181–191 (2018)
44. Yuan, Y., et al.: A hybrid self-attention deep learning framework for multivariate sleep stage classification. BMC Bioinform. **20**, 1–10 (2019)
45. Zhang, J., Yao, R., Ge, W., Gao, J.: Orthogonal convolutional neural networks for automatic sleep stage classification based on single-channel EEG. Comput. Methods Programs Biomed. **183**, 105089 (2020)

Network Security Threats Detection Methods Based on Machine Learning Techniques

Bikash Kalita and Pankaj Kumar Deva Sarma(✉)

Department of Computer Science, Assam University, Silchar, Assam 788011, India
pankajgr@rediffmail.com

Abstract. The world is now interconnected to share information and resources in a significant manner. The network must support its integrity and network security, and it must be scalable to sustain its performance and capacity. The network system is subjected to attack and due to the greater extent of applications, development of technology, added users and IoT (Internet of Things), the network system is vulnerable to different security threats. The network security threats are framed and launched with highly effective techniques targeting the network resources. The computer network system must be protected against any attack to save its resource. Machine Learning techniques are preferably being used by researchers as a solution to these problems. The development of machine learning shows effective capabilities to analyse and detect, sometimes also to prevent network security threats. In the last few years, various machine learning algorithms and applications have been designed and experimented with network traffic analysis, including Data Mining and Network-based techniques in the field of Data Science. In this work, we have conducted a detailed analysis of various machine learning techniques. Some Machine Learning approaches and related algorithms are reviewed in connection with their applications to analyse and detect network threats and the prospective solutions proposed in various research works.

Keywords: Machine Learning Techniques · Network Security · Threats · Intrusion Detection System

1 Introduction

Along with the constant evolution of the internet, users are increasing rapidly. At present all sectors of activity including banking, health, education, entertainment etc. are using the internet for their routine uses. Due to different kinds of added users and technologies, the volume of network traffic and the complexity of internet services continue to increase. Also, the information that is carried out through the network system is vulnerable [1]. The computer network system requires prevention from threats and anomalous activities. To solve the problems of detection and prevention of network security threats, many researchers have explored it from different perspectives. Several methods are proposed to solve these issues. Network security experts have developed throughout the years various network security threats detection systems to shield information and other resources from malicious activity [2].

P. Das et al. (Eds.): AMRIT 2023, CCIS 1953, pp. 137–156, 2024.
https://doi.org/10.1007/978-3-031-47224-4_13

The kind of users and the protocols that are followed by the network packets are not under specific control. Hence the error rate may increase day by day. There is no exact rule to determine what packets can be considered as anomalous or normal. There are many problems occurring on the internet. Network traffic detection, cyber security threats detection, and fault management are significant problems on the internet. The internet traffic needs to be monitored on a regular basis for improving security challenges to support confidentiality and integrity. Use of Machine learning (ML) algorithms to analyze the network traffic data are helpful in detecting and solving issues related to network security.

Machine Learning Techniques is widely being used to prevent intrusion activity and to improve system performances. Network intrusion is malicious activity, i.e., it is an unauthorized act over the internet that accesses important data [3]. The network security threats detection system is a remarkable prevention technique that is being used to detect unauthorized intrusion in a computer network system. An intrusion detection system allows to collect and analyse network data packets to detect and to respond to the security threats of the network. Today's cybercriminals are highly skilled to attack any network system. So, it is essential to monitor network traffic to provide proper security on network operations. Application of intrusion detection system can help in detecting and analysing network traffic data patterns in order to detect and prevent network security threats.

Network traffic analysis is mainly performed to detect data type, the data flowing over a network, and the data source. Some machine mechanisms can be used in finding malicious activity patterns and to identify criminal activities while it is occurring. Increased network traffic demands new ways to detect intrusions, analyse malware behaviour, discover data patterns, users' behaviour and other security aspects. For this purpose, many classification algorithms, such as Decision Tree (DT), Random Forest (RF), Naïve Bayes (NB), Support Vector Machine (SVM), Neural Network etc. can be used. In these methods different log data are used as any activities performed during a network security breach will mostly result in different log entries are recorded in one or more log files.

The paper is organised as follows: Sect. 1 introduces the background knowledge of some ML techniques that are widely used in the detection of network security threats, giving also some background of the same. Section 2 reviews other related works, mainly the application of machine learning techniques, illustrating how these techniques are used in the analysis and detection of network security threats. Section 3 focuses on datasets used by other researchers, pre-processing and selection of the data features. Section 4 overviews machine learning techniques and their accuracy. Finally, Sect. 5 describes the outcomes of the review and related open issues.

1.1 Machine Learning Techniques

Machine learning is being used for traffic classification and is essential for the detection of network security threats and resource management [4]. Some researchers used some classic algorithms, while others used some hybrid techniques to detect and solve various issues with high accuracy. Though there are different phases of the models and which vary depending on the models, these phases can be categorized into three primary components: dataset, training the model, and testing.

Depending on the working principle of the model, the methodologies for network security threats detection can be classified into three categories: Anomaly-based Detection, Stateful Protocol Analysis and Signature-based Detection [5].

Different algorithms have different uses and working principles. The ML techniques that are mostly used for network security threats detection system be can be categorized into three categories. They are- Supervised Learning, Unsupervised Learning and Reinforcement Learning. Out of all the learning algorithms, some supervised learning algorithms namely SVM, DT, NB, KNN and RF algorithms are considered for this work based on knowledge about some known datasets.

Supervised Learning

For a supervised machine learning algorithm, a set of known datasets, as well as known results, are used to train the model to get reasonable predictions on new data. The model learns from the labelled dataset by making a prediction on the dataset and generates accurate results. Since the data are known or labelled, hence these algorithms have more potential to give accurate results. In these algorithms, both classification and regression techniques are used to develop a model.

Classification

The classification technique involves in predicting a class label. This technique takes each instance of data to classify it into a different class. In network traffic analysis, the classification technique is useful to categorize all traffic into normal or malicious traffic. The limitation of the classification technique in the implementation on network traffic analysis is in order to lessen the incidence of false positives and false negatives [6].

Regression

The regression technique involves in predicting a numerical label. This technique is useful when the dependent and independent variables have a linear or non-linear relationship between them, and the dependent variables have continuous values.

Some classification and regression techniques that are used by many researchers in their works for network traffic analysis are –

a) Support vector machine (SVM): SVM algorithms are supervised ML algorithms that are mostly used for analysis of data for classification and regression problems. However, SVM algorithms are mostly used in classification problems. This model designs the training points in a high-dimensional feature space, and labels every point by its class [7]. This algorithm is capable to handle both continuous and categorical variables. Some used points in the classification of data are-

- Support Vector: Support vectors are the data points that are closest to the decision plane.
- Hyper plane: This is the decision line or decision plane that segregates different data points of different classes.
- Margin: This is the area between the decision lines on the support vectors.

SVM classifies dataset by determining its support vectors from the training dataset that determines the hyperplane. For this, a kernel function (e.g., linear, Gaussian, polynomial, or sigmoid) can be used for the SVM in training the model. Other reasons for using SVMs are speed and scalability. This algorithm can potentially run fast and can learn on a large dataset containing more different patterns and is also capable to scale better.

In [8] a SVM technique is proposed in which a generic mechanism was used to fit the hyperplane using a kernel function. In the experiment, the dataset is partitioned into two binary classes: normal traffic and anomalous traffic. Here, the objective was to separate normal traffic and malicious traffic patterns. The training of the dataset had been conducted using the RBF (radial bias function) kernel option and found 99.8% accuracy in testing as a result.

b) Decision Tree (DT): This algorithm is a supervised ML algorithm. This algorithm is broadly being used to resolve classification and regression problems analysis [9]. A decision tree uses a divide-and-conquer strategy in implementation. In a DT, each node is classified into two nodes, either as a Decision Node or a Leaf Node. This algorithm makes a decision based on the Decision nodes and according to the specified conditions, divide the node into sub-nodes., whereas the leaf nodes contain the results of those decisions, which makes a tree-like hierarchical data structure. Each decision is made based on the data features of the dataset.

c) Naive Bayes (NB): This method of categorization uses the Bayesian theorem as its foundation. The naive Bayes model is useful for very large data sets. It analyses and defines the connections between each data feature in the dataset's class.. "From the computational results, it can derive a conditional probability of an attribute and the class. In the classification process, the classifier must estimate the probabilities of the unknown sample instance as a class by combining the prior knowledge with the actual value of the unknown sample instance. Moreover, the classifier must estimate the probabilities of the feature having a certain value." [10].

In [11] it is stated that the NB technique is especially used when the dimensionality of the input data is high. This technique is adequate for calculating every possible result based on the given input. It can also be used to add new raw data at runtime which may result in an improved probabilistic classifier. The general effectiveness of the Nave Bayes technique is found to be 77.8% in that research work.

d) Random Forest (RF): This is an ensemble classifier that can be used for better accuracy. It is a combination of many decision trees. Random forest is a better classification algorithm other than traditional classification algorithms as it has a low classification error [11].

Some advantages of Random Forest classifier are listed below.

I. The generated forests could be useful for future reference [1].
II. RF is accurate and can run efficiently on large data sets [12].
III. In this classifier, variable importance is automatically generated [11]

e) K-Nearest Neighbor (KNN): Kneighbors is a supervised ML algorithm, Fix and Hodges first used the word in 1951 [13]. The model's output is produced by calculating k, after which it can choose the closest observations by using the Euclidean Distance notion. Non-parametric classifier, this approach makes no assumptions on the dataset [14]. The KNN classifier has the following benefits [15]:

a) KNN is very effective and useful on a very large dataset.
b) It repeatedly evolves over the time and adapts to any new environment.
c) Easy for implementation on a multi-class problem.

2 Classification of Network Security Threats

The basic dimension for classifying network security threats is the goal of cyber threats. The goal of the threats falls into one of the following categories [16].

- Stealing information: this kind of activity is usually performed by spyware.
- Monitoring users' information: this type of action is generally performed by mobile malware.
- Taking control of the system: this kind of attack is usually done by rootkit, trojan and botnet [17].

Another dimension for classifying a cyber threat is the threat vector which implies the vulnerability detected by an intruder to hack the system to perform its malicious activity. Cyber threat vectors can be seen on mainly three different layers: application, network and hardware. [2] Some common cyber security threats are-

- **Black Hole Attacks:** Black holes in a network are areas where malevolent nodes trash network traffic without alerting the originator that the packet did not arrive at its intended destination.Every node between the source and destination is a possible blackhole attacker, regardless of the mobile routing protocol [18].
- **Gray Hole Attacks:** A gray hole is a network attack where a compromised network element selectively drops or modifies network packets [19]. Unlike a black hole that drops all packets, a gray hole attack selectively chooses which packets to drop or manipulate. Communication breakdowns, data corruption, or even unauthorised access, depending on the situation, may result.
- **Brute Force Attack:** It is a trial-and-error method used by hackers to obtain passwords, encryption keys, or sensitive information. The attacker systematically tries all possible combinations until the correct one is found [20]. For example, in the case of password cracking, an attacker attempts different combinations of characters until they discover the correct password.
- **Denial of Service (DoS) Attack**: By flooding a network, system, or service with unauthorised requests or by taking advantage of weaknesses, a DoS attack seeks to make it unavailable to its intended users to exhaust its resources [21]. This can result in a loss of service, downtime, or performance degradation.

Machine learning techniques are highly relevant in the analysis and detection of network security threats. Machine learning techniques are broadly used in finding unforeseen threats and malicious activity created by malicious software [22, 23].

The dataset used for our experiment, NSL-KDD, consists mostly of four types of attacks. [24].

- DoS: An attacker attempts to stop authorised users from using a service.

- Remote to Local (R2L): The attacker attempts to get access even if they do not have a user account on the victim's machine.
- User to Root (U2R): The attacker attempts to achieve super user privileges while having local access to the target computer.
- Probe: An attacker uses this technique to learn more about the target host.

Despite the fact that there are many different attack types, our primary attention will be on the detection of DoS, Probe, U2R, and R2L attacks and how the models perform in terms of accuracy, precision, recall, and f-measure.

3 Related Works

Network Traffic Analysis analyses the communication patterns to detect network traffic and security breaches, resource requirements and management, and other different problems. For a better understanding of these problems and solutions to these, to find the trends and the steps to perform network traffic classification, several related works of literature are reviewed.

Network traffic analysis using machine learning started being used since 2005, Even though, several problems are occurring due to the network traffic evolution and scalability [24].

A generic DoS detection model that makes use of the biologically inspired Random Neural Network and numerous Bayesian classifiers has been suggested [4]. Their study investigated the proper detection, missed detection, and false alarm rates for a technique for diverse input data in a big test network. They chose different statistical features which are "bitrate, increase in bitrate, entropy, Hurst parameter, delay and delay rate" [4]. After selecting the input features, they computed likelihood ratios and they were first-level judgements, and later they were merged with averaging or with RNNs.

A new hybrid model by combining the Covariation Orthogonal (CO) and Artificial Neural Networks (ANNs) was proposed [25] to classify the bursty and self-similar data pattern. In their study, the CO method gave a very good accuracy in the prediction implemented on real traffic traces.

In [26] it is proposed that "an IoT device can be accurately identified based on characteristics of the network traffic it generates. Machine Learning Approach can be used to automatically and accurately recognise connections of IoT devices to an enterprise's computer network, and thus mitigate violations of operational policies.

In a work [27] Stated that Machine learning could be used in different areas of cyber security to provide analytical approaches to detect and respond to an attack. It can also enhance safety operations. Intrusion detection is usually achieved using a model developed by analysing big data sets of security events and identifying the pattern of malicious activities. As a result, when similar activities are detected, they are automatically dealt with. Another mechanism, i.e. Network-based defence method, protects the network by attempting to reduce the risk of attacks by providing an additional layer of security because each layer has established policies and controls in place to define users who are authorised to access the network. As a first step, a framework was developed to collect data and filter traffic. This basis then uses ML techniques in defence against attacks. The results can help in determining the probability and impact of the attack in relation to a specific network area. Therefore, it can help different organisations to reduce the risk of victim exposure.

A machine learning model running in Real-time [28] was experimented with to detect malicious data flow traffic based on the ISOT-CID dataset. They extracted six features of the dataset to implement the machine learning techniques. They used six models i.e. Decision Tree, K-nearest Neighbor, Support Vector Machine, Artificial Neural Network, Random Forest and Naive Bayes. But out of those, Decision Tree and Random Forest have given optimal accuracy (100%) results when evaluated by cross-validation and split-validation methods. There are some limitations in the model that Intrusion Detection System for computer networks was a little bit slow, this can be very fast to deploy in real-time to extract the communication traffic characteristics and to give its response in real-time. And the deployment of this model in real networks will affect the speed required.

The paper [1] compared the performances of three machine learning models to detect security threats. They used frequently used and benchmark datasets to analyse results in terms of accuracy, precision and recall. They came to the conclusion that for every cyber security threat detection, a particular machine learning technique can not be implemented. Different machine learning techniques are being used for different categories of cyber threats.

In [5] stated that each machine learning model has its advantages and limitations, so everyone must be cautious while selecting the approaches. The pattern-based intrusion detection system is very effective to inspect known attacks, which could hardly detect unknown attacks. Rule-based approaches are good in detecting unknown attacks but still have some limitations in creating and updating the knowledge for a particular attack. So, a comprehensive review of the intrusion detection system and application requirements could help for better practical usage.

[23] propose a model that uses a correlation-based feature selection method. The model shows better accuracy in classifying R2L (Remote-to-Local) and U2R (User-to-Root) attacks than DoS (Denial of Service) and Probe attack. [50] proposed a system to detect the malicious behavior of nodes by intrusion detection system with fuzzy logic technique and also to identify the type of attacks. The system is robust enough to detect attacks such as black hole attacks and gray hole attacks and is also able to prevent those kinds of attacks by using efficient node blocking mechanism such that the proposed system provides a secure communication between nodes.

[20] propose a new mechanism for mitigating the impact of smart gray hole attacks. Utilising the NS-2.35 simulator, the efficacy of the mechanism is validated. The simulation findings demonstrate that, when subjected to a smart grey hole attack, the suggested method outperforms the current scheme marginally.

4 Working Process

The working of machine learning model for analysis and detection of network security threats can be categorised into four parts: Collection of data, Feature selection, Analysing the data, and the last part is results and response. Figure 1 shows the pictorial representation of the working process.

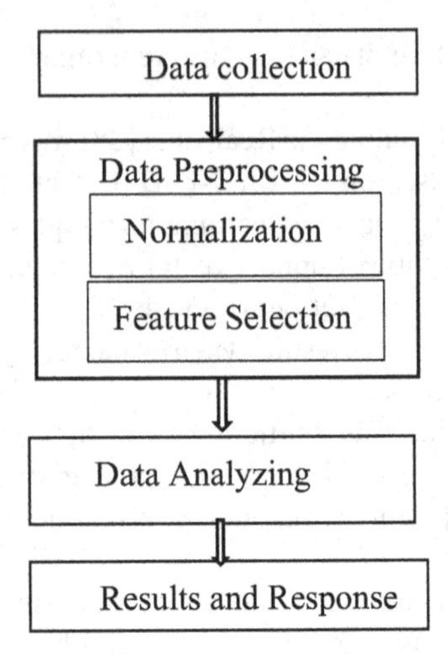

Fig. 1. Steps involved in the analysis and detection of network security threats.

5 Dataset

In-network traffic analysis, to evaluate the effectiveness of the model, different researchers used different data sets from different sources in recent years. Some important data sets that are being used by researchers for network traffic analysis are DARPA, KDD cup, Wireshark software, peer-peer Bit Torrent, UNSW-NB15, from the internet service provider etc. Some datasets and their sources are stated below in table 1-

The "NSL KDD" is a benchmark dataset for the detection of network security threats. The NSL-KDD dataset is the improved version of the KDD'99 dataset. The benefits of the NSL-KDD dataset over KDD'99 [13] are –

I. It excluded redundant data so that the model will not be biased for frequent records in the training phase.
II. It also excluded duplicate records, so that the performance of the model will not be biased for the methods which have better results on the repeated records.
III. It has a reasonable number of records in both the train and test datasets, which may expect a better result instead of selecting a small portion randomly.
IV. The records selected from each difficulty level group are inversely proportional to their percentage of the original KDD dataset, which helps in the accurate evaluation of the models [14].

Table 1. Different datasets used in developing network security threats detection systems and their sources.

Sl No	Dataset	Source
1	DARPA 1998	https://www.ll.mit.edu/r-d/datasets/1998-darpa-intrusion-detection-evaluation-dataset [7]
2	KDD cup	https://www.kaggle.com/datasets/galaxyh/kdd-cup-1999-data?select=corrected.gz [8]
3	CIC-IDS2017 (Intrusion Detection Evaluation Dataset)	https://www.unb.ca/cic/datasets/ids-2017.html [9]
4	LEDBAT+BitTorrent Dataset	https://nonsns.github.io/dataset/ledbat+bittorrent/ [10]
5	UNSW-NB15	https://research.unsw.edu.au/projects/unsw-nb15-dataset [11]
6	CSE-CIC-IDS 2018 (A collaborative project between the Communications Security Establishment (CSE) & the Canadian Institute for Cybersecurity (CIC))	https://www.unb.ca/cic/datasets/ids-2018.html [1]
7	ADFA IDS Datasets	https://research.unsw.edu.au/projects/adfa-ids-datasets [12]

6 Normalization

Normalization is a useful scaling technique in Machine Learning applications which can be applied during data preparation to change the faulty values of numeric columns in the dataset to use a common scale [15]. It is not compulsory to use every dataset in a model. It is generally used only when features of datasets for machine learning models have different ranges. Normalization of a data set can be done with help of the following formula [16].

$$Z\prime = \frac{Z - min_A}{max_A - min_A}$$

$$\begin{aligned} &\text{Where } Z\prime = \text{Value of Normalization} \\ &max_A = \text{Maximum value of a feature} \\ &min_A = \text{Minimum value of a feature} \end{aligned}$$

7 Feature Selection

In network traffic analysis, feature selection of the dataset plays a major role to improve accuracy for anomaly detection. Removing redundant and irrelevant features and adopting easily derivable and low-cost observations help in improving the performance. Using the filter method which uses statistical approaches for providing a ranking to the features, which is very useful for the selection of features of the datasets for the experiments.

Features in the NSL-KDD dataset are classified into three categories according to their relevance for anomaly detection: Negligible, low contribution and strong contribution [17]. Table 2 shows features with high relevance, Table 3 shows features with low relevance, and Table 4 shows features with minimal significance.

Table 2. A table with strong contribution (high relevance) features.

ID	TYPES/SCALE	FEATURE	GROUP
f3	nominal	Service	Basic
f4	nominal	Flag	Basic
f6	integer	dst_bytes	Basic
f8	integer	wrong_fragment	Basic
f23	integer	Count	Traffic
f25	Real	serror_rate	Traffic
f26	Real	srv_serror_rate	Traffic
f28	Real	srv_rerror_rate	Traffic
f29	Real	same_srv_rate	Traffic
f32	integer	dst-host_count	Traffic
f33	integer	dst_host_srv_count	Traffic
f34	Real	dst_host_same_srv_rate	Traffic
f35	Real	dst_host_diff_srv_rate	Traffic
f38	Real	dst_host_serror_rate	Traffic
f39	Real	dst_host_srv_serror_rate	Traffic
f40	Real	dst_host_rerror_rate	Traffic

A brief explanation of the relevance of the data features [18] is given below:

Land: The land attack can be identified by considering land features. In this type of attack, the attacker sends a spoofed SYN packet where the source and destination addresses are the same. If the land value in the data feature is 1, it can be considered as a land attack.

Wrong Fragment: The feature 'Wrong Fragment' provides some clues about malformed IP packets if it has a sum of bad checksum. Here, the attacker sends fragment packets in which the subsequent packets overlap with each other. Hence it is important to detect such a kind of attack.

Service, Duration, dst_host_same_srv_rate, same_srv_rate: These features are most important to detect smurf like Denial of Service (DoS) attack where the attacker sends a very large number of Internet Control Message Protocol (ICMP) echo messages to

Table 3. A table with low contribution (low relevance) features:

ID	TYPES/SCALE	FEATURE	GROUP
f2	nominal	protocol_type	Basic
f5	integer	src_bytes	Basic
f7	binary	Land	Basic
f10	integer	Hot	Content
f12	binary	logged_in	Content
f13	integer	num_compromised	Content
f14	binary	root_shell	Content
f15	binary	su_attempted	Content
f17	integer	num_file_creations	Content
f18	integer	num_shells	Content
f22	binary	is_guest_login	Content
f27	Real	rerror_rate	Traffic
f41	Real	dst_host_srv_rerror_rate	Traffic

Table 4. A table with negligible/ no relevance features:

ID	TYPES/SCALE	FEATURE	GROUP
f1	Integer	Duration	basic
f9	Integer	Urgent	basic
f11	Integer	num_failed_logins	content
f16	Integer	num_root	content
f19	Integer	num_access_files	content
f20	Integer	num_outbounds_cmds	content
f21	Binary	is_hot_login	content
f30	Real	diff_srv_rate	traffic
f31	Real	srv_diff_host_rate	traffic
f36	Real	dst_host_same_src_port_rate	traffic
f37	Real	dst_host_srv_diff_host_rate	traffic

a broadcast IP address with the source address of the spoofed address of victim's IP address. So this kind of attack can be detected by considering the features such as 'service', 'duration', 'dst_host_same_srv_rate' and 'same_srv_rate' (used to determine the percentage of connections) that are used to find the total number of ICMP echo packet sent to a particular machine within a limited duration.

Dst_bytes, Flag: These features are most useful to detect Ping of Death like Denial of Service (DoS) attacks where the attacker sends IP packets larger than 65,536 bytes that are allowed by the IP protocol. This attack can be detected by considering dst_bytes and flagging those packets.

Dst_host_same_srv_rate: 'Dst_host_same_srv_rate' Feature (used to find the connections to the same service) and 'flag' (represents connection status) are useful to examine the total number of ping messages during a short period of time. These ping packets are sent by an ipsweep attack to check which hosts are listening. If any host replies, the reply reveals the target's IP address.

Dst_host_diff_srv_rate: 'dst_host_diff_srv_rate' is important to detect synchronised (SYN) scan types of attacks. These attacks can be detected by checking large half-open connections that use 'flag' and 'dst_host_diff_srv_rate' (used to find the percentage of connections to different ports and the same destination IP).

However, the features presented here are not sufficient to consider to detect any attack, but a few features such as 'num_compromised', 'num_file_creations', 'Service', 'Count', 'dst-host_count', 'dst_host_srv_count', 'num_failed_logins" provides sufficient information to indicate about abnormal behaviour of the traffic.

8 Results and Discussion

For evaluation of the models, some performance metrics [36] that are popularly used have also been defined below. Metric is a performance evaluation method of ML algorithms towards a specific dataset. It provides a better way to compare, to determine which model performed better.

TP: The proposed model identifies an attack instance as an attack.

FP: The suggested model classifies an ordinary case as an attack.

FN: According to the suggested model, an attack incident is categorised as normal.

TN: The proposed model classifies a normal occurrence as normal.

ACCURACY: Accuracy is the most used metric. It is the ratio of correctly predicted output compared to total observed output. Accuracy is evaluated as

$$\text{Accuracy} = \frac{TP + TN}{TN + TP + FP + FN}$$

PRECISION: It can be defined from the confusion matrix as

$$\text{Precision} = \frac{P}{FP + TP}$$

RECALL: Recall is defined as

$$\text{Recall} = \frac{TP}{TP + FN}$$

F-MEASURE: f-measure can be explained as

$$\text{F} = \frac{2 * recall * precision}{recall + precision}$$

On various datasets, several machine learning models have been applied to assess their efficacy. Based on the literature analysis, the effectiveness of different machine learning methods on the NSL-KDD dataset and its parent dataset KDDcup99 is shown in Table 5.

Table 5. Summary of the implementation of various machine learning techniques done by other researchers and its accuracy.

Model	Dataset	Cyber Threat	Accuracy (%)	Published Year	Reference
Decision Tree (DT)	KDDCup99	DoS	91.5	2021	[19]
		R2L	93.6		
		U2R	97.61		
		Probe	99.88		
			99.50	2015	[20]
	NSL-KDD	DDoS	99.80	2022	[21]
		Intrusion	93.40	2019	[22]
			90.30	2017	[23]
Support Vector Machine (SVM)	KDDCup99	DoS	97.2	2021	[19]
		R2L	93.56		
		U2R	98.64		
		Probe	99.82		
			98.11	2015	[20]
	NSL-KDD	DDoS	89.18	2022	[21]
Naïve Bayes (NB)	KDDCup99	DoS	90.06	2021	[19]
		R2L	95.26		
		U2R	98.67		
		Probe	99.11		
			93.89	2015	[20]
	NSL-KDD	DDoS	92.12	2022	[21]
Random Forest (RF)	KDDCup99	DoS	99.99	2021	[19]
		R2L	94.44		
		U2R	98.51		
		Probe	99.90		
			85.06	2015	[20]
	NSL-KDD	DDoS	99.97	2022	[21]
Logistic Regression	KDDCup99	DoS	92.44	2021	[19]

(*continued*)

Table 5. *(continued)*

Model	Dataset	Cyber Threat	Accuracy (%)	Published Year	Reference
		R2L	99.98		
		U2R	98.68		
		Probe	99.89		
K-Nearest Neighbor (KNN)	KDDCup99	DoS	94.2	2021	[19]
		R2L	98.97		
		U2R	98.61		
		Probe	99.90		
			91.73	2015	[20]
	NSL-KDD	DDoS	98.57	2022	[21]

To assess how well the aforementioned literary work performed, we implemented some of them algorithms for classification like DT, KNN, SVM, NB, and RF classifiers. The algorithms are implemented on the NSL KDD data set. It has 42 features and 1,48,517 instances and the results are mapped into five main classes, i.e., DoS attack, Probe, U2R, R2L and normal. The work has four steps: data collection, preprocessing, classification and evaluation.

In the first step of the model i.e., in the data collection phase, the dataset "NSL KDD" has been collected from Kaggle which is available for free. In the data preprocessing phase, the data features are mapped and selected using the Random Forest algorithm. In the classification step, DT, NB, SVM, KNN and RF classifiers are used. The dataset is divided into two sets, one is the training set and another one is the test set using the 80–20 rule. In the evaluation step, the metrics like accuracy, precision, recall and f-measure are used to analyze the performance of the models. The models were implemented on Jupiter Notebook using Python. Table 6 displays the classification outcomes of the proposed machine learning models, and Fig. 2 is the graphical representation.

In Table 7, and Fig. 3 which is a graphical depiction of Table 7, we compare how the suggested model outperforms certain previous studies.

After analysis of the models, it can be concluded that the RF model got the highest accuracy to detect security threats, and the Decision Tree model also has a good accuracy among them. But the models that detect security threats should be trained continuously due to rapid changes in the types and features of network attacks.

Our experimental results are displayed in Fig. 3. The above results show that the RF, DT and Logistic Regression algorithms yield better accuracy than NB and SVM algorithms. In our experiment also, the Random Forest algorithm has better accuracy. From our experimental results and other literature that has been surveyed, it is observed that Random Forest and Decision Tree algorithm gives better results for classifying network security threats.

Some issues that have been identified after the literature review are discussed below-

Table 6. Performance evaluation of the suggested model.

Model	Results	DoS	Probe	R2L	U2R
Random Forest	Accuracy	99.82	99.63	98.05	99.74
	Precision	99.89	99.62	97.58	96.16
	Recall	99.67	99.35	97.09	85.21
	F-measure	99.83	99.42	96.97	90.78
Support Vector Machine (SVM)	Accuracy	99.37	98.45	96.79	99.63
	Precision	99.11	96.91	94.85	91.06
	Recall	99.45	98.37	96.26	82.91
	F-measure	99.28	97.61	95.53	84.87
K-Nearest Neighbors (KNN)	Accuracy	99.72	99.08	96.74	99.70
	Precision	99.68	98.61	95.31	93.14
	Recall	99.67	98.51	95.49	85.07
	F-measure	99.67	98.55	95.39	87.83
Decision Tree	Accuracy	99.64	99.61	97.90	99.67
	Precision	99.47	99.37	97.19	88.53
	Recall	99.68	99.31	96.90	91.70
	F-measure	99.60	99.25	96.96	90.11
Naïve Bayes	Accuracy	86.73	97.99	93.56	97.26
	Precision	98.82	97.32	89.10	60.16
	Recall	70.31	96.05	95.51	97.91
	F-measure	82.15	96.65	91.62	66.09

I. Lack of a standard and updated benchmark dataset: To analyse the performance of the in the realm of machine learning, sophisticated ML models for network security threat detection and for better accuracy of any ML model, the dataset must have representative information and the latest content. The available datasets contain very less diversity and advanced attacks and contain null values [24].
II. Lack of appropriate ML algorithms:

Most of the problems in the detection of network security threats are lack of sufficient data. Hence it can be considered that the ensemble of both supervised and unsupervised machine learning techniques may be a better solution to get better performance [24].

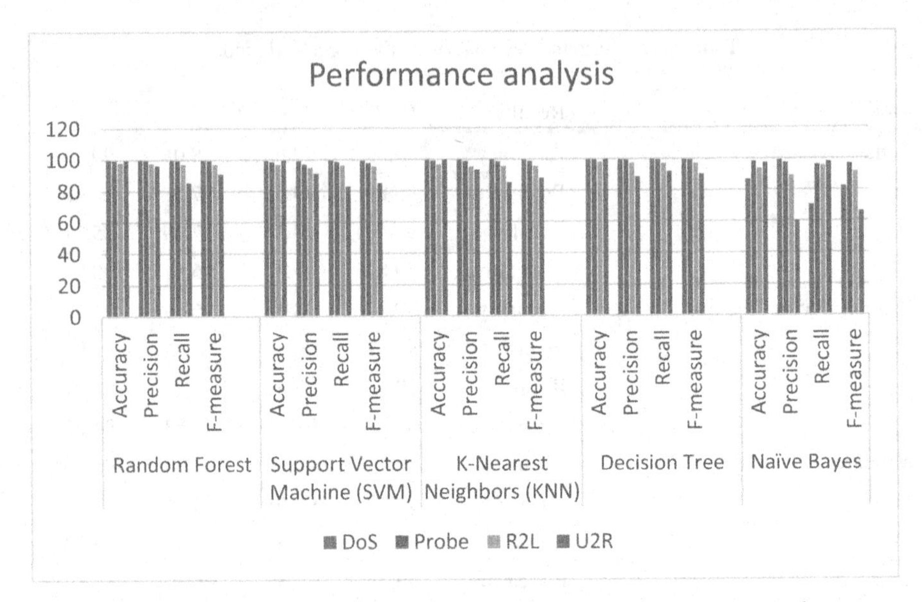

Fig. 2. An summary of different ML techniques that have been compared.

Table 7. An overview of the comparison of proposed model with some others' works.

Model	Accuracy in proposed model	Accuracy in [43]	Accuracy in [44]	Accuracy in [45]
Decision Tree	99.64	93	X	81.05
Naïve Bayes	96.82	91	83.63	X
Support Vector Machine	99.55	X	98.23	83.09
K-Nearest Neighbor	99.53	X	95.13	94.17
Random Forest	99.81	94	99.81	99.0

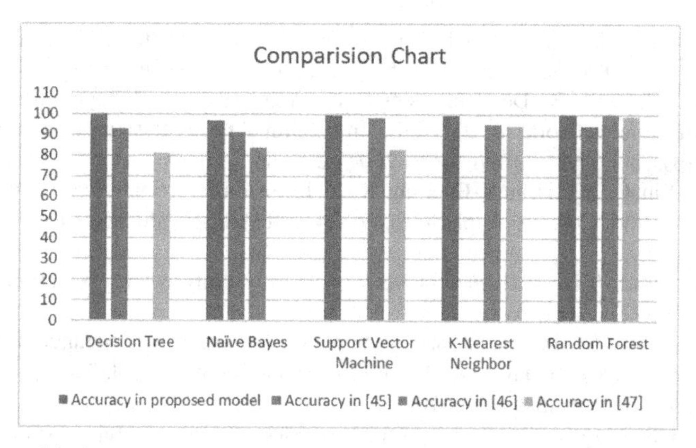

Fig. 3. The graphical representation of the comparison of proposed model with some others' works.

9 Conclusion

Here a comprehensive review of works of literature has been undertaken in the area of ML techniques used in the analysis and detection of network security threats. Different approaches are used to detect network security threats and to improve system performance and security. Today's cybercriminals are highly skilled to attack any organization and adapt new techniques every day to access network resources, and many new protocols and IoT are being added to the network system leading the system to be more complex. So improved network security techniques are still required to improve security management. We have reviewed the literature on various supervised ML techniques used to identify cyber security concerns in this work. Recent ML tools and methods are focused and solutions to the different categories of cyber threats are discussed and summarized. This work would provide insight into the topic to the new researchers.

The following can be used as an overview of the area's potential for future research:

- Comparison of performance of classification and cyber threats detection methods for detection of network security threats.
- Studying other types of attacks available in today's computer networks.
- Application of several ML and Deep Learning (DL) methods for the detection of cyber security threats.
- Developing a high-performance ML and DL model for analysing a large amount of network traffic data.

References

1. Alqudah, N., Yaseen, Q.: Machine learning for traffic analysis: a review. Procedia Comput. Sci. **170**, 911–916 (2020). https://doi.org/10.1016/j.procs.2020.03.111
2. Geetha, R., Thilagam, T.: A review on the effectiveness of machine learning and deep learning algorithms for cyber security. Arch. Comput. Meth. Eng. **28**(4), 2861–2879 (2021). https://doi.org/10.1007/s11831-020-09478-2

3. Rakshe, T., Gonjari, V.: Anomaly based network intrusion detection using machine learning techniques. Int. J. Eng. Res. Technol. **6**(5), 216–220 (2017). www.ijert.org
4. Loukas, G., Gelenbe, E.: Detecting denial of service attacks with bayesian classifiers and the random neural network. In: 2007 IEEE International Fuzzy Systems Conference, pp. 1–6 (2007). https://doi.org/10.1109/FUZZY.2007.4295666
5. Liao, H.-J., Lin, C.-H.R., Lin, Y.-C., Tung, K.-Y.: Intrusion detection system: a comprehensive review. J. Network Comput. Appl. **36**(1), 16–24 (2013). https://doi.org/10.1016/j.jnca.2012.09.004
6. Joshi, M.R., Hadi, T.H.: A Review of Network Traffic Analysis and Prediction Techniques. arXiv:1507.05722 (2015)
7. Hasan, M., Nasser, M., Pal, B., Ahmad, S.: Support vector machine and random forest modeling for Intrusion Detection System (IDS). J. Intell. Learn. Syst. Appl. **06**(01), 45–52 (2014). https://doi.org/10.4236/jilsa.2014.61005
8. Mukkamala, S., Janoski, G., Sung, A.: Intrusion detection using neural networks and support vector machines. In: Proceedings of the International Joint Conference on Neural Networks, vol. 2, pp. 1702–1707 (2002). https://doi.org/10.1109/ijcnn.2002.1007774
9. Maseer, Z.K., Yusof, R., Bahaman, N., Mostafa, S.A., Foozy, C.F.M.: Benchmarking of machine learning for anomaly based intrusion detection systems in the CICIDS2017 dataset. IEEE Access **9**, 22351–22370 (2021). https://doi.org/10.1109/ACCESS.2021.3056614
10. Ewards, V.S.: Survey of traffic classification using machine learning. Int. J. Adv. Res. Comput. Sci. **4**(4). www.ijarcs.info
11. Farnaaz, N., Jabbar, M.A.: Random forest modeling for network intrusion detection system. Procedia Comput. Sci. **89**, 213–217 (2016). https://doi.org/10.1016/j.procs.2016.06.047
12. Meeragandhi, G.: Machine learning approach for attack prediction and classification using supervised learning algorithms. Int. J. Comput. Sci. Commun. **1**(2), 247–250 (2010)
13. Dhanabal, L., Shantharajah, S.P.: A Study on NSL-KDD dataset for intrusion detection system based on classification algorithms. Int. J. Adv. Res. Comput. Commun. Eng. **4** (2015). https://doi.org/10.17148/IJARCCE.2015.4696
14. Nikhitha, M., Jabbar, M.A.: K nearest neighbor based model for intrusion detection system. Int. J. Recent Technol. Eng. **8**(2), 2258–2262 (2019). https://doi.org/10.35940/ijrte.b2458.078219
15. Harb, H.M.: Selecting optimal subset of features for intrusion detection systems. Adv. Comput. Sci. Technol. **4**(2), 179–192 (2011). http://www.ripublication.com/acst.htm
16. Lala, C., Panda, B.: Evaluating Damage from Cyber Attacks: A Model and Analysis (2001)
17. Cristalli, S., Pagnozzi, M., Graziano, M., Lanzi, A.: Micro-virtualization memory tracing to detect and prevent spraying attacks, p. 48 (2005)
18. Schweitzer, N., Stulman, A., Margalit, R.D., Shabtai, A.: Contradiction based gray-hole attack minimization for Ad-Hoc networks. IEEE Trans. Mob. Comput. **16**(8), 2174–2183 (2017). https://doi.org/10.1109/TMC.2016.2622707
19. Gurung, S., Chauhan, S.: A novel approach for mitigating gray hole at-tack in MANET. Wireless Netw. **24**(2), 565–579 (2018). https://doi.org/10.1007/s11276-016-1353-5
20. Bosnjak, L., Sres, J., Brumen, B.: Brute-force and dictionary attack on hashed real-world passwords. In: 41st International Convention on Infor-mation and Communication Technology, Electronics and Microelectronics, pp. 1161–1166. Institute of Electrical and Electronics Engineers Inc. (2018). https://doi.org/10.23919/MIPRO.2018.8400211
21. Ait Tchakoucht, T., Ezziyyani, M.: Building a fast intrusion detection system for high-speed-networks: probe and dos attacks detection. In: Procedia Computer Science, Elsevier B.V., pp. 521–530 (2018). https://doi.org/10.1016/j.procs.2018.01.151
22. Wu, Y., Wei, D., Feng, J.: Network attacks detection methods based on deep learning techniques: a survey. Security and Communication Networks, vol. 2020. Hindawi Limited (2020). https://doi.org/10.1155/2020/8872923

23. Hatcher, W.G., Yu, W.: A survey of deep learning: platforms, applications and emerging research trends. IEEE Access **6**, 24411–24432 (2018). https://doi.org/10.1109/ACCESS.2018.2830661
24. Huang, H.H., Liu, H.: Big data machine learning and graph analytics: current state and future challengesWashington DC, USA. In: IEEE International Conference on Big Data (2014)
25. Muralidharan, R., Gifty Jeya, P., Ravichandran, M., Ravichandran, C.S.: Efficient classifier for R2L and U2R attacks. Int. J. Comput. Appl. **45**(21), 27–32 (2012). https://www.researchgate.net/publication/267558979
26. Pacheco, F., et al.: Towards the deployment of machine learning solutions in network traffic classification: a systematic survey. Commun. Surv. Tutor. **21**(2), 1988–2014 (2018). https://doi.org/10.1109/COMST.2018.2883147
27. Xiang, L., Ge, X.-H., Liu, C., Shu, L., Wang, C.-X.: A new hybrid network traffic prediction method. In: 2010 IEEE Global Telecommunications Conference GLOBECOM 2010 (2010)
28. Meidan, Y., et al.: ProfilIoT: a machine learning approach for IoT device identification based on network traffic analysis. In: Proceedings of the ACM Symposium on Applied Computing, vol. Part F128005, pp. 506–509 (2017). https://doi.org/10.1145/3019612.3019878
29. Alshammari, A., Aldribi, A.: Apply machine learning techniques to detect malicious network traffic in cloud computing. J. Big Data **8**(1) (2021). https://doi.org/10.1186/s40537-021-00475-1
30. Shaukat, K., Luo, S., Chen, S., Liu, D.: Cyber threat detection using machine learning techniques: a performance evaluation perspective. In: 1st Annual International Conference on Cyber Warfare and Security, ICCWS 2020 – Proceedings (2020). https://doi.org/10.1109/ICCWS48432.2020.9292388
31. MIT Lincoln Laboratory. https://www.ll.mit.edu/r-d/datasets/1998-darpa-intrusion-detection-evaluation-dataset. Accessed 3 Jan 2023
32. Kaggle dataset. https://www.kaggle.com/datasets/galaxyh/kdd-cup-1999-data?select=corrected.gz. Accessed 3 Jan 2023
33. University of New Brunswick (UNB) datasets. https://www.unb.ca/cic/datasets/ids-2017.html. Accessed 3 Jan 2023
34. Github dataset. https://nonsns.github.io/dataset/ledbat+bittorrent. Accessed 3 Jan 2023
35. UNSW Sydney dataset. https://research.unsw.edu.au/projects/unsw-nb15-dataset. Accessed 3 Jan 2023
36. Javapoint homepage. https://www.javatpoint.com/normalization-in-machine-learning. Accessed 3 Jan 2023
37. Gautam, R.K.S., Doegar, Er.A.: An ensemble approach for intrusion detection system using machine learning algorithms. In: 8th International Conference on Cloud Computing, Data Science & Engineering (Confluence) (2018)
38. Iglesias, F., Zseby, T.: Analysis of network traffic features for anomaly detection. Mach. Learn. **101**(1–3), 59–84 (2015). https://doi.org/10.1007/s10994-014-5473-9
39. Mishra, P., Varadharajan, V., Tupakula, U., Pilli, E.S.: A detailed investigation and analysis of using machine learning techniques for intrusion detection. IEEE Commun. Surv. Tutorials **21**(1), 686–728 (2019). https://doi.org/10.1109/COMST.2018.2847722
40. Afolabi, H.A., Aburas, A.: Comparison of single and ensemble intrusion detection techniques using multiple datasets. Int. J. Adv. Trends Comput. Sci. Eng. **10**(4), 2752–2761 (2021). https://doi.org/10.30534/ijatcse/2021/161042021
41. Srikanth, U., Priyadharsini, S.: Prediction of network attacks using machine learning techniques. Int. J. Eng. Appl. Sci. Technol. **5**(10) (2021). https://doi.org/10.33564/IJEAST.2021.v05i10.017
42. Ambedkar, C., Kishore Babu, V.: Detection of Probe Attacks Using Machine Learning Techniques (2015). www.arcjournals.org

43. Ingre, B., Yadav, A., Soni, A.K.: Decision tree based intrusion detection system for NSL-KDD dataset. Smart Innov. Syst. Technol. **84**, 207–218 (2018). https://doi.org/10.1007/978-3-319-63645-0_23
44. Nadeem, M.W., Goh, H.G., Ponnusamy, V., Aun, Y.: Ddos detection in sdn usingmachine learning techniques. Comput. Mater. Continua **71**(1), 771–789 (2022). https://doi.org/10.32604/cmc.2022.021669
45. Ahmim, A., Maglaras, L., Ferrag, M.A., Derdour, M., Janicke, H.: A novel hierarchical intrusion detection system based on decision tree and rules-based models. In: Proceedings - 15th Annual International Conference on Distributed Computing in Sensor Systems, DCOSS 2019, pp. 228–233. (2019). https://doi.org/10.1109/DCOSS.2019.00059
46. Mohammed Qasim, O., Hashim Kraidi Al-Saedi, K., Hashim Al-Saedi, K.: Malware detection using data mining naïve bayesian classification technique with worm dataset. Int. J. Adv. Res. Comput. Commun. Eng. ISO **3297**(11) (2007). https://doi.org/10.17148/IJARCCE.2017.61131
47. Alqahtani, H., et al.: Cyber intrusion detection using machine learning classification techniques. In: Chaubey, N., Parikh, S., Amin, K. (eds.) Computing Science, Communication and Security: First International Conference, COMS2 2020, Gujarat, India, March 26–27, 2020, Revised Selected Papers, pp. 121–131. Springer Singapore, Singapore (2020). https://doi.org/10.1007/978-981-15-6648-6_10
48. Sumaiya Thaseen, I., Poorva, B., Ushasree, P.S.: Network intrusion detection using machine learning techniques. In: International Conference on Emerging Trends in Information Technology and Engineering, ic-ETITE 2020 (2020). https://doi.org/10.1109/ic-ETITE47903.2020.148
49. Devi, R.R., Abualkibash, M.: Intrusion detection system classification using different machine learning algorithms on KDD-99 and NSL-KDD datasets - a review paper. Int. J. Comput. Sci. Inform. Technol. **11**(03), 65–80 (2019). https://doi.org/10.5121/ijcsit.2019.11306
50. Vishnu Balan, E., Priyan, M.K., Gokulnath, C., Usha Devi, G.: Fuzzy based intrusion detection systems in MANET. Procedia Comput. Sci. **50**, 109–114 (2015). https://doi.org/10.1016/j.procs.2015.04.071

Optimized Traffic Management in Software Defined Networking

M. P. Ramkumar[1](✉), J. Lece Elizabeth Rani[1], R. Jeyarohini[2], G. S. R. Emil Selvan[1], and S. Arun Karthick[1]

[1] Department of Computer Science and Engineering, Thiagarajar College of Engineering, Madurai, India
{ramkumar,emil}@tce.edu, arunkarthicks@student.tce.edu

[2] Department of Electronics and Communication Engineering, PSNA College of Engineering and Technology, Dindigul, India
rjreee2008@gmail.com

Abstract. In recent years, the data center network has improved its rapid exchanging abilities. Software Defined Network (SDN) architecture adds functionality to the node globally by means of separation of the control and data plane. Switches and routers are network devices that are inflexible to deal with different loads of traffic in a network and experience limitations on routing and load balancing. Data centers are utilized whenever high user traffic is encountered. Network operators manage the network by controlling the traffic and providing optimal resource allocation strategies to their applications. A path optimization algorithm is proposed to increase the performance of SDN. The proposed model's efficiency is demonstrated and simulated.

Keywords: Datacenter Networks · SDN · Traffic Management · Path optimization · Load balancing

1 Introduction

A tremendous increase in cloud computing and big data technologies has impacted a higher demand for scaling the data center networks toward communication and management capabilities [1]. Yet many techniques have been arising to enhance the performance of Data Center Networks (DCNs). Managing and Configuring Computer networks are dynamic and complex [2]. The key role is played by the network Operator who is responsible for the configuration of the Network [3] with a huge number of switches, routers, and many firewalls [4]. The Networks today provide more automated mechanisms with wide scope [5]. The entire network is monitored by the controller of the control plane. The control plane provides control logic for the proper network functioning and it acts as a brain [6, 7] of the network. The SDN can integrate with the applications of big data, provide some of the services of data center, and to an extend it can support rich multimedia content [8]. Contrary to traditional methods, SDN adds an open interface layer that runs on hardware, between packet forwarding hardware and the network operating system

P. Das et al. (Eds.): AMRIT 2023, CCIS 1953, pp. 157–168, 2024.
https://doi.org/10.1007/978-3-031-47224-4_14

[9]. As today's data centers require different needs, SDN network administration brings in a more dynamic approach to adjust the network-wide traffic flow [10]. Open Flow protocol (OF) is one of the standardized protocols used for network communication in SDN architectures [11]. It has been proposed a framework in software-defined and load-balanced data center, where a huge variety of trials were accomplished for the function evaluation [12]. Moreover, this proposed method is assessed with major data centers. In this, the OpenFlow (OF) [13] controllers and switches are executed using floodlight and OVS [14, 15]. The outcome shows that the average delay and consumption of resources are enhanced [16] Gleaned from SDN technology, they proposed a novel method which is a Top-Down Parsing Language (TPDL) based method [17, 18], and look forward to functioning on various topologies. A* algorithm is used for the path selections to speed up, whereas distance formulas are the heuristic functions. The outcome shows that the proposed method outperforms in calculating the routing rendition and error forbearance [19]. To overcome some issues in the network a novel hybrid algorithm called Genetic Algorithm and Ant Colony Optimization (G-ACO) algorithm was proposed. It finds the ideal path for the large flow of overcrowding links and also redirects that [20]. Many trials were carried out to test this Method and the outcome shows that this method minimizes the link utilization and enhances the bandwidth [21, 22] Various methods are proposed for load balancing using an open daylight controller and Ryu controller with ant colony optimization algorithm. Since the bandwidth performance of the Open Network Operating System (ONOS) controller is better than these controllers [23]. Local optimization is the major issue in the ACO algorithm. It is not fully solved and that leads to poor network performance and secondary collision on large flows [24]. SDN is a software-based approach that controls and monitors the network by designing, building, and managing the network. The control plane is separated from the data plane and is an abstraction through which a single control plane can manage several data planes. Control planes are software-based system that decides in which path the packet should be transmitted while a data plane is a packet forwarder that forwards those packets in that desired path [16]. The SDN network administrators can easily manage the services in the network from a centralized location without manually configuring the route or a switch. The concept of centralizing the control plane is by allowing the forward decision to be merged globally across the SDN domain rather than at each Hop [25]. In SDN too many relationships exist to control and manage a thousand devices with a single command. It enables centralized management of networking devices. Since administrators can adjust the network-wide traffic flow dynamically to satisfy every service, SDN architecture is agile. The most awesome thing about SDN is an abstraction, the centralized controller abstracts the Network so any changes to be made can be done in one location that is, a controller [26]. The SDN based controller aids the benefits like scalability, efficiency and flexibility by means of automation of networking devices.

2 Material and Method

2.1 SDN Architecture

The SDN architecture of any controller comprises various components, and a few of the fundamental features that exist in the controllers irrespective of the vendors are CPU, RAM, Storage, Networking (Infrastructure layer), topology and device manager (Control layer), and Java, Python API (Application layer) [27].

Network Devices. They can be either physical or virtual. The primary function is forwarding in the network. Here the network devices have a data plane only.

SDN Controller. The network environment of SDN is functioning intelligently with the support of a logical component is called the SDN controller. Apart from that it also establishes an effective communication between the networking devices involved in the environment.

Application and Services. A software application known as an SDN application is created to carry out activities in an SDN environment [13].

Southbound Interface. It is an OpenFlow protocol specification that enables the communication between the network component and lower-level details.

Northbound Interface. The network program that is fetching the service from the interface for allowing the components to communicate with higher-level features.

Network Operating System. The network service provider builds a carrier-grade SDN to enhance high scalability, availability, and performance in the network based on service providers' requirements [22].

2.2 Planes in SDN Architecture

- Forwarding plane
- Control plane
- Operational plane
- Management plane
- Application plane

Forwarding Plane/Data Plane. Using the instructions given by the control plane, it is in charge of forwarding the data packets. Packet dumping, packet modifying, and packet forwarding are the three fundamental responsibilities of forwarding planes [28].

Control Plane. Packet forwarding and sending commands to the network are carried out by the control plane. The objective is to construct the infrastructure layer's flow table based on the topology.

Operational Plane. The status of the networking device is managed by the operational plane.

Management Plane. The management plane is liable for configuring, monitoring and maintenance of network devices. It is also make decision about the state of the network.

Application Plane. The applications are basically programs that instructs the controller to perform specific functions like network virtualization, controlling the data flow and monitoring the network.

2.3 Layers in SDN Architecture

- Infrastructure layer
- Control layer
- Application layer

Infrastructure Layer. The processing and forwarding of data packets are carried out by the infrastructure layer [29].

Control Layer. The SDN has the intelligence by means of keeping the control planes that is used to control the network infrastructure. The various networking information is created, fetched and maintained.

Application Layer. It is an open area to develop as much innovation as possible by leveraging network information. It enables the user to use the network services.

3 Proposed Method

The Proposed methodology, G-ACO is explained along with the limitations of the previous Method ACO.

3.1 LB with ACO

Ant leaves pheromone when it travels in a route searching for food, and other ants follow the pheromone trail. Through these, ants can figure out the shortest path. ACO Algorithm used in networks, the ants are assumed as data packets, the way is simulated using network topology, and the pheromone is used to find the optimal path [16, 24]. The probability to select switch j by an ant k, sitting at the place i given by Eq. (1).

$$P_{ij}^{k} = \frac{\left[T_{ij}\right]^{\alpha}\left[\eta_{ij}\right]^{\beta}}{\sum_{\delta \in allowed_k}\left[T_{ij}\right]^{\alpha}\left[\eta_{ij}\right]^{\beta}}, j \in allowed_k \tag{1}$$

$$P_{ij}^{k} = 0|otherwise, j \in allowed_k \tag{2}$$

The Regulating factors for T_{ij} are α and β, respectively. The pheromone values between each pair of switches are represented as T_{ij}. The Distance between buttons is defined as d_{ij}, and the clear path is measured using the formula $\frac{1}{d_{ij}}$. $allowed_k$ and k is the number of switches that have not been visited yet [25]. The pheromones are updated so that the shorter path gets a more intense pheromone than the longer path, and the updated pheromone is calculated using Eq. (2). The emulator and Ryu controller for enhanced Efficiency [28]. Ryu controller is used in managing the OF protocol, which is a southbound protocol [29].

3.2 Initial Population

The integral numbers are used in the path-searching function. The number of paths formed is exactly equal to the total number of ants. The packet's code is generated based on the switches that have been approached by a packet where the packets are assumed as ants. For example, the switches closed are S9, S5, S1, S3, and S7 then, the packet code is (9,5,1,3,7), and it is valid for that iteration.

3.3 Fitness Function

The factors deciding fitness functions are path length, energy consumption for sending and receiving a packet, and network energy [30]. In Our methodology, the path length is taken for the fitness function.

$$F(p) = W * l(p) \tag{3}$$

The length-weight and path length is represented by W and l(p), respectively, as shown in Eq. (3). The smallest value of the fitness function is chosen later.

3.4 Selection

In every iteration, the code of the path can be obtained. The Weight is obtained from previous steps, and the optimal value is passed on to the next iteration. By doing this process for each iteration, the optimal path can be obtained.

3.5 Crossover

The algorithm 1 explains the crossover method that is used to create more paths and then finds optimal and sub-optimal paths. Let's consider B1 and B2 as two parents.

B1= (2,3|4,5,6|7,8,9) B2= (5,2|9,6,3|8,4,7)
The offspring by crossover operation, b1= (**|4,5,6|***)
b2= (**|9,6,3|***)
b1= (2,9|4,5,6|3,8,7)
b2= (2,4|9,6,3|5,7,8)

let's take B1 = optimal path and B2 = sub-optimal path.

ALGORITHM 1: CROSSOVER Algorithm

```
1.  Let's take B1 is (a1, b1, c1) and path B2 is
    (a2,b2,c2)
2.  Crossover: b1 and b2
3.  New path B3 :(a1, b2, b1, c1)
4.  Delete the duplicate switch, and the new path B3 is
    determined.
5.  In the same crossover operation, path B3 is formed.
6.  FF (B1, B2, B3, B4) and selected the optimal path.
```

3.6 Mutation

The algorithm 2 explains the mutation from which the predefined values are taken. Let's take two paths for mutation operation.

$$h1 = (2, 9|3, 8, 5|7, 6, 4)$$

After the process, 9 and 5 get changed, h1 is changed as h' = (2,5|3,8,9|7,6,4).

```
ALGORITHM 2: MUTATION Algorithm
1.  M: frequency (simulation part, number of switches
    from the optimal path)
2.  Random numbers m1 and m2(m2<m1<A)
3.  Exchange operation m2 and m1 in h1 gives
4.  Fitness (h1, h') and select a smaller one
```

The algorithm 3 explains the load balancing with the G-ACO algorithm.

```
ALGORITHM 3: LB with G-ACO
1.  Mc+=1; the number of iterations
2.  Yk=1;
3.  Yk: switch selection and updation of pheromone.
4.  K+1
5.  If k<=A go to step 3, or else continue to proceed GA;
6.  Get the FF value and if it satisfies the rule, end
    the process. Otherwise, proceed from step 1.
```

The packets get easily transmitted through the optimal path after the G-ACO algorithm.

4 Experimental Analysis

Ryu controller and Mininet are utilized for the controller and switches setup. The Ryu controller provides routing with more flexibility and increases the network's agility. The advantage of Ryu is that, it is an open SDN controller which provides software components and a clear API. The developers can easily design the applications required for the network management and control. Also, it helps to achieve maximum network agility. The developers can easily tweak the existing components or create new ones to guarantee the underlying Network with this strategy, which helps companies adapt deployments to fit particular needs. The Ryu controller interacts with the forwarding plane using OpenFlow or other protocols to change the traffic flow on the network.

For more adaptable custom routing and SDN, the switches enable OpenFlow. A less expensive and simple testbed for creating OpenFlow applications is provided by Mininet. Additionally, the more complicated topology can be tested even without the physical network, making it possible to create networks and conduct experiments using an extendable Python API. Fat tree topology is used in this Method which is the most commonly used topology. The Sample topology is given in Fig. 1, where C0: Controller S1- S4: Switches and h1- h5: Hosts.

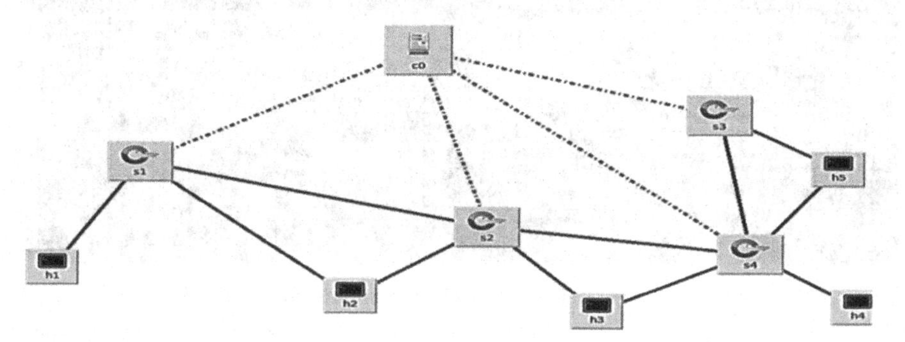

Fig. 1. Sample topology

The Sample topology configuration is given in Fig. 2.

```
dell@dell-Vostro-15-3568: ~
File Edit View Search Terminal Help
dell@dell-Vostro-15-3568:~$ sudo ~/mininet/examples/miniedit.py
[sudo] password for dell:
topo=none
Getting Hosts and Switches.
Getting controller selection:ref
Getting Links.
*** Configuring hosts
h1 h5 h4 h2 h3
**** Starting 1 controllers
c0
**** Starting 4 switches
s3 s4 s2 s1
No NetFlow targets specified.
No sFlow targets specified.
*** Stopping 1 controllers
c0
*** Stopping 11 links
...........
*** Stopping 4 switches
s3 s4 s2 s1
*** Stopping 5 hosts
h1 h5 h4 h2 h3
*** Done
```

Fig. 2. Sample topology Configuration

The Fat tree topology configuration is given below in Fig. 3.

In the proposed G-ACO method, the ant selects the path based on two main parameters such as α and β. The α regulates the pheromone dependency. The β maximizes the possibility of finding the following random way for the ant. To find out a proper value for the α and β, the simulations were accomplished.

In this method, the major parameter is the quantity of ants(A) in numbers. When this parameter is very few, the entire network search will be minimized, leading to a local optimum. In the case of very large, it might provide a path by random selection. Thus $\alpha = 1$, $\beta = 2$, $m = 8$, $\rho = 0.4$ are the parameter values whereas the mutation

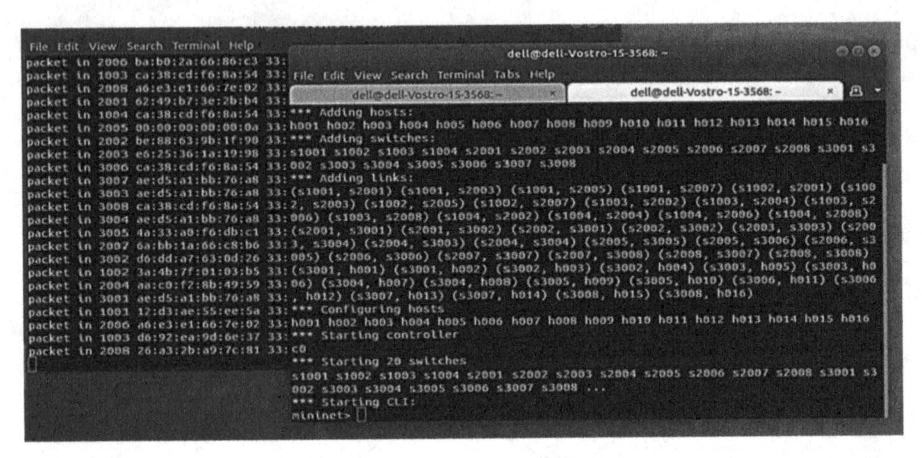

Fig. 3. Fat tree topology Configuration

and crossover probabilities are 0.04 and 0.9. At least 0.1 is the pheromone difference between the largest and smallest path.

The scapy tool is utilized for the 5 min and 10 min simulation run. Scapy, a packet manipulation tool that can decode packets and match the requests and replies. Additionally, handles tasks like scanning, probing, tests, and attacks. The results of 5 min' simulation time are revealed in Figs. 4 and 5. The ACO, Round Robin (RR), and G-ACO are simulated based on Round-Trip Time (RTT) and packet loss rate. It is observed that the ACO and G-ACO have almost the same RTT and packet loss rates. The computational overlooking of the Genetic algorithm in starting phase is more. In contrast, the data needed for finding an ideal path for the proposed Method is not enough to decrease the simulation time by less than 5 min. That makes the G-ACO and ACO have identical performance.

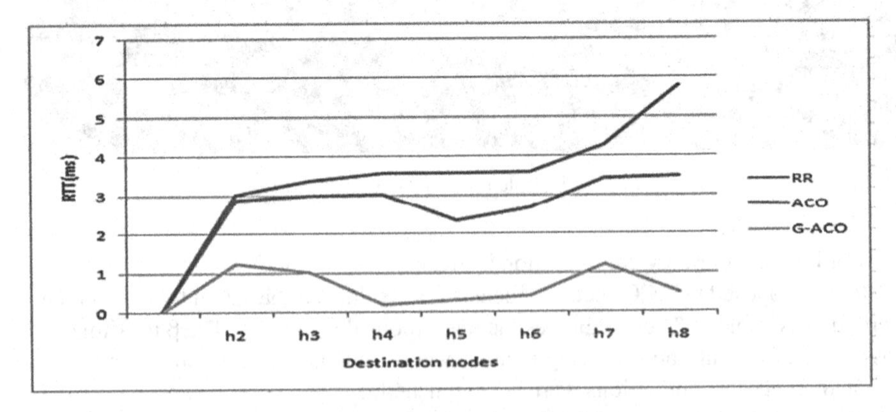

Fig. 4. RTT comparison (5 min)

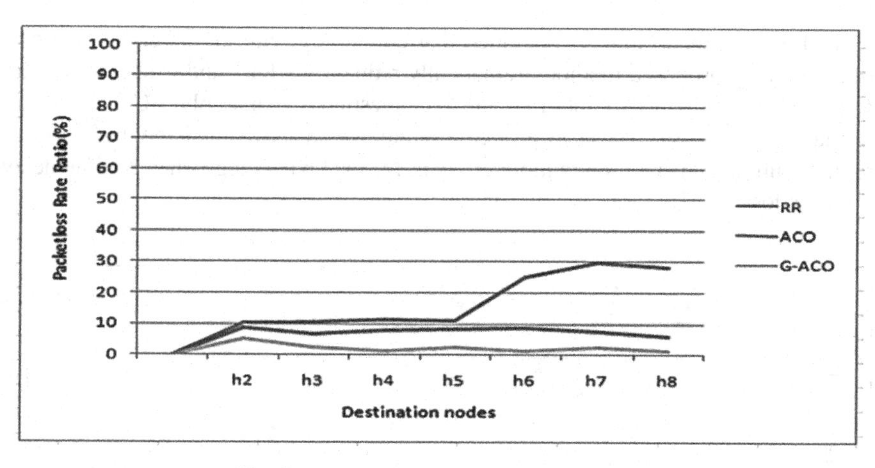

Fig. 5. Packet loss rate comparison (5 min)

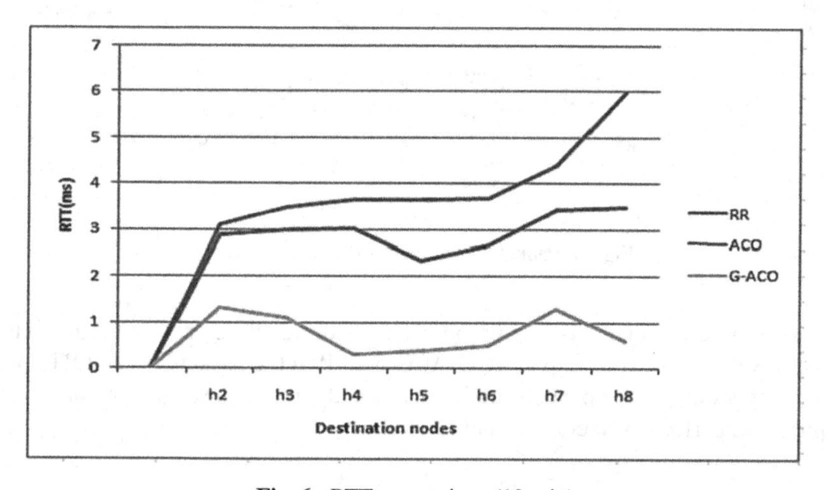

Fig. 6. RTT comparison (10 min)

The Figs. 6 and 7 shows the results of 10 min' simulation time of RR, ACO and C-ACO algorithms. It has been observed that, if the network operations are carried out longer, additional information can be obtained so that the packet loss and RTT of the G-ACO method turn down compared to existing methods. Recollecting that the proposed method has benefits of GA and ACO, which involves the quick and ideal search.

By observing the graph, it concluded that comparing with the other two existing methods and the proposed method dramatically reduces the RTT and packet loss. With RR, h1-h6 has a more packet loss rate due to congestion. Comparably, ACO has higher RTT than RR whereas it has a low packet loss rate, which shows the benefit of ACO on load distribution. This method is more efficient for load balancing, whereas time delay and packet loss rate are more or less similar.

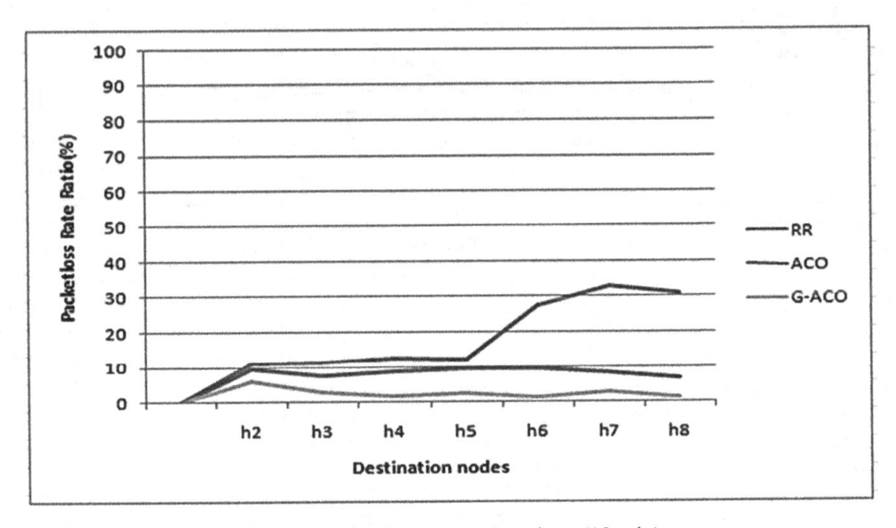

Fig. 7. Packet loss rate comparison (10 min)

Comparing other methods, the proposed method broadly gives out the packets in the entire Network. While combining G-ACO with Ryu the packet loss and RTT were significantly reduced. The packets are levelly issued with the help of more nodes, and that promotes performance enhancement.

5 Conclusion and Future Work

SDN is more useful in the fields of IoT, Cloud Integration, and Cloud services. SDN Control plane controls the Network in a centralized and automated manner. In SDN, OF protocol supports the SDC-CO in the control plane. Data Center Networks are said to benefit from the potential of SDN, which can offer centralized network management and traffic control. The OF protocol establishes the communication between control and data layer. Future research on energy efficiency might extend the scope of SDN. The optimization algorithm is that improves the performance may provide high throughput and balanced load in Software-Defined Data Centers.

References

1. Jia, Z., et al.: cRetor: an SDN-based routing scheme for data centers with regular topologies. IEEE Access **8**, 116866–116880 (2020)

2. Chahlaoui, F., Dahmouni, H.: A taxonomy of load balancing mechanisms in centralized and distributed SDN architectures. SN Comput. Sci. **1**(5), 268 (2020)
3. Madhukrishna, P., et al.: An energy-efficient load distribution framework for SDN controllers. Computing **102**, 2073–2098 (2020)
4. Li, J., et al.: A load balancing approach for distributed SDN architecture based on sharing data store. In: 2020 21st Asia-Pacific Network Operations and Management Symposium (APNOMS). IEEE (2020)
5. Pooja, M., Wilfred Godfrey, W., Kumar, N.: A green computing-based algorithm in software defined network with enhanced performance. In: 2021 International Conference on Computing, Communication, and Intelligent Systems (ICCCIS). IEEE (2021)
6. Hodaei, A., Babaie, S.: A survey on traffic management in software-defined networks: challenges, effective approaches, and potential measures. Wireless Pers. Commun. **118**(2), 1507–1534 (2021)
7. Lei, J., et al.: Energy-saving traffic scheduling in backbone networks with software-defined networks. Cluster Comput. **24**, 279–292 (2021)
8. Chiang, M.-L., et al.: SDN-based server clusters with dynamic load balancing and performance improvement. Cluster Comput. **24**, 537–558 (2021)
9. Bavani, K., Ramkumar, M.P., Selvan, E.: GSR. Statistical approach based detection of distributed denial of service attack in a software defined network. In: 2020 6th International Conference on Advanced Computing and Communication Systems (ICACCS). IEEE (2020)
10. Mohammed, S.H., Jasim, A.D.: Evaluation of firewall and load balance in fat-tree topology based on floodlight controller. Indon. J. Electric. Eng. Comput. Sci. **17**(3), 1157–1164 (2020)
11. Iyer, N., et al.: Load balancing using openday light SDN controller: case study. Int. Res. J. Adv. Sci. Hub **2.9**, 59–64 (2020)
12. Kadim, U.N., Mohammed, I.J.: A hybrid software defined networking-based load balancing and scheduling mechanism for cloud data centers. J. Southwest Jiaotong Univ. **55**(3) (2020). https://doi.org/10.35741/issn.0258-2724.55.3.3
13. Mohammad Riyaz, B., et al.: A systematic review of load balancing techniques in software-defined networking. IEEE Access **8**, 98612–98636 (2020)
14. Muthuperumal Periyaperumal, R., et al.: Deep maxout network for lung cancer detection using optimization algorithm in smart Internet of Things. Concurr. Comput. Practice Exper. **34.25**, e7264 (2022)
15. Deeban Chakravarthy, V., Amutha, B.: Path based load balancing for data center networks using SDN. Int. J. Electric. Comput. Eng. **9**(4), 3279 (2019). https://doi.org/10.11591/ijece.v9i4.pp3279-3285
16. Torkzadeh, S., Soltanizadeh, H., Orouji, A.A.: Energy-aware routing considering load balancing for SDN: a minimum graph-based Ant Colony Optimization. Clust. Comput. **24**, 2293–2312 (2021)
17. Ali, T.E., Morad, A.H., Abdala, M.A.: Traffic management inside software-defined data centre networking. Bull. Electric. Eng. Inform. **9**(5), 2045–2054 (2020). https://doi.org/10.11591/eei.v9i5.1928
18. Lu, L.: Multi-path allocation scheduling optimization algorithm for network data traffic based on SDN architecture. IMA J. Math. Control. Inf. **37**(4), 1237–1247 (2020)
19. Muhamad, H., et al.: SDN mininet emulator benchmarking and result analysis. In: 2020 2nd Novel Intelligent and Leading Emerging Sciences Conference (NILES). IEEE (2020)
20. Mosab, H., et al.: A comprehensive survey of load balancing techniques in software-defined network. J. Network Comput. Appl. **174**, 102856 (2021)
21. Montazerolghaem, A.: Software-defined load-balanced data center: design, implementation and performance analysis. Clust. Comput. **24**(2), 591–610 (2020). https://doi.org/10.1007/s10586-020-03134-x

22. Li, H., Hailiang, L., Xueliang, F.: An optimal and dynamic elephant flow scheduling for SDN-based data center networks. J. Intell. Fuzzy Syst. **38**(1), 247–255 (2020)
23. Mosab, H., et al.: Flow-aware elephant flow detection for software-defined networks. IEEE Access **8**, 72585–72597 (2020)
24. Ramkumar, M.P., et al.: FACVSPO: Fractional anti corona virus student psychology optimization enabled deep residual network and hybrid correlative feature selection for distributed denial-of-service attack detection in cloud using spark architecture. Int. J. Adaptive Control Signal Process. **36.7**, 1647–1669 (2022)
25. Shreya, T., et al.: Ant colony Optimization-based dynamic routing in Software defined networks. In: 2020 11th International Conference on Computing, Communication and Networking Technologies (ICCCNT). IEEE (2020)
26. Yang, J., et al.: An ant colony optimization method for generalized TSP problem. Progress Natl. Sci. **18.11**, 1417–1422 (2008)
27. Anil, S., Gupta, N., Niazi, K.R.: Efficient reconfiguration of distribution systems using ant colony optimization adapted by graph theory. In: 2011 IEEE Power and Energy Society General Meeting. IEEE (2011)
28. Mutaz Hamed Hussien, K., et al.: Detection and classification of conflict flows in SDN using machine learning algorithms. IEEE Access **9**, 76024–76037 (2021)
29. Li, Y., et al.: Performance analysis of floodlight and ryu SDN controllers under mininet simulator. In: 2020 IEEE/CIC International Conference on Communications in China (ICCC Workshops). IEEE (2020)
30. Xue, H., Kyung, T.K., Hee, Y.Y.: Dynamic load balancing of software-defined networking based on genetic-ant colony optimization. Sensors **19.2**, 311 (2019)

Information Extraction for Design of a Multi-feature Hybrid Approach for Pronominal Anaphora Resolution in a Low Resource Language

Shreya Agarwal(✉), Prajna Jha, Ali Abbas, and Tanveer J. Siddiqui

Department of Electronics and Communication, University of Allahabad, Prayagraj, India
agarwal.shreya1994@gmail.com

Abstract. In this paper, we present a hybrid approach for anaphora resolution in Hindi Discourse which is a low resource language. We propose a rule-based module for resolving reflexive anaphoric references and a data driven approach for resolving demonstrative, relative pronouns in Hindi data. The combined module tends to resolve all three types of pronominal anaphors using syntactic features, semantic-disambiguator (named-entity and animacy), linguistic features and statistical metrics, to resolve both inter-sentential and intra-sentential anaphoric references. We evaluated our approach on Hindi tourism data and developed a corpus for developing statistical model for the same. We obtained encouraging results which holds promise and achieves accuracy of 82.9% (0.829) comparable to that of state of art approaches developed so far. The dataset used in this work contains complex, compound sentences which makes information extraction pipeline designing a challenging task, while most of the learning and non-learning approaches developed so far in Hindi have been tested and evaluated mostly on simple short stories.

Keywords: Natural Language Processing · ensemble learning technique · discourse analysis · contextual knowledge · stanza library

1 Introduction

Collection of interrelated sentences is a discourse which requires contextual knowledge for correct semantic interpretation [1]. Reference in a text to an entity that has been previously introduced into the discourse is called the anaphora. The referring expression is called an antecedent. The process of determining the antecedent of an anaphor is called anaphora resolution (abbrev AR). Given a Discourse T, Anaphora resolution tends to find referents to items mentioned earlier in the discourse or pointing back [Mitkov, 1999] [2].

For example - 'अंटार्कटिका पहुंचना आसान नहीं है पर ये इसके आकर्षण का हिस्सा है |'.

In the above sentence 'ये'refers to 'पहुंचना आसान नहीं है'and 'इसके'refers to 'अंटार्कटिका'. 'ये'and 'इसके'are the anaphors, 'पहुंचना आसान नहीं है'and 'अंटार्कटिका'are their respective antecedents.

P. Das et al. (Eds.): AMRIT 2023, CCIS 1953, pp. 169–180, 2024.
https://doi.org/10.1007/978-3-031-47224-4_15

Anaphora resolution is an important subtask for correct semantic interpretation of text. Applications like Automatic Text Summarization that extract important information given a document are efficient when the generated summaries are coherent. However such applications suffers from problem of fragmented anaphoric references whose context and antecedent is not included in the summary making summary less coherent [19]. Resolution of AR as a pre-processing step or post-processing step in Automatic Text Summarization is proved to improve ROGUE metric and coherence of generated summaries [19]. Extrinsic evaluation of Anaphora resolution systems alludes to improvement in other Natural language processing task like Machine translation, document similarity identifier, Information extraction metrics and hence reserach in AR is gaining traction. AR is a discourse phenomenon. It is a challenging subtask in the field of NLP which acquires new dimension in Hindi language as Hindi is a free word order language and is morphologically rich.

Our objective is to identity correct antecedent-anaphora pair in our Hindi tourism discourse. The anaphors are different forms of pronouns (demonstrative, reflexive and relative) while potential antecedents are concrete noun phrases (head of noun phrase) and not abstract entities or events. The research in Anaphora resolution in English domain has achieved unprecedented attention and accuracy while AR in Hindi is still at a nascent stage due to lack of resources in Indian languages like tagged corpora, efficient named entity taggers, efficient morphological analyzers, which are readily available for English language.

We have built a annotated corpus with different factors discussed in Sect. 3 formulated as features that are specific to anaphora resolution. We have used random forest classifier for correct anaphora-antecedent pair whose efficiency is discussed in Sect. 6. Instead of using phrase structure grammar for anaphora resolution which is used in most of existing approaches, our hybrid module uses dependency structures for resolving anaphoric references in rule based as well as statistical data model. The dependency relation helps in determining the salience of potential noun phrases by determining the syntactic role of potential antecedents. The rest of the paper is organized as follows:

The following section briefly discusses earlier work done in Anaphora resolution in Hindi. Section 3 offers a brief introduction to factors, key to pronominal anaphora resolution. Section 4 gives some important statistical data regarding our dataset, explains the details of the proposed system and describes the hybrid model in detail. Section 5 includes the experimental details conducted for anaphora resolution. Results of the proposed system and its interpretation is discussed in Sect. 6. Conclusion and future scope regarding our methodology is discussed in Sect. 7.

2 Related Study

Approaches to Anaphora resolution can be divided into rule based model, data driven model, hybrid based model and deep learning based model. Literature suggests earlier work done in Anaphora resolution in Hindi, using hybrid approach for resolving pronominal anaphoric references [3]. The authors used rule based setting for resolving simple anaphoric references and more ambiguous references were resolved using decision tree classifier. They used dependency structure and prepared a list of potential antecedent based on karaka relations for determining the salience of noun phrases.

Detection of potential antecedents is important for an efficient anaphora resolution system. Utpal and Asif in [4] experimented with Conditional Random field for mention detection and used the same classifier for anaphora resolution in Hindi and Bengali language highlighting the importance of part of speech tagging, correct noun phrase extraction and detection of probable mentions for building an efficient anaphora resolution system. Sobha and Vijay in [5] developed a generic algorithm exploring the morphological richness of the Hindi language. They used a heuristic based model for finding the potential antecedent candidates and machine-learning algorithm for filtering the correct antecedents from the potential candidate list using CRF's. The preprocessing tool Indian Language–Indian Language Machine Translation (IL-ILMT) consortium was used for detailed morphological analysis of given word.

In [6] role of discourse salience and grammatical function is investigated for anaphora resolution system. They proposed a novel ranking method for forward looking centers and observed that grammatical role helps in determining the salience of potential noun referents. In [8] Bhargav and Uppalapu proposed a variant of S-LIST algorithm which developed a previous list of antecedent candidates and a present list of probable discourse entities.An important theory in discourse analysis is the centering theory which was used for anaphora resolution in English language.Centering theory approach to anaphora resolution in any language is based on coherence of different discourse entities in a discourse segment that attempts to capture the perceived local coherence present in a segment of a discourse. Salience ranking for S-LIST was obtained using dependency relation from Paninian framework (karaka relations). The author proposed a language independent approach for resolving first person and second person pronouns and observed that they were generally intra-sentential while third person pronouns were inter-sentential. Compound sentences were broken into multiple utterances for improving intra sentential anaphoric references. A rule based approach for reflexive and possessive pronouns using semantic information and gender agreement is presented in [10] where vibhakti and karaka relations are studied and approach adapted to Hobbs's algorithm. A statistical approach using WEKA random forest implementation is investigated using number, gender and verb semantics as factors for anaphora resolution in Hindi, Bengali and Tamil. Priyamani in [13] had detailed the constraints of number, gender, recency, and its contribution as factors for an efficient anaphora resolution system. They observed that recency was a baseline criteria for anaphora resolution in Hindi language with agreement feature improving the overall accuracy. This was verified by our learned model. A more comprehensive review of anaphora resolution in Hindi text can be studied through [15, 17].

3 Factors Affecting Anaphora Resolution

Approaches to anaphora resolution rely on certain set of features that are termed as anaphora resolution factors. Most of the existing work in anaphora resolution in hindi divides the factors into constraints and preferences. Constraints are used to filter the list of potential antecedent candidates for an anaphor in a text. Preferential factors are employed on the filtered list and the most probable antecedent is stated as the antecedent for a given anaphor. We now explain the factors used in development of our feature set by designing an information extraction pipeline using stanza library [7]. They are as follows:

3.1 Number Agreement

This constraint requires that anaphor and its antecedent should agree in their number information. For example: "सिटी पैलेस महाराजा जयसिंह द्वितीय के द्वारा बनवाया गया था और यह एक मुगल और राजस्थानी स्थापत्य कला का मिश्रण है". Both सिटी पैलेस and यह are singular and can co-refer.

3.2 Gender Agreement

This constraint requires that the anaphor and its antecedent to agree in their gender information. In Hindi, we cannot explicitly determine the gender of pronominal anaphors as in English. The gender agreement in Hindi is achieved through nearest verb phrase of the concerned sentence.

"जैसलमेर की सरहद पार स्थित प्रसिद्ध पर्यटन स्थलों में से एक लोदुर्वा प्रसिद्ध जैन मंदिर जाना जाता है, जो साल भर विशाल संख्या में तीर्थयात्रियों को आकर्षित करता है|".

Relative pronoun "जो" is Masculine in the above sentences, decided through "आकर्षित करता है" and not "आकर्षित करती है". We in our dataset have not used this feature but various methodologies have used this agreement feature.

3.3 Animacy Agreement

This constraint works on animistic knowledge of the anaphor and antecedent, referring to living and non-living things. Animistic knowledge contributes significantly to the overall accuracy of anaphora resolution in Hindi Discourse [13]. An animate noun phrase cannot be the antecedent of an inanimate anaphor and an inanimate noun phrase cannot be the antecedent of an animate anaphor. If an anaphor cannot be determined as animate or inanimate, then this constraint cannot be used for anaphora resolution.

"सिटी पैलेस महाराजा जयसिंह द्वितीय के द्वारा बनवाया गया था और यह एक मुगल और राजस्थानी स्थापत्य कला का मिश्रण है".. In this sentence "यह" have two potential antecedent "सिटीपैलेस" and "महाराजा जयसिंह द्वितीय". "यह" is inanimate. "सिटी पैलेस" is inanimate and "महाराजा जयसिंह द्वितीय" is animate. The potential antecedent list is reduced to "सिटीपैलेस" as it matches the animistic constraint. The animistic agreement feature act as a semantic disambiguator.

3.4 Grammatical Role

Role of subject and object in sentences are prominent in Hindi discourse for resolving anaphoric references [6]. When an anaphor can refer to more than one antecedent, the antecedent that play the role of subject are more likely to be the antecedent than the antecedent that play the role of object. Consider the following sentence.

"जैन मंदिर का प्रमुख आकर्षण एक दिव्य वृक्ष है|". In this sentence "जैन मंदिर" is the subject and "दिव्यवृक्ष" is the object.

3.5 Principal Noun(Head Noun)

The first noun (noun phrase) occurring in a sentence is called the head noun or principal noun which is more preferable to be considered an antecedent of an anaphor in a list of potential antecedents. The remaining noun(s) in the same sentence are termed as modifier noun (noun phrases).

3.6 Named Entity

This feature classifies potential candidates into semantic classes of person, organization, time, money, percent, nature, events or others. This feature is used to determine the class of antecedent and can be used as an agreement feature between the antecedent and anaphor in case the pair belongs to same semantic classes. We have added the named entity of the potential antecedent list in our tagged corpus as a feature.

3.7 Referring Type

This feature checks whether the anaphor is a demonstrative pronoun, relative pronoun or a reflexive pronoun.

3.8 Term Frequency

This statistical metric determines the count of occurrence of a potential candidate. It is an important metric for information extraction applications and is widely used in combination with Inverse Document Frequency as a metric in various ranking and information retrieval problem statement. In our dataset we have used the occurrence of a potential antecedent frequency in previous three sentences from the place of occurrence of an anaphor.

3.9 Part of Speeh of Context Word

Context words are words before and after a potential antecedent. We used the part of speech of the context words in our feature set which were generally found to be case markers after the potential occurrence. The intent is to investigate the effect of case markers after nouns along with grammatical role in correct antecedent determination for a given anaphor.

A combination of different text features that can be employed for effective automatic anaphora resolution is discussed extensively by Kusum lata., et.al.[22].

4 Data and Methodology

4.1 Dataset

Our Dataset is a tourism data provided by CFLIT, Bombay that was converted into a discourse. The dataset statistics is shown in Table 1. We focus our study on identification and resolution of correct explicit concrete antecedent to anaphors.

Table 1. Dataset Pronoun statistics

Document	Demonstrative	Relative	Reflexive
Doc-1	30	22	10
Doc-2	48	36	11
Doc-3	59	45	20

4.2 Methodology

The task of finding the correct antecedent of an anaphora encountered while parsing the sentence from left to right is modelled as a binary classification problem in case of demonstrative and relative pronouns.

Resolution of Reflexive Pronoun: If the anaphora was a reflexive pronoun then after number and animacy agreement, it was resolved to the subject of the same clause as given by the dependency tree as its referent or the subject within its governing category. We obtained an accuracy of 80% in case of reflexive pronoun.

For demonstrative and relative pronouns the work was a two stage pronoun reference resolution. In first stage we identified the potential antecedents, anaphors and in the second stage we identified the correct antecedent-anaphora pair.

4.3 Proposed Architecture

The first stage includes Mention Detection and second stage include feature extraction and anaphora resolver module as shown in Fig. 1.

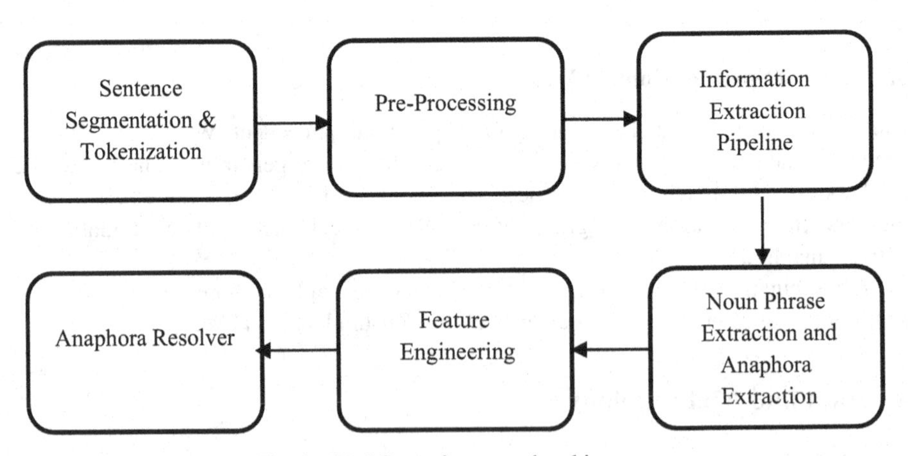

Fig. 1. Workflow of proposed architecture

Preprocessing. Preprocessing tasks involves sentence segmentation of each document, tokenization and morphological analysis of text. The preprocessing tasks included tokenization, part of speech tagging, Noun phrase extraction and identification, anaphora identification, dependency parsing and morphological analysis of text data using stanza library [7].

Information Extraction. The information extraction pipeline was designed with morphological analyzer, Part of speech tagger, dependency parser. The morphological analysis and Part of speech was done using stanza library [3] The POSProcessor labels words with their universal POS (UPOS) tags, and tree specific POS(XPOS) tags· The dependency relationship in a sentence was identified using dependency parser available in stanza library, which represents the syntactic dependency relations between words.

Mention Detection. The antecedents which are the referring expressions in a discourse are called mention or markables. We considered all noun (head of noun phrase) as potential candidate for antecedent from within the same sentence in which the anaphora occurred and noun (head of noun phrases) from previous three sentences (not valid in case of anaphor present in the first sentence of the discourse).

Feature Extraction and Feature Engineering. The problem was modeled as a binary classification task. The feature vector was prepared for testing and training data. The training data consists of sample where each sample was a pair of antecedent and anaphora and a set of features. Every anaphor was paired with all mention detected within its sentence and previous three sentences.

Training samples were developed using the framework proposed by soon et al. [11]. The features include number agreement, animacy agreement, sentence distance, markable distance,part of speech of the context words, head noun or modifier noun, grammatical role of the antecedent, named entity of the antecedent,Anaphora type(demonstrative, relative),Term frequency of antecedent within and before the considered anaphora sentence, and Target feature alluding to whether the antecedent- anaphora is the correct pair or not. The correct pair(s) were the positive sample and pairing of an anaphor with all other antecedents formed the negative samples in our dataset.

5 Experiment

The dataset consists of numerical and categorical values which cannot be directly used in machine learning algorithms. This paper attempts to explore the H20.ai Framework which is an open access, fully implemented machine learning algorithm framework and can deal with categorical data in a dataset without explicitly encoding the categorical data into its numerical equivalent. The developed tagged corpus was divided into 70:30 ratio for training and testing purposes. In order to identify correct anaphora resolution we experimented with Random forest classifier (Table 2).

$$nij = WjCj - Wleft(j)Cleft(j) - Wright(i)Cright(j) \tag{1}$$

where,

nij = importance of node nij

Wj = weighted number of samples reaching node j
Cj = impurity value of node j
left(j) = child node from left split node j
right(j) = child nodef from right split node j
The importance for each feature on decision tree is calculated as:

$$fii = \sum_{j:node\ j\ splits\ on\ feature\ i} nij \tag{2}$$

where,
fii = importance of feature i
n_j = importance of node j

Table 2. Distributed Random Forest Model Summary

Number of trees	Number of internal trees	Min and Max depth	Min and Max leaves	Mean depth
50	300	(1,20)	(2,86)	9.583
100	600	(1,20)	(2,90)	9.383
200	1200	(1,20)	(2,93)	9.448

Methodology developed using decision tree classifier in WEKA implementation was explored which is a knowledge rich approach and used only agreement features and recency for anaphora resolution [12]. The learned model was then applied on test data. To determine anaphora resolution in a new test document, all markables and potential pairs of antecedent-anaphora are presented to the classifier, which decides whether the antecedent-anaphora pair actually co-refer. The number of tree estimator fed for model development was 50,100 and 200 respectively to check the effect of number of trees estimator on the classification accuracy.

6 Results and Discussion

In order to assess the performance of the proposed method, we conducted a test run on Tourism test dataset.

Table 3. Model Evaluation Metrics

Number of trees estimators	Accuracy(relative and demonstrative)	MSE	RMSE	Log loss
50	0.849	0.123	0.351	0.636
100	0.858	0.123	0.351	0.490
200	0.844	0.124	0.353	0.500

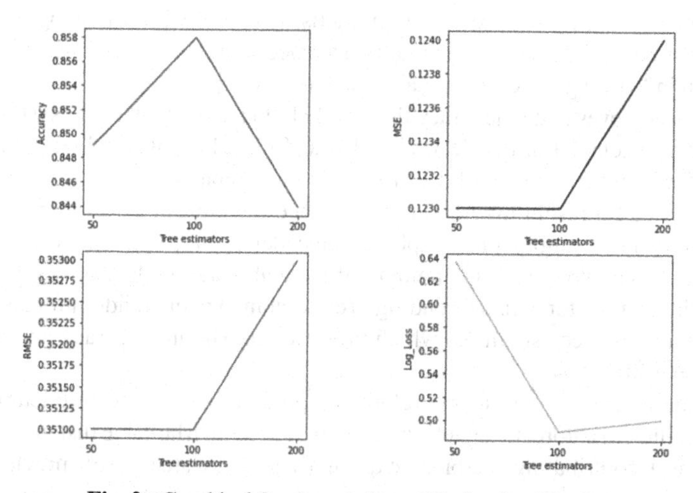

Fig. 2. Graphical Representation of Evaluation Metrics

Table 4. Learned Feature Salience Score (n = 100)

Feature	Scaled importance
Markable Distance	1.0
Part of Speech of Context Word after antecedent	0.959
Part of Speech of Context Word before antecedent	0.801
Sentence Distance	0.628
Grammatical role	0.586
Named Entity	0.529
Anaphora type	0.303
Principal Noun	0.247
Term Frequency	0.194
Number Agreement	0.168
Animacy Agreement	0.045
Direction of Reference	0.049

Table 3 shows the result of evaluation in case of relative and demonstrative anaphors.The methodology showed 85.8% for demonstrative and relative pronouns when the number of tree estimators in random forest is 100 and and an accuracy of 80% is achieved in case of reflexive pronouns with overall accuarcy of 82.9%. Figure 2 depicts the optimal accuracy, RMSE,MSE and Log loss achieved at number of tree estimators being 100. The y axis has the Evaluation metric as shown in Table 3. And the number of tree estimators used are represented on x axis. A model with lower log loss is considered better. Table 4 delineates the feature salience score, in descending order of their contribution in learning the correct anaphora antecedent pair.

Our work achieves better accuracy than [8, 13]. It is better to existing systems developed so far as discussed in the Sect. 2. In [8] and [13], both worked on general short stories in Hindi, with no standard data for absolute comparison. In [3], first rule based module was used for resolution for simple sentences and machine learning module was used in case of more ambiguous anaphora antecedent pair. Moreover rule based approach are tedious and require large amount of manual work. Table 4 shows the salience score of individual features in descending order obtained using random forest classifier with the number of tree estiamtors which produced maximum accuracy and minimum log loss (n = 100).

The data in Table 4 suggests the relative importance of each feature learned while building an efficient automatic anaphora resolution system. Markable distance is verified to be the most contributing factor as expected and suggested by all previous works [6, 13]. Grammatical role plays an important role in building an efficient anaphora resolution system as seen in Table 4. The number and animacy agreement feature low relative importance is attributed to posing the problem statement as a binary classification problem where anaphoric ambiguity is present. A set of antecedent agreeing in number and animacy information among which only single candidate is appropriate results in them contributing lesser, while a majority of rule based approach rely on these two factors primarily to filter the probable candiates and then apply rule for resolution, along with sentence distance for antecedent anaphora resolution in simple Hindi sentences.

7 Conclusion and Future Work

We present a hybrid approach for pronominal anaphora resolution in Hindi data. The focus of this paper is to build an automated anaphora resolution system for a resource poor language namely Hindi. We explore the use of random forest classifier in its efficacy to learn constraints and preferences between potential antecedents for a given anaphor automatically.The classifier resulted in improved accuracy from existing learning approaches. While investigating the correct antecedent and anaphora pair we observed that Demonstrative pronouns, Relative pronouns have both inter-sentential and intra-sentential antecedent while reflexive pronouns are generally resolved to the subject of the same clause (as given by the dependency tree) as its referent or the subject within its governing category.Dependency structure is more reliable for a free word order language like Hindi. We also observed that maximum antecedents are nouns (head of noun phrases) in context of demonstrative pronouns.

Future scope include use of deep learning methodology for anaphora resolution in Hindi discourse. Anaphora resolution in English language using deep learning models have been developed extensively while little effort is explored in Anaphora resolution in hindi data containing complex sentences [20, 21]. We aim to investigate the resolution of anaphors to entities which may be verb clauses and adjectival clauses apart from noun phrases. We also aim to investigate the improvement in document similarity identification and automatic text summarization after anaphora resolution to access its role in Natural Language processing applications.

References

1. Tiwary, U.S., Siddiqui, T.: Natural Language Processing and İnformation Retrieval. Oxford University Press, Inc. (2008)
2. Mitkov R.: Anaphora resolution: state of the art, Technical report, School of Languages and European Studies, University of Hampton (2000)
3. Dakwale, P., Mujadia, V., Sharma., D.M.: A Hybrid approach for anaphora resolution in Hindi. In: International Joint Conference on Natural Language Processing, pp. 977–981. Japan (2013)
4. Sikdar, U.K., Ekbal, A., Sanha, S., Uryupina, O., Poesio, M.: Adapting state of art anaphora resolution approaches for resource poor langauge. In: International Joint Conference on Natural Language Processing, pp. 815–921. Japan (2013)
5. Devi, S.L., Ram, V.S., Rao, P.R.K.: A generic anaphora resolution engine for Indian languages. In: 25th International Conference on Computational Linguistics, pp. 1824–1833. COLING, Dublin (2014)
6. Prasad, R., Strube, M.: Discourse salience and Pronoun resolution in Hindi, vol. 6, article 13. University of Pennsylvania (2000)
7. Qi, P., Zhang, Y., Bolton, J., Manning, C.D.: Stanza: a python natural language processing toolkit for many human languages. In: 58th Annual Meeting of the Assoiation for Computational Linguistics: System Demonstration, pp. 101–108. ACL (2020)
8. Uppalapu, B., Sharma, D.M.: Pronoun Resolution in Hindi. Language Technologies Research Center. International Institute of Information Technology, Hyderabad (2009)
9. Dutta, K., Prakash, N., Kaushik, S.: Machine learning approach for the classification of demonstrative pronouns for indirect anaphora in Hindi news items. In: Prague Bulletin of Mathematical Linguistics, pp. 33–50 (2011)
10. Dutta, K., Prakash, N., Kaushik, S.: Resolving pronominal anaphora in Hindi using Hobbs algorithm. Web J. Formal Comput. Cogn. Linguist. **1**(10S) (2008)
11. Soon, W.M., Lim, D.C.Y., Ng, H.T.: A machine learning approach to co-reference resolution of noun phrases. In: Associationa of Computational Linguistics (ACL) (2001)
12. Chatterji, S., Dhar, A., Barik, B., Moumita, P.K., Sarkar, S., Basu, A.: Anaphora resolution for bengali, hindi and tamil using Random tree algorithm in WEKA. In: ICON NLP Tool Contest (2011)
13. Lakhmani, P., Singh, S.: Anaphora resolution in Hindi language. Int. J. Inform. Comput. Technol. **3**, 609–616 (2013)
14. Singh, P., Dutta, K.: Sensitivity analysis of feature set employed for anaphora resolution. Int. J. Comput. Appl. **128** (2015)
15. Mahato, S., Thomas, A., Sahu, N.: A relative study of factors and approaches for hindi anaphora resolution. Int. J. Manage. IT Eng. **7**(12) (2017)
16. Sikdar, U.K., Ekbal, A., Saha, S.: A generalized framework for anaphora resolution in Indian languages. Knowl. Based Syst. 147–159 (2016)
17. Yadav, D.S., Dutta, K., Singh, P., Chandel, P.: Anaphora resolution for Indian languages: the State of Art. In: National Conference on Recent Innovations in Science and Engineering, pp. 01–07 6 (2016)
18. Lalitha, S., Patnaik, B.N.: Vasisth: an anaphora resolution system for indian languages. In: Artificial and Computational Intelligence for Decision,Control and Automation in Engineering and Industrial Applications, Tunisia (2000)

19. Batista, J., Lins, R.D., Lima,R., Riss, M., Simiske, S.J.: Automatic ccohesive summarization with pronominal anaphora resolution. In: Computer Speech and Language (2018)
20. Singh, S., Patel, K., Bhattacharya, P.: Attention based anaphora resolution for code mixed social media text for Hindi language. In: Forum for Information Retrieval Evaluation(Fire) (2020)
21. Volta, E., Serdyukov, P., Sennrich, R., Titov, I.: Context aware neural machine translation learns anaphora resolution. In: 56th Annual Meeting of the Association for Computational Linguistics, pp. 1264–1274. Australia (2018)
22. Lata, K., Singh, P., Dutta, K.: A comprehensive review on feature set used for anaphora resolution. Artific.l İntell. Rev. 2917–3006 (2021)

Signature-Based Batch Auditing Verification in Cloud Resource Pool

Paromita Goswami[1,3](✉), Munmi Gogoi[1], Somen Debnath[2], and Ajoy Kumar Khan[3]

[1] Department of Computer Engineering and Applications, GLA University, Mathura, India
{paromita.goswami,munmi.gogoi}@gla.ac.in
[2] Department of Information Technology, Mizoram University, Aizawl, India
somen@mzu.edu.in
[3] Department of Computer Engineering, Mizoram University, Aizawl, India

Abstract. Online cloud resource pool provides cloud users with a multitude of appealing developments in highly sought-after online scalable storage services for them to eager new creative and investment business benefits. Today, the majority of cloud data security research focuses on ways to increase the accuracy of outsourced data audits rather than internal and foreign foes who could hack a cloud server. Even though the data owner enviously envies the data auditing work of stored data to a trustworthy Third Party Auditor(TPA) to avoid communication as well as computational overhead of data when outsourcing to a reputable Cloud Service Provider, TPA is unreliable in a realistic setting. In this research, we offer a model for an effective data integrity system based on the ZSS signature, which secures data privacy in cloud storage. Finally, the actual results of the performance of the proposed prototype of the system model were tested and indicated greater efficiency of provided model in comparison to BLS signature.

Keywords: Data auditing · Cloud resource pool · Data confidentiality · Auditing correctness · Cloud customer · ZSS Signature

1 Introduction

Due to pay-per-services (IaaS, SaaS, PaaS), flexibility, decentralization, and long list of features, both academic and industrial fields are paying increased attention to the current internet-based era which is known as Cloud Computing. Users can now gain virtual access to cloud computing resources more easily by moving their data to online cloud storage and benefiting from lower cloud service costs without having to deal with the managerial difficulties of handling devices [1,2,26]. In the modern era of technology, cloud computing has become a very common practice for users to use it as a backup repository. In real-world scenarios, cloud computing poses a wide variety of security threats in terms of data privacy, data integrity, and data availability because it facilitates online shared

P. Das et al. (Eds.): AMRIT 2023, CCIS 1953, pp. 181–193, 2024.
https://doi.org/10.1007/978-3-031-47224-4_16

data storage via unreliable cloud service providers [22,27]. On the flip side of this observation is the possibility that a cloud service provider is actually a malicious attacker who is always trying to make a profit by either deleting data in order to make more space available in the cloud or intentionally sharing important data belonging to data owners with other parties in order to make a profit for their own company. Sometimes, CSPs are trustworthy; however, they may lose data coming from the side of the data owner either as a result of a lack of managerial capabilities [3–5,13] or as a result of replacement attacks by authorised others in the same shared storage space [6]. Therefore, cloud service providers conceal this loss on purpose from the owners of the data in order to protect their professional reputation [5]. Because the owners of the data do not have physical control over the data that is outsourced, they are unable to determine who is responsible for improperly handling the data [6,7,25]. Previously, data owners used the traditional MAC scheme for data security, but it creates a massive communication overhead. Due to the public key-based nature of HLA, a third-party auditor is able to verify the data integrity of outsourced information without the need for a duplicate copy [8,9]. To ensure data correctness, data intactness, and auditing correctness, researchers are constantly working on the data integrity concept by improving data integrity verification techniques and by enhancing the data privacy-preserving technique [1,2,10]. In data integrity auditing schemes, several researchers have used the provable data possession (PDP) method to shorten the amount of time needed to compute very long files and to shorten the amount of time needed to transfer a very large hash value [10]. Not supported by the PDP scheme are two crucial parameters of the auditing scheme: error localization and data recovery method. Since then, researchers have proposed the PoR method to address these issues [12]. The PoR method guarantees 100% retrieval probability of corrupted blocks of an original data file and also facilitates dynamic data operations [12,16]. However, the DO must always keep vital data for the data integrity test and the authorization test in order to pass. It makes the computation more expensive. To create signatures for data blocks and maintain DO data authorization in cloud storage, some researchers proposed a bilinear pairing concept. The BLS cryptographic signature was developed by a few studies and is the shortest signature currently in use [9]. This signature relies on a one-of-a-kind hash function that is probabilistic rather than deterministic. It's more work because it requires exponential and hash calculations. Therefore, in light of the above issues, we propose a public data auditing scheme based on ZSS signature-based outsourced data auditing strategy for ensuring data confidentiality and privacy while doing data integrity audits and it consumes less overhead computation than BLS signature.

Taking the existing scenario into account, and in order to address the aforementioned difficulties, we offer an efficient public data auditing method based on ZSS signatures that maintains data confidentiality, privacy, auditing correctness, and resists collusion and forgery attacks. We describe in this paper a modified ZSS signature-based outsourced data auditing scheme that can protect data security and privacy while performing data integrity batch audits.

2 Related Work

Using the tailing method, Gokulakrishnan S. et al. [13] developed a methodology for spliting data and generating tokens for data split-up by adding the cloud address or locations of cloud storage. Thus, every problematic node's missing segment is immediately pointed out within a slender range of bounds and receives data support from adjoining nodes. Vineela A. et al [14] established a data integrity auditing scheme for cloud-based medical big data to assist reduce the danger of unwanted data access. Multiple copies of the data are kept to guarantee that it may be restored rapidly in the event of damage. This approach can also be used to allow clinicians to readily track changes in patient's conditions using a data block. The simulation results demonstrated the effectiveness of the proposed method. Wang Y. et al [15] created a Mapping-Trie tree-based data integrity checking scheme. To prevent the leaking of private information, remove the TPA platforms. To audit the data, the method of random sampling is utilized, which considerably reduces the calculation and conveying of resources of cloud users and servers. Both the Hash and Trie schemes have been improved in terms of audit efficiency and storage space savings. Traditional data auditing has issues with data security, processing speed, and communication efficiency. To address these issues, Bian J. et al. [16] suggested a data integrity audit scheme based on data blinding. To reduce transmission time, this strategy employs edge devices in the coveying node to construct a fog computing layer between the CSP and the data owner. This paper presented a security model and proof based on computational Diffie-Hellman (CDH) assumptions. The experimental results suggest that including a fog computing layer and a blind element into the data auditing process can effectively minimize data connection delays and increase data audit security. Zhao T. et al. [17] introduced a cloud storage data auditing scheme based on Dynamic Array Multi-branch Tree (DA-MBT) to minimise the height of the tree, enhance node utilisation, simplify the dynamic update process, reduce data block query time, and improve verification efficiency. Finally, the suggested DA-MBT technique allows full dynamic operation, protects user privacy, can successfully verify the integrity of a large number of data, and can significantly reduce communication and processing overhead throughout the verification process. Ali A. S. et al [18] suggested a new encryption paradigm as a two-person integrity scheme for cloud-outsourced data is offered here. This strategy employs a trusted third-party manager who serves as a go-between for the data owner and the cloud service provider. The suggested model employs a two-stage encryption process. The key is acquired from the cloud user and data manager in this first stage. The key is significant in the second stage by the data user and the CSP. The suggested system's experimental findings outperform existing cloud security techniques.

Shen W. et al. [20] conceived and designed a new trend termed data auditing short of private key storage. To avoid using the hardware token, biometric data is used as the user's fuzzy private key in this technique. It used a linear sketch with coding and error correction methods to certify the user's identity. Furthermore, a novel signature method is proposed here that not only allows blockless verifiability but is also consistent with the linear sketch. This suggested approach delivers desirable security and efficiency, as demonstrated by the security proof and performance analysis.

3 Preliminaries

3.1 BLS Signature

This section describes the bilinear pairing concept [21], which is used in the proposed approach. Allow two cyclic groups G1 and G2 to generate two group generators g1 and g2. Both cyclic groups have additive properties. Both generators have the same order, q, where q is a prime number. G2 is a multiplicative cyclic group with the same q order as G1. A bilinear map is a type of mappe: $G_1 * G_2 \rightarrow G_t$ such that $x_1 \in G_1$, $x_2 \in G_2$ and $q_1, q_2 \in Z_q$, $e(x_1^{p_1}, x_2^{p_2)} = e(x_1, x_2)^{p_1, p_2}$. Of course, an efficiently computable algorithm exists for computing e, and the map should be nontrivial, indicating that e is non-degenerate: $e(g_1, g_2) \neq 1$

3.2 ZSS Signature

Bilinear pairing based ZSS signature contains four algorithms parameters generation P_{gen}, key generation k_{gen}, signature generation S_gen and signature verification S_{ver} are introduced in ZSS public key cryptographic signature [5]. It is defined as: P_{gen}: A set of public parameters are e, G_1, G_2, g_1, q, H

K_{gen}: K is a positive integer number, randomly selected to compute $P_{pub} = k * g1$. Here P_{pub} is a public key and K is a secret key where $k \in Z_q^*$.

S_{gen}: If S be a ZSS signature and m be a message, then

$$S = 1/(H(m) + K).g_1 \tag{1}$$

S_{ver}: Given a message m, signature S and public key P_{pub}, verify if

$$e(H(m) \cdot g_1)P_{pub}, S) = e(g_1, g_1) \tag{2}$$

4 Proposed System Framework

The suggested data auditing architecture is based on a shared cloud resource environment and contains three entities: Cloud Customer(CC), Cloud Resource Pool(CRP), and Auditing Expertise(AE), as shown in Fig. 1.

Cloud Customer(CC): CC hires storage services from CRP and uploads a considerable volume of data to the shared resource pool to achieve the goal of storing data remotely. CC could be a user inside an organisation or an individual user.

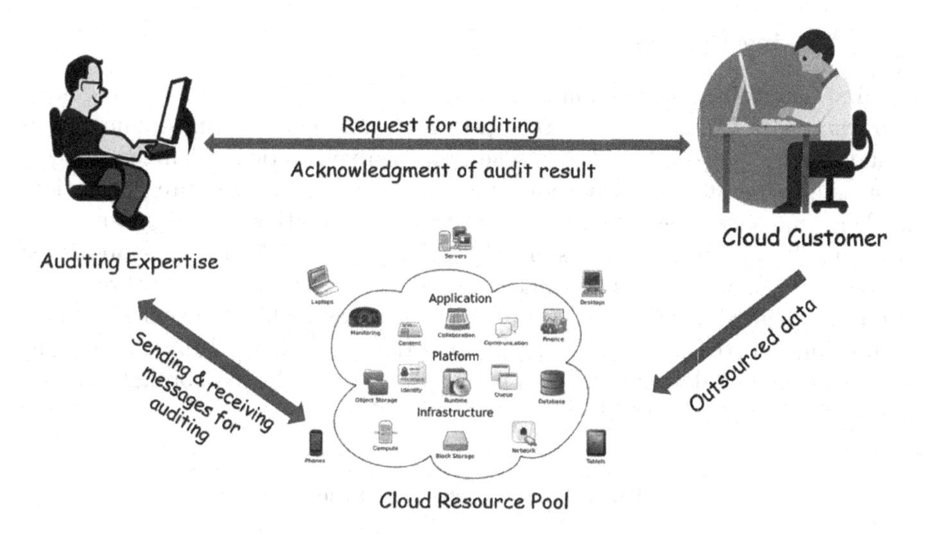

Fig. 1. Proposed System Model

Cloud Resource Pool(CRP): Public and private cloud storage servers are combined to provide a shared cloud resource pool. The versatility of private and public cloud computing is combined with the efficiency of a regular dedicated server in new cloud storage server technology. CRP receives the outsourced data that CC has uploaded via wireless transmission and provides cloud computing and storage services to CC.

Auditing Expertise(AE): Apart from users and the cloud, AE is an independent third party. The AE is qualified and equipped to conduct cloud data analysis. The AE will create a challenge for the data auditing process after CC files an audit request to them. The AE will then forward the challenge to CRP. The CRP will give proof in accordance with the challenge information. The AE can generate an audit result after reviewing the proof. The AE can be used to audit user data in a cloud system, saving CC compute and storage resources while ensuring the verification process's integrity.

The major goals are to maintain the data integrity of CC's outsourced data while it is being stored on an insecure remote cloud server and to confirm the audit verification results given by AE on the CC end, as summarised above. To earn money, CC generates data before uploading it to any remote cloud storage service. CRP is an independent company that supplies CC with storage services. CRP stores all CC's outsourced data in a resource pool, an online shared data storage. AE relieves CC of the responsibility of managing its data by confirming the correctness and integrity of outsourced data. We assume in this prototype system model that each entity has limited personal storage and processing power to process data.

4.1 Working Process

CC first determines the maximum file size F to be w. The file F is then divided into chunk size variable chunks $B = \{B1, B2, ...Bj\}$, with a maximum size of w and $1 < l < cel(\frac{w}{d})$, $2 \leq d \leq k$. The proposed verification technique consists of six steps: parameter generation, message encryption, signature generation, challenge message generation, response message generation, and signature verification. Please note that our essential architecture is made up of six processes: keyGen, MsgEncrypt, SignGen, QGen, QRes, and SignVeri. The first three operations are located in the Block Data Processing phase, while the next three procedures are located in the Data auditing verification phase. Table 1 shows the nomenclature and descriptions used in our proposed model. A thorough description of the two phases as follows:

Table 1. Definitions of Notations

Notation	Meaning
G_1	Multiplicative Cyclic group with q prime order
r_1	Random no.s
r_2	Secret key
g_1	Group generator
Z_q	Ring of integer modulo q
e	Billinear paring map $e : G_1 \times G_1 = G_2$
F	Original File
ω	File Tag
$B = \{B_1, B_2, ...B_j\}$	Original blocks
$B_t[M_i]$	Block Elements of tth block where $t \in j$, $i \in k$
$B_t[M_i']$	Encrypted Block Elements of tth block
w	Maximum size of a filer
H	Cryptographic Hash Function: $H : \{0,1\}^* \rightarrow \{0,1\}^\lambda$
q	Prime number
k	Total size of a particular block
j	Total no. of block
B_τ	Particular challenged block no where $i \in \tau$
$S = \{S_1, S_2, .., S_j\}$	Signature of all blocks B_j
$B_t[M_i']$	Encrypted blocks where $t \in j$ and $i \in k$
ϕ	Collude value
CC_{pri}	Private key of CC
CC_{pub}	Public key of CC

i. Block Data Processing Phase

- $KeyGen(g, r_1, r_2) \rightarrow (CC_{pub}, CC_{pri})$: CC produces a pair of public and private keys accordingly CC_{pub} and CC_{pri} by generating a group generator g_1 from G_1 using a bilinear map and two random numbers r_1, r_2 to generate a pair of private and public key CC_{pri} and CC_{pub} respectively where $g_1 \in G_1$ and $r_1, r_2 \in \mathbb{Z}_q$.
- $MsgEncrypt(CC_{pri}$, M=$(\sum_{t=1}^{j}\sum_{i=1}^{k} B_t[M_i])]) \rightarrow EncryptMessage(M' = (\sum_{t=1}^{j}\sum_{i=1}^{k} B_t[M_i'])]$: CC breaks the original file F into j no. of variable-size blocks $B = \{B_1, B_2, ..., B_j\}$ and encrypts each block elements using the CC_{pri} where $t \in j$.
- *SignGen*: $(B_t[M_i'], g_1, r_1) \rightarrow (S)$: Input file F blocks B and generator parameter g_1, random number r_1, output the signature of data block set $S = \{S_1, S_2, ..., S_j\}$ where

$$S_i = [1/H(B_t[M_j'] + r_1)].g_1 \tag{3}$$

CC generates a message as $detail\_msg = \{S, B_j', \omega, g_1, r_1\}$ and send it to CRP.

ii. Data Auditing Verification Phase
Few data blocks are questioned for often outsourced data accuracy over the stored original encrypted data blocks at the resource pool in block-level auditing. In this case, CC sends a query message to AE for auditing, using $query\_msg = \{B_j, g_1, CC_{pub}, \omega\}$ for verification.

- *QGen*: $(query\_msg) \rightarrow chal\_pro(\omega, B_\tau)$: Input is an question as a message and output is *chal_pro* message and send it to CRP.
- *QRes*: $(\omega, \tau, S_{1,2,..,\tau}) \rightarrow chal\_res(\omega, sk, \gamma)$: When CRP receives a *chal_pro* message, it produces a *chal_res* message before sending it to AE where

$$\gamma = \sum_{t=1}^{\tau} r_1/[sk + g1 * H(B_t[M_i])] \tag{4}$$

and $sk = r1 * g1$.
- *SignVer*: $(S_\tau, \gamma, sk, g1) \rightarrow Result(T/F)$: After receiving $chal_r es$ message, AE verify the message as follows:

$$e(g1 * \sum_{t=1}^{\tau} S_t, sk/\gamma) = e(g_1, g_1) \tag{5}$$

The output will be true if the above equation is true; otherwise, the output will be false. After verification, the verification result is sent to CC.

4.2 Designing Features

Desire design goals of outsourced data auditing scheme can be summarized as follow:

- Public Auditing: According to our suggested public auditing approach, CC outsources critical data to CRP for storage purposes. Later, CC delegated the outsourced data auditing work to a third-party auditor known as AE, who is capable of doing similar activities. This model provides this feature like AE-based partial data verification. This technique, like [10], does not contain additional computing costs for data preprocessing performed by CC throughout the auditing process.
- Auditing Correctness: The evidence of CRP can only pass the AE validation test if both CRP and AE are truthful and CRP follows the pre-defined data storage mechanism as described in [5].
- Data Confidentiality: This crypto parameter ensures that original data must not disclose in front of either CRP or AE during the auditing process [9].
- Unforgeability: If the LCSP properly maintains outsourced data at SCS, then only *chal_res_pro* message can pass the verification test i.e. our system model can resist internal and external forgery attack [24].
- Collision Resistance: If stored data at the resource pool is not colluded by CRP, then the signature verification test shows the data intactness [23].

5 Security Analysis

This section describes the security parameters of the proposed auditing scheme. The following concepts are introduced: auditing correctness, data confidentiality, data privacy, unforgeability, and collusion resistance.

- Auditing Correctness: The evidence of CRP can only pass the AE validation test if both CRP and AE are truthful and CRP follows the pre-defined data storage mechanism. The auditing process is defined as follows using Eqs. (4) and (5):

$$
\begin{aligned}
& e(g1 * \sum_{t=1}^{\tau} S_t, \frac{sk}{\gamma}) \\
&= e(g_1 * \frac{g_1}{H(B_1[M_i']) + r_1} + \frac{g_1}{H(B_2[M_i']) + r_1} + ... + \frac{g_1}{H(B_\tau[M_i']) + r_1}, \frac{sk}{\gamma}) \\
&= e(g_1 * \frac{g_1}{H(B_1[M_i']) + r_1} + \frac{g_1}{H(B_2[M_i']) + r_1} + ... + \frac{g_1}{H(B_\tau[M_i']) + r_1}, \\
&\quad \frac{sk}{\frac{r_1}{r_1*g_1+g_1*H(B_1[M_1'])} + \frac{r_1}{r_1*g_1+g_1*H(B_2[M_1'])} + .. + \frac{r_1}{r_1*g_1+g_1*H(B_\tau[M_i'])}}) \\
&= e(g_1 * \frac{g_1}{H(B_1[M_i'] + r_1} + \frac{g_1}{H(B_2[M_i'] + r_1} + ... + \frac{g_1}{H(B_\tau[M_i'] + r_1}, \\
&\quad \frac{g_1}{\frac{1}{r_1*g_1+g_1*H(B_1[M_1']} + \frac{1}{r_1*g_1+g_1*H(B_2[M_1']} + .. + \frac{1}{r_1*g_1+g_1*H(B_\tau[M_i']}}) \\
&= e(g_1, g_1)(\frac{g_1}{H(B_1[M_i'] + r_1} + \frac{g_1}{H(B_2[M_i'] + r_1} + ... + \frac{g_1}{H(B_\tau[M_i'] + r_1},
\end{aligned}
$$

$$\frac{1}{\frac{1}{r_1*g_1+g_1*H(B_1[M_1'])}+\frac{1}{r_1*g_1+g_1*H(B_2[M_1'])}+..+\frac{1}{r_1*g_1+g_1*H(B_\tau[M_i'])}})$$
$$= e(g_1, g_1) \tag{6}$$

- Data Privacy: For verification, AE verifies signatures of challenged blocks using the information from the *query_message* from CC and the *chal_res* message from CRP. The true beauty of this proposed paradigm is that AE checks the signature of encrypted block elements that are completely unknown to both AE and CRP. They never know the CCpri or the random number r2. r2 conceals CC's confidential information. As a result, our proposed strategy effectively protects CC's privacy.

Table 2. Comparison of Computational overhead

Entity	Public Auditing using BLS signature [21]	Our Proposed Auditing Model using ZSS signature
CC	$jHash_{g_1} + jMul_{g_1} + jExpz_q + jExpz_q$	$WEncryptz_q + jhashz_q + Addz_q + Invz_q$
CRP	$\tau MulExp_{g_1} + Hashz_q + Exp_{g_1} + CAddz_q + (\tau+1)Mulz_q$	$\tau(Mulz_q + Invz_q + Addz_q)$
AE	$\tau MulExp_{g_1} + \tau Hash_{g_1} + Hashz_q + 3Exp_{g_1} + 2Pair_{g_1,g_1} + Mul_{g_1} + Mulz_q$	$Pair_{g_1,g_1} + Mulz_q + Invz_q + \tau Addz_q$

- Data Confidentiality: The original block elements cannot be obtained by AE from the *chal_res* message. CRP stores *detail_msg* of all block data pieces from CC. AE is aware of the value of g1, but no one [CRP or AE] is aware of the value of r2. As a result, it is demonstrated that AE never discovers the original CC block elements, and data confidentiality is appropriately maintained on both the CRP and AE sides i.e.,

$$MsgEncrypt(g_1 * r_2, M = \sum_{t=1}^{j}\sum_{i=1}^{k} B_t(M_i)) = (\sum_{t=1}^{j}\sum_{i=1}^{k}(g_1 * r_2) \oplus B_t[M_i']))$$
$$= (\sum_{i=1}^{n}\sum_{j=1}^{w} B_i[M_j']))$$

- Unforgeability: The ZSS signature mechanism is used in this proposed approach to establish signatures on each block. AE verifies the signature S_τ of all requested blocks B_τ after receiving the *chal_res* message sent by CRP. Because of the security nature of the ZSS signature scheme, the integrity verification cannot pass the AE's audit test if the specific signature is not properly maintained by CRP at resource pool created by CC. As a result, this entire strategy is resistant to both internal and external forgery attempts. The accuracy of a valid signature is defined by the bilinear property as follows:

$$\begin{aligned}
&e(g_1 * \sum_{t=1}^{\tau} S_t, g_1) \\
&= e(g_1 * \frac{g_1}{H(B_1[M_i']) + r_1} + \frac{g_1}{H(B_2[M_i']) + r_1} + ... + \frac{g_1}{H(B_\tau[M_i']) + r_1}, g_1) \\
&= e(g_1 * \frac{g_1}{H(B_1[M_i']) + r_1} + \frac{g_1}{H(B_2[M_i']) + r_1} + ... + \frac{g_1}{H(B_\tau[M_i']) + r_1}, g_1) \\
&= e(g_1 * \frac{g_1}{H(B_1[M_i']) + r_1} + \frac{g_1}{H(B_2[M_i']) + r_1} + ... + \frac{g_1}{H(B_\tau[M_i']) + r_1}, g_1) \\
&= e(g_1, g_1)(\frac{g_1}{H(B_1[M_i']) + r_1} + \frac{g_1}{H(B_2[M_i']) + r_1} + ... + \frac{g_1}{H(B_\tau[M_i']) + r_1}) \\
&= e(g_1, g_1)(S_1, S_2, .., S_\tau)
\end{aligned} \quad (7)$$

- Collision Resistance: The auditing verification phase of the challenge response message *chal_res* from CRP can only pass if CRP does not change the values of genuine block elements $B_t[M_i']$ with corrupted block elements $B_t[\phi_i]$. According to the collision resistance property of the hash function, $H(B_t[M_i']) \neq H(B_t[\phi_i'])$ and $e(H(B_t[M_i']).g_1 + CC_{pub}, S_t) \neq e(H(B_t[\phi_i']).g_1 + CC_{pub}, S_t)$.

6 Performance Analysis

The primary goal of this section of the study is to calculate the computational cost of signature generation with block numbers, signature verification with block numbers, and block size (KB). The test conditions are as follows: It is equipped with a 12-GB RAM configuration, an Intel(R) Core(TM) i5-9300H processor, and Windows 10. This section of the research focuses mostly on calculating the computational cost of signature creation time with multiple block numbers at the cloud customer side, signature verification time with multiple block numbers

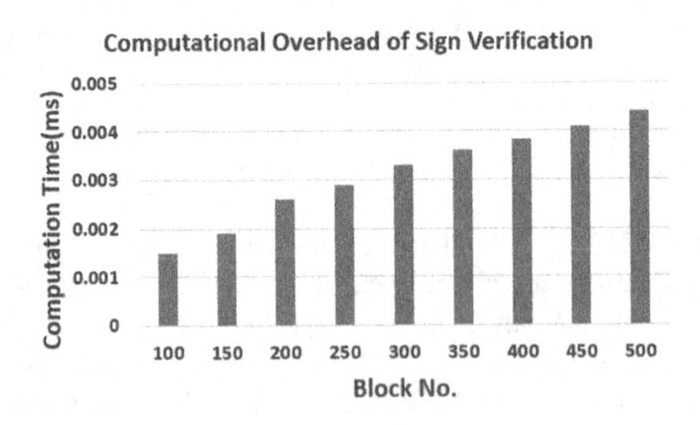

Fig. 2. Computational Overhead of AE for Signature Verification

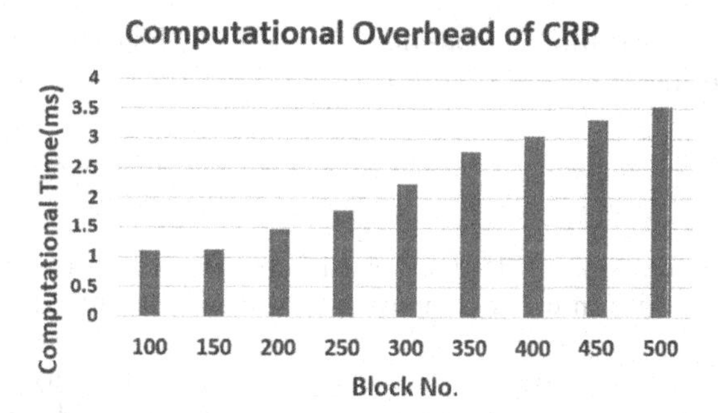

Fig. 3. Computational Overhead of CRP for preparation of *chal_res* message

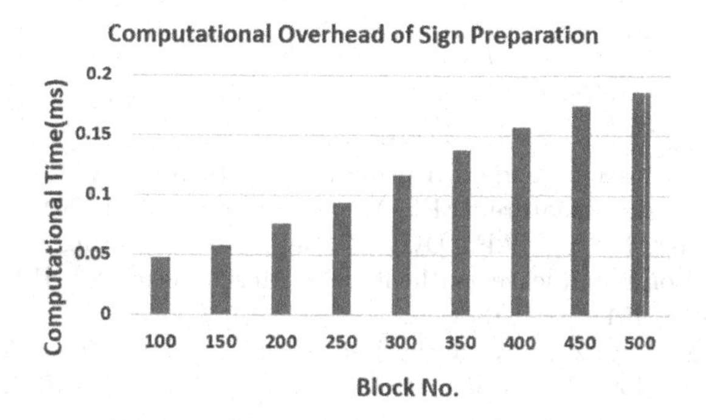

Fig. 4. Computational Overhead of CC for Signature preaparation

100, 150, 200, 250, 300, 350, 400, 450, and 500 at the auditing expertise side in Fig. 2 and Fig. 4 consequently. Figure 3 also depicts the computational overhead of preparing the *chal_res* message by CRP. Table 2 summarises a comparison of computational overheads of CC, CRP, and AE that aids understanding of the proposed model's efficiency over BLS signature where Mul_{g_1} denotes multiplication operation in g_1, Inv_{Z_q} denotes inversion operation in Z_q,Mul_{Z_q} denotes multiplication operation in Z_q,$Hash_{g_1}$ is used to map Strings which are consistently assigned to the group G1, Exp_{g_1} denotes exponential operation, $Pair_{g_1,g_1}$ denotes pairing operation to verify the signatures, Add_{Z_q} denotes addition operation in Z_q.

7 Conclusion

The use of ZSS signature and original data encryption method based on random number generation ensures the consistency of outsourced data for CC at cloud resource pool. It helps to prevent forgeability attacks, and resist collusion. Furthermore, AE can detect and inform CC of block corruption during auditing. Furthermore, while completing the audit, this user-friendly auditing architecture secures the privacy of CC's data from all entities, including CRP and AE. In this research study, we plan to extend this auditing methodology for CC-side audit message verification.

References

1. Singh, A.P., Pasupuleti, S.K.F.: Optimized public auditing and data dynamics for data storage security in cloud computing. Procedia Comput. Sci. **2**(93), 751–759 (2016)
2. Shen, W., Qin, J., Yu, J., Hao, R., Hu, J., Ma, J.F.: Data integrity auditing without private key storage for secure cloud storage. IEEE Trans. Cloud Comput. **9**(4), 1408–1421 (2019)
3. Thangavel, M., Varalakshmi, P.F.: Enabling ternary hash tree based integrity verification for secure cloud data storage. IEEE Trans. Knowl. Data Eng. **32**(12), 1408–1421 (2019)
4. Shen, W., et al.: An efficient data integrity auditing protocol for cloud computing. Future Gener. Comput. Syst. **109**, 306–316 (2020)
5. Zhu, H., et al.: A secure and efficient data integrity verification scheme for cloud-IoT based on short signature. IEEE Access **7**, 90036–90044 (2019)
6. Rao, L., Zhang, H., Tu, T.F.: Dynamic outsourced auditing services for cloud storage based on batch-leaves-authenticated Merkle hash tree. IEEE Trans. Serv. Comput. **13**(3), 451–463 (2017)
7. Yu, H., Lu, X., Pan, Z.F.: An authorized public auditing scheme for dynamic big data storage in cloud computing. IEEE Access **8**, 151465–151473 (2020)
8. Zhu, Y., et al.: Dynamic audit services for outsourced storages in clouds. IEEE Trans. Serv. Comput. **6**(2), 227–238 (2011)
9. Shen, W., Qin, J., Yu, J., Hao, R., Hu, J.F.: Enabling identity-based integrity auditing and data sharing with sensitive information hiding for secure cloud storage. IEEE Trans. Inf. Forensics Secur. **14**(2), 331–346 (2018)
10. Zhang, J., Lu, R., Wang, B., Wang, F.: Privacy-preserving public auditing protocol for regenerating-code-based cloud storage. IEEE Trans. Inf. Forensics Secur. **16**(2), 1288–1289 (2020)
11. Ogiso, S., Mohri, M., Shiraishi, Y.F.: Transparent provable data possession scheme for cloud storage. In: International Symposium on Networks, Computers and Communications (ISNCC), pp. 1–5. IEEE (2020). https://doi.org/10.10007/1234567890
12. Bian, G., Chang, J.F.: Certificateless provable data possession protocol for the multiple copies and clouds case. IEEE Access **8**, 102958–102970 (2020)
13. Gokulakrishnan, S., Gnanasekar, J.M.: Data integrity and recovery management in cloud systems. In: Fourth International Conference on Inventive Systems and Control (ICISC), pp. 645–648 (2020). https://doi.org/10.1109/ICISC47916.2020.9171066
14. Vineela, A., Kasiviswanath, N., Bindu, C.S.: Data integrity auditing scheme for preserving security in cloud based big data. In: 6th International Conference on Intelligent Computing and Control Systems (ICICCS), pp. 609–613 (2022). https://doi.org/10.1109/ICICCS53718.2022.9788365
15. Wang, Y., Chen, Z., Wang, K., Yang, Z.: Education cloud data integrity verification based on mapping-trie tree. In: International Conference on Machine Learning, Big Data and Business Intelligence (MLBDBI), pp. 155–158 (2019). https://doi.org/10.1109/MLBDBI48998.2019.00036

16. Bian, G., et al.: Data integrity audit based on data blinding for cloud and fog environment. IEEE Trans. Cloud Comput. Access **10**, 39743–39751 (2022)
17. Zhao, T., Zhang, Z., Zhang, L.: A novel cloud data integrity verification scheme based on dynamic array multi-branch tree. In: International Conference on Machine Learning, Big Data and Business Intelligence (MLBDBI) 6th International Conference on Intelligent Computing and Signal Processing (ICSP), pp. 377–380 (2021). https://doi.org/10.1109/ICSP51882.2021.9408776
18. Ahamed Ali, S., Justindhas, Y., Lakshmanan, M., Hari Kumar, P., Baskaran, P.: Secured cloud data outsourcing model using two party integrity scheme. In: 2nd International Conference on Advance Computing and Innovative Technologies in Engineering (ICACITE), pp. 509–512 (2022). https://doi.org/10.1109/ICACITE53722.2022.9823507
19. Sushma, S., Srilatha, P., Srinivas, G.: Efficient integrity checking and secured data sharing in cloud. In: International Conference on Innovative Computing, Intelligent Communication and Smart Electrical Systems (ICSES), pp. 1–4 (2022). https://doi.org/10.1109/ICSES55317.2022.9914256
20. Shen, W., Qin, J., Yu, J., Hao, R., Hu, J., Ma, J.: Data integrity auditing without private key storage for secure cloud storage. IEEE Trans. Cloud Comput. **9**(4), 1408–1421 (2021)
21. Wang, C., Chow, S.S.M., Wang, Q., Ren, K., Lou, W.F.: Privacy-preserving public auditing for secure cloud storage. IEEE Trans. Comput. **62**(2), 7362–375 (2013)
22. Singhal, S., Sharma, A.: Load balancing algorithm in cloud computing using mutation based pso algorithm. In: 4th International Conference of Advances in Computing and Data Sciences, pp. 224–233 (2020). https://doi.org/10.1007/978-981-15-6634-9_21
23. Shen, W., Qin, J., Yu, J., Hao, R., Hu, J.: Enabling identity-based integrity auditing and data sharing with sensitive information hiding for secure cloud storage. IEEE Trans. Inf. Forensics Secur. **14**(2), 331–346 (2018)
24. Yan, H., Gui, W.: Efficient identity-based public integrity auditing of shared data in cloud storage with user privacy preserving. IEEE Access **9**, 45822–45831 (2021)
25. Singhal, S., Sharma, A.: Mutative BFO-based scheduling algorithm for cloud environment. In: Proceedings of International Conference on Communication and Artificial Intelligence (ICCAI), pp. 589–599 (2020). https://doi.org/10.1007/978-981-33-6546-9_56
26. Singhal, S., Sharma, A.: Mutative ACO based load balancing in cloud computing. Eng. Lett. **29**(4), 1297 (2021)
27. Singhal, S., Sharma, A.: A job scheduling algorithm based on rock hyrax optimization in cloud computing. Eng. Lett. **103**(9), 2115–2142 (2021)

Genetic Algorithm Based Anomaly Detection for Intrusion Detection

Sushilata D. Mayanglambam[1,2](✉), Nainesh Hulke[3], and Rajendra Pamula[1]

[1] Department of Computer Science and Engineering, Indian Institute of Technology (ISM), Dhanbad 826004, Jharkhand, India
sushilatamayanglambam@gmail.com

[2] Department of Computer Engineering, Mizoram University, Aizawl 796004, Mizoram, India

[3] Arista Networks, S.No. 7, Pinnac House II, Kothrud, Pune 411038, Maharashtra, India

Abstract. This article describes a method for detecting anomalies in network intrusion data. The method utilizes a clustering-based strategy to efficiently cluster the data. Here, Genetic Algorithm (GA)-based clustering was employed, with chromosomes represented by strings of real integers encoding the cluster centres. We have adopted the Davies-Bouldin (DB) index as a fitness function in this study. After clusters have been constructed, points that are closer to the cluster centroids are deemed inliers and are eliminated. Pruning strategies considerably minimize the computing effort and complexity required to identify outlier scores. To generate outlier scores, we've used a technique based on local distance and the outlier factor. The n points with the highest value were deemed to be outliers. Our results were compared to those obtained utilizing the explicit local distance-based outlier factor algorithm.

Keywords: Genetic Algorithm · Davies-Bouldin Index · Local Distance-Based Outlier Factor

1 Introduction

The Internet has become one of the most important services in our modern society, which is undergoing a period of rapid change. Consequently, network security has also become crucial for the protection of personal information, financial transactions in online shopping, stock markets, and bank accounts, etc. The objective of intrusion detection is to detect computer intrusions by examining various information records found in network operations [1, 2]. Intrusion detection plays a vital part in addressing network security challenges. IDS (Intrusion Detection System) [3] evaluation criteria include detection stability and detection precision. IDS is a network security system that identifies vulnerable or malicious network activity and delivers notifications to the monitoring device. IDS has played an important role in the current digital world. IDS also examines the system for policy violations. The performance of an intrusion detection system is dependent on its ability to identify network threats. As the internet and cyberspace have matured, computer and internet security have become more vulnerable. Therefore, traditional IDS

P. Das et al. (Eds.): AMRIT 2023, CCIS 1953, pp. 194–203, 2024.
https://doi.org/10.1007/978-3-031-47224-4_17

will soon become obsolete. In the early stages of IDS research, statistical approaches and rule-based expert systems were dominated [4]. However, neither of the two techniques were particularly accurate for large datasets. Traditional IDS-based security methods were founded on signature-based concepts [5], making it difficult to identify new kinds of irregularities and threats [6]. The greatest drawback of signature-based IDS was the sluggish rate at which the signature database was updated with newly emerging threats [6]. Numerous IDS-related studies have recommended a variety of strategies for enhancing the precision and reliability of intrusion detection systems [7]. Nevertheless, researchers continue to develop new methods for effective threat detection. Particularly, anomaly-based intrusion detection techniques [8] for defending against modern cyber-attacks are becoming more challenging. In the real world, an intrusion detection model's dependability is determined by its capacity to detect current threats in real-time. In this regard, machine learning-based intrusion detection system (IDS) approaches are gaining popularity and are commonly used for anomaly detection in network intrusion detection devices [2]. Various machine learning techniques, such as Random Forest (RF), Decision Trees (DT), k-Nearest Neighbor (kNN), Linear Genetic Programming [9], Neural Networks (NN) [10], Bayesian networks, Support Vector Machines (SVM), Fuzzy Inference Systems [11], Multivariate Adaptive Regression Splines (MARS) [12], etc., have been proposed and implemented in IDS to detect intrusions. Even optimization strategies have been implemented in IDS and intrusion detection feature extraction. Additionally, clustering algorithms have been implemented in intrusion detection. Various clustering algorithms are described in the literature [13, 14]. K-means is one of the most popular and commonly used algorithms [14]. In general, the ideal distance criterion for data clustering was determined by either maximising the distance between clusters or minimising the dispersion within each cluster. In this study, we have used the Genetic Algorithm to cluster data based on minimising the within-cluster spread. Our Genetic clustering technique divides the entire dataset into k clusters and estimates the radius of each cluster using the elbow approach. Furthermore, we calculated the ratio between inter-cluster separation and intra-cluster dispersion using the Davies-Bouldin index [15] to obtain the optimal clustering. It is determined the distance between each cluster point and the cluster's centroid. The point is eliminated if its distance is less than the cluster's radius. To identify outliers, LDOF values are computed for every unpruned point in every cluster.

2 Related Work

IDS can be broadly categorized into Anomaly based-IDS, Signature based-IDS, and Specification based-IDS. The traditional classification algorithm, such as Support vector machine (SVM), was widely used in IDS [16–19]. In addition, the k-nearest neighbour (kNN) method was employed extensively for unsupervised applications in IDS [13]. Earlier papers [14, 20, 21] have adopted the Random Forest (RF) method as a machine learning algorithm to overfit and manage such unbalanced data in IDS. In the real-time intrusion detection system, rule-based pattern matching was proposed to identify security violations by tracking aberrant computer usage patterns [22]. However, the evolution of the current digital world has put us at risk from new content-based attacks that cannot be identified by existing machine learning models. Consequently, numerous academics have recently developed new IDS solutions employing deep learning methodologies. IDS has embraced techniques such as stacked autoencoders, convolutional, and recurrent neural networks for monitoring novel attack types [23, 24]. Intrusion detection mechanism using time-series, Marchov chains, and other statistical methods have been reported in the literature. Meanwhile, the first distance-based outlier detection technique was introduced by Knorr and Ng [22]. Subsequently, Angiulli and Pizzuti have proposed an outlier detection algorithm to find out the outliers in the dataset by accounting the whole neighborhood of the objects [25–27]. The reported method does not solely consider the distance between the cluster center to the kth nearest neighbor. Instead, they have taken all the points that were ranked based on the sum of the distances from the k-nearest neighbors. Breunig et al. further reported that the Local Outlier Factor (LOF) for each element in the dataset indicates its degree of the outlier [28]. Here, the outlier factor was considered by limiting the analysis to just each object's immediate surroundings. Later, Zhang et al. proposed a local distance-based outlier detection method. First they determined the degree to which an object deviates from its neighborhood. Then, those objects which deviated from its neighborhood were considered as outliers or anomalies in the data set [29]. Meanwhile, Clustering methods such as Clustering Large Applications based on RANdomized Search (CLARANS) [30], Density-Based Spatial Clustering of Applications with Noise (DBSCAN) [31], Balanced Iterative Reducing and Clustering using Hierarchies (BIRCH) [32], and Clustering Using Representatives (CURE) [33] were also adopted to detect outliers. Further, optimized clustering based on Genetic-based clustering technique was proposed in [34]. They showed that the GA-clustering performed superiorly to the K-means algorithm. In [35], a clustering-based method was adopted to capture outliers. They carried out pruning of possible inliers to reduce computations required to calculate outlier scores.

3 Background

Genetic Algorithm (GA) is a heuristic search optimization algorithm that adopts the principles of natural evolution to solve various optimization problems. Here, population refers to a set of all the probable solutions that can solve the given problem. In the GA, the chromosome represents one of the solutions in the population. The degree of goodness of every chromosome was evaluated using an objective or fitness function. The principle of GA algorithm relies on genetics and natural selection [i.e. survival for the fittest], and chromosomes are chose based on their objective or fitness value. A certain number of copies of each chromosome are designated for the mating pool. To create a new generation of chromosomes, these chromosomes are subjected to biologically inspired operations including crossover and mutation. Until a termination condition is met, the process of selection, crossover, and mutation continues for a specific number of generations. In this study, we employed GA to generate optimum centroids by data clustering. Then, in order to identify the outliers, we used a local distance-based outlier detection algorithm. Here, Zhang et al. [29] developed a local distance-based outlier detection method to locate outliers from the data set. They determined the degree to which an object deviates from its surroundings using local distance-based outlier factor

(LDOF). The overall LDOF calculation complexity for the entire data set is O (N2), where N denotes the dimension of the dataset. Therefore, we have adopted the pruning strategy to prune the possible inliers and finding the LDOF values for only those unpruned data points. Higher the value of LDOF, the higher the probability of the point being outlier.

4 Proposed Method

In this work, we have put forth an approach to finding anomalies that is based on GA-clustering. The clustering analysis seeks to classify a set of data into distinct groups. The objects having similar features are grouped into same clusters and those objects with no similarities can be placed into different clusters. Many measures for assessing cluster validity have recently been proposed [15, 36, 37]. The fitness function is a way of assessing the solution. Here, we have adopted Davies-Bouldin index as a fitness function to obtain optimized centroids. The LDOF technique to detect outliers has one major flaw as it is computationally expensive. Because LDOF values must be computed for each data point in the dataset. We are only interested in a few outliers, hence it is unnecessary to compute the LDOF values for all data points. Therefore, points whose distance from the centroid is smaller than the cluster radius are discarded, as they are unlikely to be outliers. After that, we carry out the outlier detection procedure for each unpruned point in every cluster.

Pseudo code of Genetic Algorithm for data clustering

```
Initialize a population of 100 individuals with randomly generated real value genes
Calculate the fitness value of solution using proposed fitness function
  i ← 0
  REPEAT
    Selection is performed using steady-state selection
    Single-Point crossover is performed
    Random mutation is used. 10% of the total genes are mutated. The minimum
    value of the mutated gene is -1, and the maximum value is +1
    Calculate the fitness value of the obtained population using proposed fitness
    function (Davies-Bouldin index)
  i ← i + 1
  UNTIL i < 300
STOP
```

Pseudo code of LDOF for Outlier Detection

```
Input: Dataset X
Output: top-o outlier in X
Use GA algorithm to obtain the optimized centroids
For each x belongs to X do
   Group the data into clusters using the optimized centroids
   Compute radius of each cluster centroid
   Prune those data points for a distance less than its radius
   Compute LDOF of unpruned points
End
LDOF result: Sort LDOF (unpruned points) in descending order.
Output top-o outlier results
```

5 Results and Discussion

The dataset was taken from earlier reported literature [38]. It contains both benign and various attacks. The data points are labeled with various network flow features, including IPs, source and destination ports, and protocols. In this work, there are 5200 samples, among which 5002 are benign and remaining are the attacks or anomalies. There are 8 different intrusion attacks are covered and the total of 198 anomaly samples. There are described in Table 1. After feature selection, the dataset consists of 33 dimensions. Thus, it is impossible to visualize the data without feature reduction [39]. So, here we have used T-distributed Stochastic Neighbor Embedding (T-SNE) to visualize data in a two-dimensional space. It is a nonlinear dimensionality reduction technique. This reduced-feature data is illustrated in Fig. 1.

Table 1. Intrusion Attacks

Type of Intrusion	Description	Number of samples
DoS Hulk	Attacks on the web servers by generating unique and obfuscated traffic volumes	85
PortScan	Common method employed by hackers is to locate open doors or vulnerabilities in a network	63
DoS GoldenEye	Opening several parallel connections against a URL to compromise the webserver	15
FTP-Patator	Brute force attack on FTP servers	11
DoS Slowhttptest	Application Layer Denial of Service attacks by	10
DDos	Overwhelming a target website with fake traffic	8
DoS slowloris	Application layer is DDoS attack. Executed partial HTTP requests to establish connections between a single computer and a specified Web server	4
Bot	Large-scale cyber-attack by malware-infected devices and controlled remotely	2

Estimating the number of underlying clusters in the dataset is essential for GA-based clustering. This was performed utilising the elbow technique. For each number of clusters, the total of squared distances between each point and its assigned centroid (distortions) is computed and shown. The transition from a steep to a shallow slope (an elbow) is utilised to calculate the ideal number of clusters. Here, four clusters have been selected. The GA-based clustering technique requires probabilities of crossover and mutation, population size, number of generations, and other parameters. In this study, we used a mutation rate of 10%, a population size of 100, and 300 generations. Figure 2 illustrates the clustered data visualized with the aid of T-SNE.

Dataset Pre-processing Details are Given Below

- To select the important features, we calculated features importance for all features using random forest classifier.
- Features with importance greater than a threshold 0.0125 were selected.
- Selected top 1110 data points with high LOF values from a sample of 30,000 data points.
- We select anomaly data points from 1110 data points.
- Added the benign samples of 5002.
- Dropped rows which has field with NAN values
- Dropped duplicated rows
- Dropped non-numerical columns
- Finally, dataset of 5200 was prepared for the detection.
- Normalized the data.

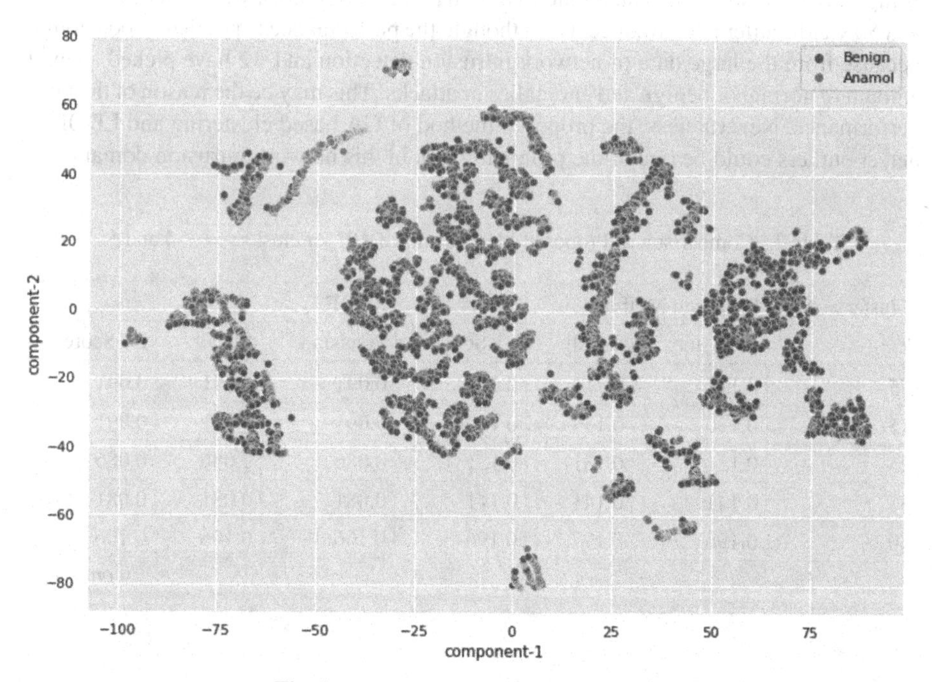

Fig. 1. Dataset Visualized using T-SNE.

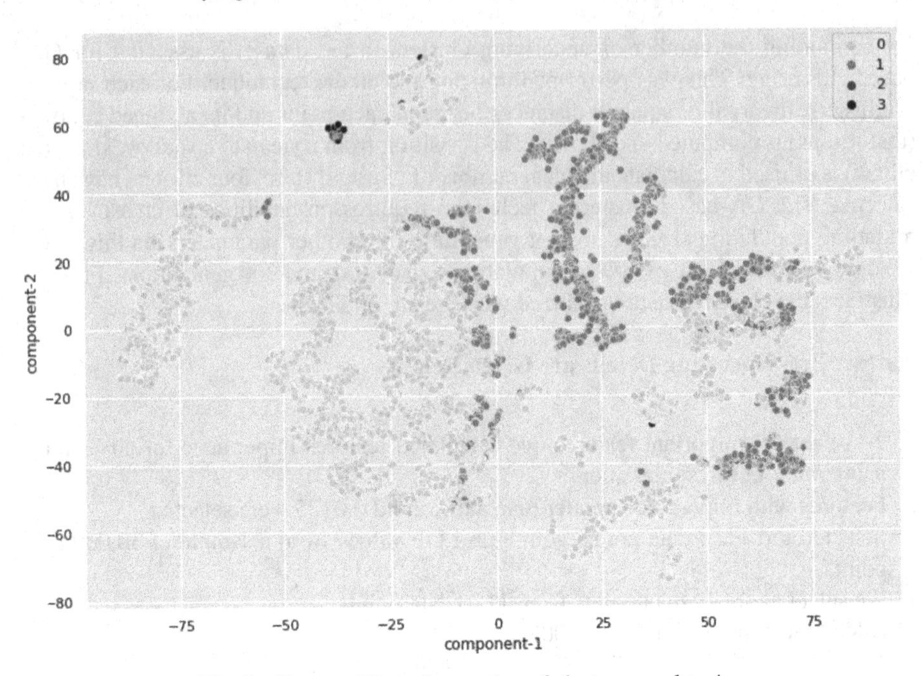

Fig. 2. Clustered Data for number of clusters equal to 4.

From the Table 2, we can see that the result of cluster = 4 was the best as compared with cluster = 3, 5. The same cluster = 4 was also given by the elbow method. While comparing the result with standalone LDOF in all the KNN, i.e. k values of 5, 15, 25, 35 and 50 yields better performance. Even though, the performance is not that good enough because from the large data of network intrusion detection and we have picked a small amount of normal or benign and anomalies or attacks. This may be the reason of the poor performance. Nevertheless, the proposed method of GA based clustering and LDOF to detect outliers could be one of the promising tool in this network intrusion domain.

Table 2. Comparison of Proposed Method and LDOF for clusters = 3, 4 and 5.

Cluster = 4	Proposed Method			LDOF		
k	Precision	Recall	F1-Score	Precision	Recall	F1-Score
5	0.111	0.111	0.111	0.091	0.091	0.091
15	0.136	0.136	0.136	0.091	0.091	0.091
25	0.131	0.131	0.131	0.086	0.086	0.086
35	0.141	0.141	0.141	0.081	0.081	0.081
50	0.196	0.197	0.196	0.106	0.106	0.106

(continued)

Table 2. (*continued*)

Cluster = 3	Proposed Method			LDOF		
k	Precision	Recall	F1-Score	Precision	Recall	F1-Score
5	0.116	0.116	0.116	0.091	0.091	0.091
15	0.141	0.141	0.141	0.091	0.091	0.091
25	0.106	0.106	0.106	0.086	0.086	0.086
35	0.091	0.091	0.091	0.081	0.081	0.081
50	0.086	0.086	0.086	0.106	0.106	0.106
Cluster = 5	Proposed Method			LDOF		
k	Precision	Recall	F1-Score	Precision	Recall	F1-Score
5	0.111	0.111	0.111	0.091	0.091	0.091
15	0.136	0.136	0.136	0.091	0.091	0.091
25	0.076	0.076	0.076	0.086	0.086	0.086
35	0.066	0.066	0.066	0.081	0.081	0.081
50	0.071	0.071	0.071	0.106	0.106	0.106

6 Conclusion

The data has samples corresponding to 8 different intrusion attacks. Feature selection using random forest helped to reduce the dimensionality of the dataset for efficient analysis. The optimal number of clusters was determined to be 4 using the elbow method. It can also be seen from the Table 2 that the result of cluster = 4 was the best among others. The parameters of the GA-based clustering algorithm were hyper-tuned for better clustering results. Here, only 32.12% of data points were left after pruning. Therefore, the time and space complexity were reduced after adopting the pruning method as compared to standalone LDOF. However, the reason of the poor performance could be that we have picked a small amount of normal or benign and anomalies or attacks from the large data of network intrusion detection. Nevertheless, the proposed method of GA based clustering and LDOF to detect outliers could be one of the promising tools in this network intrusion domain.

References

1. Anderson, J.P.: Computer security threat monitoring and surveillance. Technical report, Fort Washington, PA, USA, pp. 9–11 (1980)
2. Endorf, C., Schultz, E., Mellander, J.: Intrusion Detection and Prevention. Illustrated edn. McGraw-Hill Education, California (2004)
3. Silva, L.D.S., Santos, A.C., Mancilha, T.D., Silva, J.D., Montes, A.: Detecting attack signatures in the real network traffic with ANNIDA. Expert Syst. Appl. **34**(4), 2326–2333 (2008)

4. Manikopoulos, C., Papavassiliou, S.: Network intrusion and fault detection: a statistical anomaly approach. IEEE Commun. Mag. **40**(10), 76–82 (2002)
5. Lv, L., Wang, W., Zhang, Z., Liu, X.: A novel intrusion detection system based on an optimal hybrid kernel extreme learning machine. Knowl. Based Syst. **195**(105648), 1–17 (2020)
6. Gao, X., Shan, C., Hu, C., Niu, Z., Liu, Z.: An adaptive ensemble machine learning model for intrusion detection. IEEE Access **7**, 82512–82521 (2019)
7. Patcha, A., Park, J.M.: An overview of anomaly detection techniques: existing solutions and latest technological trends. Comput. Netw. **51**(12), 3448–3470 (2007)
8. Atefinia, R., Ahmadi, M.: Network intrusion detection using multiarchitectural modular deep neural network. J. Supercomput. **77**(4), 3571–3593 (2021)
9. Mukkamala, S., Sung, A.H., Abraham, A.: Modeling Intrusion detection systems using linear genetic programming approach. In: Orchard, B., Yang, C., Ali, M. (eds.) IEA/AIE 2004. LNCS, vol. 3029, pp. 633–642. Springer, Heidelberg (2004). https://doi.org/10.1007/978-3-540-24677-0_65
10. Mukkamala, S., Sung, A.H., Abraham, A.: Intrusion detection using ensemble of soft computing paradigms. In: Abraham, A., Franke, K., Köppen, M. (eds.) Intelligent Systems Design and Applications. ASC, vol. 23, pp. 239–248. Springer, Heidelberg (2003). https://doi.org/10.1007/978-3-540-44999-7_23
11. Shah, K., Dave, N., Chavan, S., Mukherjee, S., Abraham, A., Sanyal, S.: Adaptive neuro-fuzzy intrusion detection system. In: IEEE International Conference on Information Technology: Coding and Computing (ITCC 2004), USA, vol. 1, pp. 70–74. IEEE Computer Society (2004)
12. Mukkamala, S., Sung, A.H., Abraham, A., Ramos, V.: Intrusion detection systems using adaptive regression splines. In: Seruca, I., Filipe, J., Hammoudi, S., Cordeiro, J. (eds.) Proceedings of the 6th International Conference on Enterprise Information Systems, ICEIS, Portugal, vol. 3, pp. 26–33 (2004)
13. Li, W., Yi, P., Wu, Y., Pan, L., Li, J.: A new intrusion detection system based on KNN classification algorithm in wireless sensor network. J. Electr. Comput. Eng. **2014**(240217), 1–9 (2014)
14. Breiman, L.: Random forests. Mach. Learn. **45**(1), 5–32 (2001)
15. Davis, D.L., Bouldin, D.W.: A cluster separation measure. IEEE Trans. Pattern Anal. Mach. Intell. **1**, 224–227 (1979)
16. Heba, F.E., Darwish, A., Hassanien, A.E., Abraham, A.: Principle components analysis and support vector machine based intrusion detection system. In: 10th International Conference on Intelligent Systems Design and Applications, ISDA 2010, Cairo, Egypt, pp. 363–367. IEEE (2010)
17. Chen, R.C., Cheng, K., Chen, Y., Hsieh, C.: Using rough set and support vector machine for network intrusion detection system. In: Nguyen, N.T., Nguyen, H.P., Grzech, A. (eds.) First Asian Conference on Intelligent Information and Database Systems, ACIIDS 2009, Dong hoi, Quang binh, Vietnam, 1–3 April 2009, pp. 465–470. IEEE Computer Society (2009)
18. Jia, N., Liu, D.: Application of SVM based on information entropy in intrusion detection. In: Xhafa, F., Patnaik, S., Zomaya, A. (eds.) IISA 2017. AISC, vol. 686, pp. 464–468. Springer, Cham (2018). https://doi.org/10.1007/978-3-319-69096-4_64
19. Wang, H., Gu, J., Wang, S.: An effective intrusion detection framework based on SVM with feature augmentation. Knowl. Based Syst. **136**, 130–139 (2017)
20. Zhang, J., Zulkernine, M., Haque, A.: Random-forests-based network intrusion detection systems. IEEE Trans. Syst. Man Cybern. Part C **38**(5), 649–659 (2008)
21. Farnaaz, N., Jabbar, M.: Random Forest modeling for network intrusion detection system. Procedia Comput. Sci. **89**, 213–217 (2016)
22. Knorr, E.M., Ng, R.T.: Algorithms for mining distance based outliers in large datasets. In: Proceedings of the 24th International Conference on Very Large Data Bases, VLDB, pp. 392–403 (1998)

23. Aldweesh, A., Derhab, A., Emam, A.Z.: Deep learning approaches for anomaly-based intrusion detection systems: a survey, taxonomy, and open issues. Knowl. Based Syst. **189**(105124), 1–19 (2020)
24. Kim, K., Aminanto, M.E.: Deep learning in intrusion detection perspective: overview and further challenges. In: International Workshop on Big Data and Information Security, IWBIS 2017, Jakarta, Indonesia, pp. 5–10. IEEE (2017)
25. Angiulli, F., Basta, S., Pizzuti, C.: Distance-based detection and prediction of outliers. IEEE Trans. Knowl. Data Eng. **18**, 145–160 (2006)
26. Angiulli, F., Pizzuti, C.: Fast outlier detection in high dimensional spaces. In: PKDD 2002: Proceedings of the 6th European Conference on Principles of Data Mining and Knowledge Discovery, pp. 15–26 (2002)
27. Angiulli, F., Pizzuti, C.: Outlier mining in large high dimensional data sets. IEEE Trans. Knowl. Data Eng. **17**, 203–215 (2005)
28. Breunig, M.M., Kriegel, H.-P., Ng, R.T., Sander, J.: ACM SIGMOD Rec. **29**(2), 93–104 (2000)
29. Zhang, K., Hutter, M., Jin, H.: A new local distance-based outlier detection approach for scattered real-world data. In: PAKDD 2009: Proceedings of the 13th Pacific-Asia Conference on Advances in Knowledge Discovery and Data Mining, pp. 813–822 (2009)
30. Ng, R.T., Han, J.: Efficient and effective clustering methods for spatial data mining. In: VLDB 1994: Proceedings of the 20th International Conference on Very Large Data Bases, pp. 144–155 (1994)
31. Ester, M., Kriegel, H.-P., Xu, X.: A database interface for clustering in large spatial databases. In: Proceedings of 1st International Conference on Knowledge Discovery and Data Mining (KDD-95), pp. 94–99 (1995)
32. Zhang, T., Ramakrishnan, R., Livny, M.: Birch: an efficient data clustering method for very large databases. ACM SIGMOD Rec. **25**(2), 103–114 (1996)
33. Guha, S., Rastogi, R., Shim, K.: CURE: an efficient clustering algorithm for large databases. ACM SIGMOD Rec. **27**, 273–284 (1998)
34. Maulik, U., Bandyopadhyay, S.: Genetic algorithm-based clustering technique. Pattern Recognit. **33**, 1455–1465 (2000)
35. Pamula, R., Deka, J.K., Nandi, S.: An outlier detection method based on clustering. In: 2011 Second International Conference on Emerging Applications of Information Technology, pp. 253–256 (2011)
36. Theodoridis, S., Koutroumbas, K.: Pattern Recognition, Academic Press, 2nd edn. San Diego (2003)
37. Bezdek, J.C.: Some new indexes of cluster validity. IEEE Trans. Syst. Man Cybern. Part B **28**(3), 301–315 (1998)
38. https://www.unb.ca/cic/datasets/ids-2017.html. Accessed 05 Sept 2022
39. Akashdeep, Manzoor, I., Kumar, N.: A feature reduced intrusion detection system using ANN classifier. Expert Syst. Appl. **88**, 249–257 (2017)

Machine Learning Based Malware Identification and Classification in PDF: A Review Paper

Vaishali and Nahita Pathania(✉)

Lovely Professional University, Phagwara, India
nahita.19372@lpu.co.in

Abstract. Today's modern antivirus software fails to provide protection against malicious PDF (Portable Document Format) files, which is considered a threat to system security. This study introduces a new machine-learning-based classification approach to PDF malware to reduce PDF malware somewhat. A unique feature of this system is static and dynamic checking of provided PDF files. As a result, each of these classification algorithms may more accurately determine the right document type, negative rate (FNR), and F1 score, which is the most appropriate strategy evaluated. The proposed system is also subject to malicious attacks that obfuscate the malicious code in the PDF file by hiding it from the PDF parser during the processing stage. It was determined that the proposed technique outperformed the prior art, which had an F1 metric of 0.978, by achieving an F1 metric of 0.986 utilizing a random forest (RF) classifier. Compared to current solutions, this is very effective at detecting malware encoded in PDF files.

Keywords: Machine learning · PDF · malware classification · malware detection approaches · malware features

1 Introduction

The majority of activities in the contemporary digital world revolve around using the Internet, making it more crucial to protect our applications information and data in the sight of various online threats. Adversaries are constantly coming up with innovative hostile codes and attempts to breach resources. Analyzing Malware therefore becomes one of main issues these days in the world since different malwares are produced by hackers and attackers and even the characteristics change extremely quickly. Malware today differs from that which existed in the past because its signatures are constantly changing, making it challenging to track [1, 2]. Therefore, one of the most sought-after study fields is the identification and categorization of the most recent malware. There are primarily two methods for identifying malware: behavior-based detection and signature-based detection. The behavior-based technique is complicated, but it can recognise complex and unknown malware to certain extent utilizing other methods and machine intelligence signature-based techniques, on the other hand, are fast and effective only at identifying known malware. All virus types cannot be detected by any technology, especially given the daily rise in infection. The features of the underlying object are used to establish

P. Das et al. (Eds.): AMRIT 2023, CCIS 1953, pp. 204–212, 2024.
https://doi.org/10.1007/978-3-031-47224-4_18

a unique signature in the signature-based approach. Signature-based techniques compare suspicious programmes with patterns and signatures from the detected malware. Anomaly-based approaches examine the typical behavior of mobile devices over time, using modeling tools to identify unexpected activity brought on by malware [3, 4]. The review paper also focused on assessing the machine learning algorithms used in malware detection techniques and providing details on the feature extraction process, dataset age, and assessment metrics used in earlier studies. Static analysis is complicated by the malware's source code's code polymorphism, and dynamic analysis necessitates numerous executions to identify harmful activity. Static and dynamic analyses are the two main types of malware analysis. In static technique, malware is examined without the embedded code being executed, but in dynamic method, it is examined while the embedded code is being executed.

Therefore, one of the most important tasks in the malware analysis procedure is malware identification. The static analysis is fairly straightforward and does not involve running any malware samples. While performing static malware analysis, it is not necessary to go through each step of the procedure. Dynamic analysis includes in-depth examination to find malware [5]. While the procedure is still being carried out, a thorough analysis of the entire behavior of activities is performed. Process monitoring needs to be done in great depth for this analysis. Malware can be categorized into a number of different types, including viruses, trojan horses, worms, rootkits, ransomware, and keyloggers [6]. Malware can be categorized in a variety of ways, including using feature-based approaches and image classification, which converts binary information into images. Therefore, gathering additional data will be beneficial for improving classification. For more accurate and better classification of malware utilizing the most recent machine intelligence techniques, excellent classifiers must be constructed.PDF is generally used in formats. In spite of widespread ignorance, it swiftly became a crucial computer attack vector. A victim's PC could be hijacked by hackers via one of the several Adobe Reader security holes. The inherent complexity of PDF files and the wide range of existing obfuscation techniques make it challenging for antivirus software developers to create defenses against attacks. The majority of people transmit attachments in the portable document format (PDF), which is known for its portability and light weight. But people are not aware about what attacks these files might be utilized in or spread to [7]. Phishing, vulnerabilities, and use of the features of the PDF are the three main types of PDF malware. Exploit kits take advantage of a weakness in the PDF reader's API to let the attacker infect the system by running the arbitrary code. In most of the cases, this is accomplished by including JavaScript code in the file. However, in phishing attacks, a file that the user is unaware of is used to deceive them into clicking on a malicious link [8]. These campaigns are far trickier to spot, and they have only recently been discovered. Any of these attacks have the potential to download a malicious application or steal a user's login information for a website. In order to identify a piece of software, static analysis employs techniques including format of file checking, string extraction, disassembly, in which machine code is converted to assembly language andho fingerprint—which mostly relies on hashcode data and scanning of antivirus [9]. Static approaches take the longest and are more dependent on malware behavior analysis. However, the dynamic method observes behavior and executes files to identify viruses. It can therefore identify the malware with greater authority. Similar to this, there are mainly two methods used to identify malware: One is signature-based, which uses predefined malware signatures to detect, and the other is heuristic, which uses a number of features to contrast malicious behavior [14, 16].

Signature-based techniques may ascertain the begin or malicious character of a PDF using heuristics, code sequences, and string comparison. However, it has not been proven that technique works well against modern stealth attacks [15]. Dynamic techniques are more effective at uncovering a file's concealed dangerous behaviors because they execute the file in a supported environment while analyzing the steps it takes, the API calls it makes, and compiling an exhaustive log of its operations. The execution log contains a wealth of information about the properties of a file. This is true for all malware detection techniques because the characteristics used to identify malware differ based on the strategy.

The classification block, which compares the data collected with the algorithm's conclusions and categorizes the file as "malicious," or "correct" sends these results to the recipient [18].

The following are the paper's primary highlights:

(1) It offers a novel way to identify malware using machine learning (ML) for PDF files.
(2) It offers testing and training of the proposed model using several Machine learning approaches.
(3) It also demonstrates the effectiveness of the suggested strategy when subjected to the fictitious hostile attack.

2 Related Work

A. Malware Identification: Malware analysis focuses on determining how malware functions and how it impacts systems and programmes. In the past, identification methods based on signatures were frequently used. This method swiftly and efficiently combats known malware [5, 6] however it falls short when dealing with zero-day malware. The authors suggested a methodology for identifying malware based on genetic algorithms (GAs) [7] and signature generators. Although the study omits important information about the proposed framework, such as test results, the volume of malware inspected, and comparisons with other recent studies, the authors use this method to detect previously unknown malware. Claims to be able to find Fukushima et al. developed a behavior-based detection method [8]. On Windows OS, the described method may be used to find new, encrypted malware.

This model uses the SVM kernel base to weight the frequency of each library call to identify Mac OS X malware. Machine learning has also been significantly adopted by malware analysis (ML). Artificial neural networks are the basis for deep learning, a feature of machine learning (ML) (ANNs). Although it is a cutting-edge method, it is insufficient for virus detection and is frequently employed for image analysis and autonomous vehicles. Although it significantly and effectively lowers the amount of room that may be used for features, it does not prevent attacks from evading. Shabtai et al. established a classification for malware identification using machine learning approaches by describing certain functionalities and selecting features from the literature. The author uses the convolution neural network (CNN) in image processing with machine learning to detect malware.

B. File-Based Malware Identification: From top to bottom, authors looked at PDF design and JavaScript content contained in PDFs. They developed a wide range of design and metadata features, including bytes per second, encoding methods, object names, keywords, and human-readable strings in JavaScript. Additionally, since subtle changes have a significant impact on AI calculations, it is challenging to develop hostile models when the attributes vary [17]. To reduce the risk of malicious attacks, they developed a classification model that uses heuristic models while preserving structure and data properties. They created opposing attacks to embrace the proposed paradigm. Current attacks are used to perform on PDF malware through a naturally derived and reliable attack model, as the authors have also provided an outline of the PDF [18] (Fig. 1).

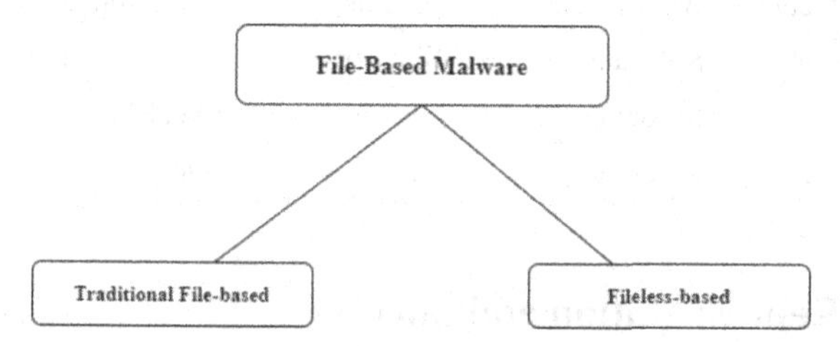

Fig. 1. File based malware types

They demonstrated how to do a quantifiable analysis of a PDF file to find evidence of malware implantation using coding techniques [15]. They looked at some of the emerging AI-powered tools for detecting PDF malware that can support computerized scientific analyses and can flag questionable documents before a more thorough, more conclusive statistical analysis is published. They looked at the PDF restrictions and other unresolved problems, paying close attention to how their flaws might be used to potentially misdirect subsequent measured studies [16]. Authors have concentrated on malware inserted into PDF files as a representative example of contemporary cyber-attacks. They began by classifying the numerous techniques 13 used to create PDF virus scientifically. They have used a proven adversarial AI structure to counter PDF malware classifiers that rely on learning. By using this technique, we can, for instance, find weaknesses in learning-oriented PDF malware locators and new attacks that could endanger such frameworks, as well as potential countermeasures [17]. In order to get beyond malware classifiers, authors have designed and implemented a revolutionary framework called AIMED. The authors have provided an example of how the worst-case scenario behavior of a malware classifier with reference to explicit vigor attributes might be assessed. Additionally, they discovered that by forgoing simple assault avoidances, the development of classifiers that satisfy officially verified energy qualities can increase the avoidance cost of unrestrained attackers [18]. In addition, they identified two classes of strength features, including subtree inclusions and erasures, and presented a new distance metric that operates on the PDF tree structure [20] (Table 1).

Table 1. Differentiation techniques used in traditional File-based and fileless based malware

Techniques	Traditional File-based malware	Fileless based malware
Source code	Yes	No
Malicious code	Yes	No
Malicious process	Yes	No
File Types	Executable files Format that executes (PDF, Word, Excel etc.)	JavaScript PowerShell Flash
Tenacity	Medium	Low
Attacks	File that can be executed with a single OS	Can target many different OS
Anti-Malware detection	An acknowledged signature	No applicable
Sandbox recognition	Substantially availability of file	No applicable
Complexity	Moderate	Very high
Detection complexity	Moderate	Very high

3 Features Representation and Success

Authors	Year	Features representation/method	Success
D. Zhu, H. Jin, Y. Yang, D. Wu, and W. Chen [21]	2017	Hybrid features, deep flow technique and DBN model	High detection F1 score of 95.05%, which performed better than conventional ML-based learning methods
Y. Ye, L. Chen, S. Hou, W. Hardy, and X. Li [22]	2018	Greedy layer-wise training operations	Superior overall performance compared to typical learning methods
A. Narayanan, M. Chandramohan, L. Chen, and Y. Liu [23]	2017	Security-sensitive behaviors	On a benchmark dataset, the method outperformed two cutting-edge approaches, achieving a 99.23% f-measure
H. Sun, X. Wang, J. Su, and P. Chen [24]	2015	Multiple hash function, bucket cross-filtering method	Performs better than competing systems while using less time and communication
A. Cimitile, F. Martinelli [26]	2017	Elective Mu-calculus logic and phylogenetic trees	Identify new and different forms of malware

(*continued*)

(continued)

Authors	Year	Features representation/method	Success
A. Azmoodeh [25]	2018	Monitors the energy consumption patterns of different processes	Higher effective than KNN, neural networks, SVM, and RF
W. Huang and J. W. Stokes [27]	2019	Raw bytes, gradient based attack	Show the inefficiency of deep learning
B. Anderson [28]	2020	The n-gram, Markov chain and diagrams	Identify different forms of malware with 95.41%

4 Existing Approach

A PDF file consists of the body, trailer, and cross-reference table (CRT). Although the document's current version is shown in the header, the document's source is mentioned in several body parts. The CRT contains tables for linking to objects. The trailer component contains both the root object and the table locations for the items in the body section [19].

The recommended ML-based malware detection method is described here. This system's main objective is to scan the Pdf version under examination, decode its internal code, and assess if it is risky or safe.

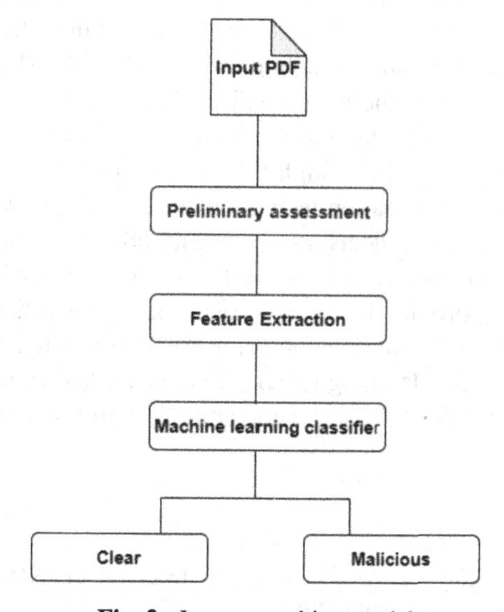

Fig. 2. Layout working model

The mechanism in place has also been found to prevent the hacker's attempts to conceal file headers. This method appropriately categorizes the file's kind but does not identify the malware family that it contains [18]. Figure 2 depicts how the planned work was classified by the system. It gives a clearer notion of how the classifier is implemented, despite the fact that this is a high-level system architecture. The PDF file document needs to be delivered to the system in order for the inspection process to start. The document is initially examined for its structure and content after submission. If it demonstrates the characteristics of files that are known to be hazardous, it is flagged for further investigation. This saves a tonne of time and speeds up the process. In other situations where it does not fit the pattern of hostile instances, the feature extraction module assesses the complete file structure and extracts the features from it. After the PDF's features have been extracted, it is then provided according to the classifier portion for evaluation. The classifier component houses the primary ML algorithm, and it is this algorithm that is in charge of carefully scrutinizing the data supplied by the feature extractor component [19]. The only method to assess whether the code contains any dubious API references is to do a fresh static analysis in addition to the dynamic code analysis. To keep a watch out for any API references, the PhoneyPDF employed the same monitoring strategy as dynamic code and the static inspections. Spider Monkey and Rhino are two examples of open-source facilitators that have been employed in the past to carry out dynamic investigations. These translators can see the JavaScript ECMA standard, but without Adobe DOM duplication, they cannot understand linkages between JavaScript and Acrobat PDF files [20].

5 Conclusion

This study presents a Machine learning model that can recognise malicious API and JavaScript calls assaults in PDF files. This study also tested the effectiveness of a variety of different classifiers on the dataset, including SGB, RF, DT, SVC, and LR. The best results in this study came from the RF classifiers. This method is 4% more successful than the other most recent PDF classifiers, according to a comparison with other PDF classifiers, which reveals that it has a high F1-score of 0.986. The system is now more resistant to harmful code obfuscation techniques thanks to the ability to examine the PDF document for unhandled objects by running an object scanner inside of it. Using this method, unparsed objects that contain the dangerous code can be quickly found and eliminated. Future plans provide for adding assistance for additional file formats. For a more in-depth understanding of documents, use a sophisticated data mining strategy. Although the use of machine learning (ML) for malware detection and classification is highly beneficial, it is useless against evasion attacks, therefore more study is required.

References

1. Alzarooni, K.M.A.: Malware variant detection, Doctoral dissertation, UCL (University College London), London, England (2012)
2. Stallings, W., Brown, L., Bauer, M.D., Bhattacharjee, A.K.: Computer Security: Principles and Practice, pp. 978–980. Pearson Education, Upper Saddle River (2012)
3. Alam, S., Horspool, R.N., Traore, I., Sogukpinar, I.: A framework for metamorphic malware analysis and real-time detection. Comput. Secur. **48**, 212–233 (2015)

4. Mehtab, A., Shahid, W.B., Yaqoob, T., et al.: AdDroid: rule based machine learning framework for android malware analysis. Mob. Netw. Appl. **25**(1), 180–192 (2020)
5. Alosefer, Y.: Analyzing web-based malware behaviour through client honeypots, Doctoral dissertation Ph.D. thesis, Cardiff University, Cardiff, Wales (2012)
6. Idika, N., Mathur, A.P.: A survey of malware detection techniques, Technical report, Purdue University, West Lafayette, IN, USA, 2007
7. Islam, R., Tian, R., Batten, L.M., Versteeg, S.: Classification of malware based on integrated static and dynamic features. J. Netw. Comput. Appl. **36**(2), 646–656 (2013)
8. Saxe, J., Berlin, K.: Deep neural network based malware detection using two dimensional binary program features. In: Proceedings of the 2015 10th International Conference on Malicious and Unwanted Software (MALWARE), Fajardo, PR, USA, pp. 11–20. IEEE, October 2015
9. Nataraj, L., Manjunath, B.S.: SPAM: signal processing to analyze malware [applications corner]. IEEE Signal Process. Mag. **33**(2), 105–117 (2016)
10. Han, K.S., Lim, J.H., Kang, B., Im, E.G.: Malware analysis using visualized images and entropy graphs. Int. J. Inf. Secur. **14**(1), 1–14 (2015)
11. Liu, D., Wang, H., Stavrou, A.: Detecting malicious JavaScript in pdf through document instrumentation. In: Proceedings of the 44th Annual IEEE/IFIP International Conference on Dependable Systems and Networks, Atlanta, Georgia, pp. 100–111. IEEE, June 2014
12. Chumachenko, K.: Machine learning methods for malware detection and classification, bachelor's thesis information technology, Xamk Kouvolan kampus, Kouvola, Finland (2017)
13. Aslan, O.A., Samet, R.: A comprehensive review on malware detection approaches. IEEE Access **8**, 6249–6271 (2020)
14. Souri, A., Hosseini, R.: A state-of-the-art survey of malware detection approaches using data mining techniques. HCIS **8**(1), 3–22 (2018)
15. Javed, A.R., Beg, M.O., Asim, M., Baker, T., AlBayatti, A.H.: Alphalogger: detecting motion-based side-channel attack using smartphone keystrokes. J. Ambient. Intell. Humaniz. Comput. **2020**, 1–14 (2020)
16. Demontis, A., Melis, M., Biggio, B., et al.: Yes, machine learning can be more secure! A case study on android malware detection. IEEE Trans. Dependable Secure Comput. **16**(4), 711–724 (2019)
17. Pektaş, A., Acarman, T.: Malware classification based on API calls and behaviour analysis. IET Inform. Secur. **12**(2) (2017)
18. Panda, P., Chakraborty, I., Roy, K.: Discretization based solutions for secure machine learning against adversarial attacks. IEEE Access **7**, 70157–70168 (2019)
19. Chen, B., Ren, Z., Yu, C., Hussain, I., Liu, J.: Adversarial examples for cnn-based malware detectors. IEEE Access **7**, 54360–54371 (2019)
20. Yuan, X., He, P., Zhu, Q., Li, X.: Adversarial examples: attacks and defenses for deep learning. IEEE Trans. Neural Netw. Learn. Syst. **30**(9), 2805–2824 (2019)
21. Zhu, D., Jin, H., Yang, Y., Wu, D., Chen, W.: DeepFlow: deep learning based malware detection by mining Android applications for abnormal usage of sensitive data. In: Proceeding of the IEEE Symposium Computer Communications (ISCC), July 2017
22. Ye, Y., Chen, L., Hou, S., Hardy, W., Li, X.: DeepAM: a heterogeneous deep learning framework for intelligent malware detection. Knowl. Inf. Syst. **54**(2), 265–285 (2018)
23. Narayanan, A., Chandramohan, M., Chen, L., Liu, Y.: Context–aware, adaptive, and scalable Android malware detection through online learning. IEEE Trans. Emerg. Topics Comput. **1**(3), 157–175 (2017)
24. Sun, H., Wang, X., Su, J., Chen, P.: RScam: cloud-based anti-malware via reversible sketch. In: Thuraisingham, B., Wang, X., Yegneswaran, V. (eds.) Security and Privacy in Communication Networks. SecureComm 2015. LNICST, vol. 164, pp. 157–174. Springer, Cham (2015). https://doi.org/10.1007/978-3-319-28865-9_9

25. Azmoodeh, A., Dehghantanha, A., Conti, M., Choo, K.-K.-R.: Detecting crypto-ransomware in IoT networks based on energy consumption footprint. J. Ambient Intell. Hum. Comput. **9**(4), 1141–1152 (2018)
26. Cimitile, A., Martinelli, F., Mercaldo, F., Nardone, V., Santone, A., Vaglini, G.: Model checking for mobile android malware evolution. In: Proceedings IEEE/ACM 5th International FME Workshop Formal Methods Software Engineering (FormaliSE), May 2017
27. Huang, W., Stokes, J.W.: MtNet: a multi-task neural network for dynamic malware classification. In: Caballero, J., Zurutuza, U., Rodríguez, R. (eds.) DIMVA 2016. LNCS, vol. 9721, pp. 399–418. Springer, Cham (2016). https://doi.org/10.1007/978-3-319-40667-1_20
28. Anderson, B., Quist, D., Neil, J., Storlie, C., Lane, T.: Graph-based malware detection using dynamic analysis. J. Comput. Virol. **7**(4), 247–258 (2020)

A Survey on Lung Cancer Detection and Location from CT Scan Using Image Segmentation and CNN

K. Hari Priya, Suryatheja Alladi, Saidesh Goje, M. Nithin Reddy, and Himanshu Nama(✉)

Department of Computer Science and Engineering, VNR VJIET, Hyderabad, TS 50090, India
namahimansh@gmail.com

Abstract. In the contemporary era, it is clear that lung cancer has always been the major cause of cancer-related patients' death. Currently, the primary method to diagnose lung cancer in a suspected patient is to examine the CT scan images of the patient's lungs. Therefore, determining it as soon as possible is the greatest strategy to prevent death. Most of the time, radiologists have a limited amount of time to examine and interpret medical images. Nevertheless, as medical technology advances, more and more imaging data is being generated. Deep learning and machine learning technologies are effective for automating the interpretation and diagnosis of medical images. In this, we used biopsy reports (symptoms including hoarseness, coughing, smoking addiction, yellow fingers, anxiety, fatigue, allergy, wheezing, alcohol consumption, breathing problem) displayed in patients', ct scans to find out the location and size of the nodule as well as the stage of lung cancer. We built an interface that requires uploading a CT scan image and selecting symptoms. In logic view, we use segmentation and a CNN model to analyze the CT scan result in order to build a patient's medical profile. Using these sophisticated models increases the ability to forecast. The majority of the image analysis effort is automated through the use of machine learning models with also including deep learning models as well. Medical professionals can then provide prophylaxis more quickly.

Keywords: Lung Cancer · Image PreProcessing and Segmentation · Boundary detection · CNN · U-NET · Computed Tomography

1 Introduction

Across the globe, it is observed that lung cancer has always been the most frequent cause of cancer-related deaths. It is estimated that in 2022, 236,740 Americans will be diagnosed with lung cancer. Lung cancer claims the lives of approximately three times as many men as prostate cancer and nearly three times as many women as breast cancer. Prostate cancer has claimed the lives of roughly three times as many men as lung cancer. The risk of dying from lung cancer has decreased by a third for women and men, respectively.

P. Das et al. (Eds.): AMRIT 2023, CCIS 1953, pp. 213–220, 2024.
https://doi.org/10.1007/978-3-031-47224-4_19

Smoking is a major risk factor for lung cancer, accounting for 80% of lung cancer fatalities. However, 20% of lung cancer fatalities occur in individuals who have never smoked. Other risk factors include exposure to asbestos, certain metals, organic compounds, pollution, and engine oil exhaust. Lung cancer development may also be influenced by genetic and familial factors. Approximately 20% of lung cancer diagnoses are made in non-smokers each year, with an estimated 47,300 cases expected in 2022.

There are two primary types of lung cancer, NSCLC and SCLC. Although symptoms may vary depending on the stage of lung cancer, chest pain and discomfort, wheezing, hoarseness, coughing up blood, loss of appetite, unexplained weight loss, and fatigue are common symptoms. A persistent cough after treatment may be the first indication of lung cancer. Hoarseness and chest discomfort are the most commonly observed symptoms.

This method involves a radiologist uploading a CT scan of a patient's lungs along with their symptoms. The CT scan undergoes preprocessing and segmentation before being passed through a VGG16 CNN model to determine the probability of lung cancer. A U-Net is used to identify the precise location of the lung nodule, while boundary detection is utilized to determine its size. Finally, using this information, a medical profile of the patient is created.

This paper presents a structured approach to reporting research findings. It includes an abstract that provides a summary of the study, an introduction that contextualizes the research problem, a related work section that reviews the literature, a proposed methodology section that outlines the research design and data collection methods, a conclusion that summarizes the key findings and their implications, an acknowledgment section that recognizes contributors, and a references section that lists the sources cited in the paper. This structure ensures clarity and organization in presenting research.

2 Related Work

[1] Azmira Krishna, CMAK, and P.C. Srinivas Rao Basha Zeelan analyzed the initial step in computer-aided identification of cancer cells, which involves the classification of CT scan images. The researchers reviewed several articles and journals to examine the concept of lung cancer detection. Along with categorizing CT images, a CNN-based classification approach was used with a filtering method, specifically Gaussian filtering, and marker-controlled Watershed algorithm in segmentation. The study collected approximately 500 CT photos of various body parts, including the kidney, neck, lung, bone, and brain, from hospitals in the oncology department.

[2] Abdalla Ibrahim, and Sergey P. Primakov have examined and assessed a for detection and volumetric segmentation f NSCLC, an automated pipeline, It is mentioned that they have validated above 1300 thoracic CT scans(1328) which were validated from 8 institutions, in the study, the CT scans for the detection and differentiation of non-small cell cancer. In addition to mathematical performance broken down by image thickness, size, difficulty of interpretation, and tumor place, they published a prospective clinical study, In which they demonstrated that the proposed method is quicker and more repeatable than the experts.

[3] Carlos Francisco Silva has conducted the study aimed to determine the performance of LCP-CNN in identifying benign nodules with more accuracy while excluding

malignancy in less than a quarter of treated individuals with small and intermediate-sized tumors. The LCP-CNN was trained using the NLST dataset to identify lung tumors, showcasing the potential of CNNs in estimating the likelihood of developing lung cancer and assisting in decision-making regarding lung nodules. By avoiding CT images with high sensitivity, a significant portion of patients may avoid unnecessary workup, including radiology and invasive treatments. However, validation in a lung cancer screening program is necessary before confirming its applicability in this case.

[3] By performing CNN classification, the researchers have been working on developing a fully automated approach for localizing and grading fluorine 18-fluorodeoxyglucose PET uptake patterns in individuals with lung cancer and lymphoma, aiming to achieve a high level of accuracy. This approach has been validated and demonstrates comparable performance to traditional methods that rely on CT and PET images. With the successful implementation of this technique, there is potential to improve efficiency and accuracy in the diagnosis of lung cancer and lymphoma.

[4] In this study, the researchers explored the segmentation of nodules after obtaining Computed Tomography Scan images from the LIDC dataset. Next, a CNN model accomplishes the diagnosis by categorizing the tissue as benign or cancerous. The suggested system is broken down into three steps, as shown in the illustration. It is recommended that the first step of the method be the acquisition of Computed Tomography Scan from the selected dataset. The segmentation of nodules is achieved in the second step using professional markers. The CNN model finalizes the diagnosis of benign or malignant tissue during the evaluation of the tests.

[5] P. Harsh has conducted a study on a new method for tumor detection using CNN and image processing algorithms. The dataset was pre-processed in the MATLAB directory before being fed into the CNN. The pooling layer reduces the dimensions by applying maximum shrinkage, while the first convolution layer creates features and the second layer identifies patterns in those features. To remove noise more effectively without distorting the margins of the image, a median filter was used.

[6] Ruchita Tekade obtained lung nodules, data from LUNA16 and Kaggle were integrated. U-Net architecture was used for segmenting the CT scan pictures. For classification, an architecture patterned by VGG was used. Lung nodules and non-nodules were classified into two categories, as were benign and malignant nodules. U-Net segmentation produced better results for detecting lung nodules and predicting the degree of malignancy. According to the results, the approach showed high accuracy.

[7] Pragya Chaturvedi1, Anuj Jhamb1 created computer technology to resolve this concern; as a response, a variety of computer-aided diagnosis (CAD) techniques and systems have been put forth, created, and/or emerged. These systems employ a variety of machine learning and deep learning techniques, and there are also a number of methods based on image processing that can be employed to predict the level of malignancy in cancer. The purpose of this study is to list, debate, contrast, and analyze several approaches to image segmentation, feature extraction, and detect lung tumors as early as possible

[8] Harshil Soni1 have developed two divisions for lung cancer diagnosis: image preprocessing and machine learning. Image preprocessing involves a collection of sequential techniques used for data normalization and segmentation. The dataset obtained from

patients is used to train CNN. The main advantage of adopting the DICOM standard is the inclusion of raw input visual data together with additional physician information, such as the number of pixels, which helps to ease computation. However, the segmented image may still have gaps, extraneous pixels, and noise, and its coherency is not guaranteed.

[9] Dr. Sooraj Hussain Nandyala has developed and reviewed a quick and accurate diagnostic procedure for lung cancer, as it does not manifest symptoms until it's progressed. The research topic is the analysis of exhaled breath to diagnose many respiratory problems, including lung cancer. VOC analysis is an effective tool for detecting lung cancer because it is non-invasive, quick, and inexpensive. This method can also detect nodules in the primary stage.This publication reviews in detail the numerous techniques used by researchers in the field of breath analysis. The publications were chosen using a variety of search engines, including Google, PubMed, Google Scholar, and EMBASE. The findings of the analyzed studies demonstrate that exhaled breath VOC analysis can efficiently diagnose lung cancer. The outcomes also show that this technique can be utilized to identify the various stages and histology of lung cancer.

[10] Wafaa Alakwa conducted research on thresholding as an initial segmentation technique for lung cancer detection. However, the approach of categorizing segmented CT images into 3D CNNs was unsuccessful. Instead, a modified U-Net trained on LUNA16 data was used, which produced a substantial number of false positives for nodules in the Kaggle CT scans. The outputs from the 3D Convolutional Neural Networks were then fed into U-Net to classify the CT scan as positive or negative for lung cancer. The algorithms Threshold, Watershed, and U-Net were used to evaluate nodules in patients, achieving a modern performance AUC of 0.93.

[11] Gaza Samy S proposed an Artificial Neural Network (ANN) in this study to determine whether a person has lung cancer. The model uses symptoms such as chest pain, anxiety, wheezing, shortness of breath, yellow fingers, difficulty swallowing, chronic illness, coughing, allergies, and fatigue, along with other personal information as inputs to the ANN. The "survey lung cancer" dataset was used to build, train, and validate the ANN. According to model reviews, the ANN demonstrated a 96.67% accuracy rate.

[12] In 2019, Nasser IM argued that lung cancer can be detected by Artificial Neural Networks (ANN), and symptoms are used to diagnose and treat the disease. However, the ANN model predicted existing lung cancer much better with a 96.67% accuracy rate and a training error rate of less than 1% after 1418105 iterations. The model also identified "Age" as the factor with the greatest influence on the results.

[13] Ying Xiea, Wei,YuMeng,Ru,Lai-Han, Leungab had developed early cancer by using machine learning methods. Patients with lung cancer have a higher chance of surviving when they receive an early diagnosis. The availability of blood-based screening could boost the awareness of patients with early lung cancer. The aim of the current investigation was to detect plasma metabolites from Chinese patients as lung cancer diagnostic biomarkers. In this study, they employ a ground-breaking interdisciplinary method to identify lung cancer early so it is easy for diagnosis. This is done by combining the metabolomics by various machine learning algorithms. This mechanism is first applied to lung cancer. In their study, we gathered a total including 43 healthy people and 110 patients with lung cancer. A targeted metabolomic analysis utilizing LC-MS/MS

examined the levels of 61 plasma compounds. The difference between stage I of lung cancer which can be detected by patients, so healthy people can be established using a particular combination of six metabolic indicators. Additionally, the top five metabolic biomarkers by relative relevance identified by the algorithm are FCBF which may be used for lung cancer early detection. It is advised to use Naive Bayes as a useful technique for early lung tumor prediction.This study will present compelling evidence in favor of the viability in which the screening is done blood-based and introduce a more sensitive, efficient, and better usage tool in order for the early identification of the lung cancer. Beyond lung disease, the proposed integrated approach may be applied to other tumors

[14] S. Shanthi et al. used a method to detect lung cancer that combines SDS with classification algorithms, including Naive Bayes, Decision Trees, and Neural Networks. They utilized 270 pictures from the TCGA dataset and extracted texture characteristics using the GLCM and shape-based characteristics using the Gabor filter. The SDS algorithm was used to select features and consists of four phases: Initialization, Evaluation, Test Phase, and Diffusion Phase. Different categorization techniques were used after SDS. When examining the accuracy of all the models, the SDS-NN method, which combines Neural Networks with the SDS algorithm, performed better than other classification models. The study reported an observation that better feature selection leads to better picture categorization.

3 Proposed Methodology

The system primarily carries out three tasks. In the first task, a web application will be developed for radiologists to upload the CT-SCAN images of lung cancer patients and select their symptoms to determine the stage of lung cancer. In the second task, image preprocessing and segmentation of the CT scan will be implemented using the watershed algorithm. The segmented images will be fed into the VGG16 CNN model, which is trained beforehand using the LIDC-IDRI dataset of CT-scan images, for detecting lung cancer. Additionally, the location and size of the nodule will be identified. Explicitly, the location of the tumor will be detected by performing u-net segmentation and boundary detection for the quantification of the nodule. This method will provide maximum accuracy for the quantification of the nodule by evaluating the ratios of the size of the tumor in each region. Then, the identification of nodules' location is established, and finally, the staging of lung cancer will be determined by considering the symptoms of patients mentioned earlier. Lastly, in the third stage, the findings will be compiled into a medical profile of the patient, which doctors can easily access through various patient reports.

3.1 Image Preprocessing

To prevent lungs from receiving a higher radiation dose during a CT scan, pre-processing is performed on the images. Gaussian filtering is used to reduce noise and higher frequency data, followed by resizing the image to 224x224x3. The resized image is then sent for the segmentation process.

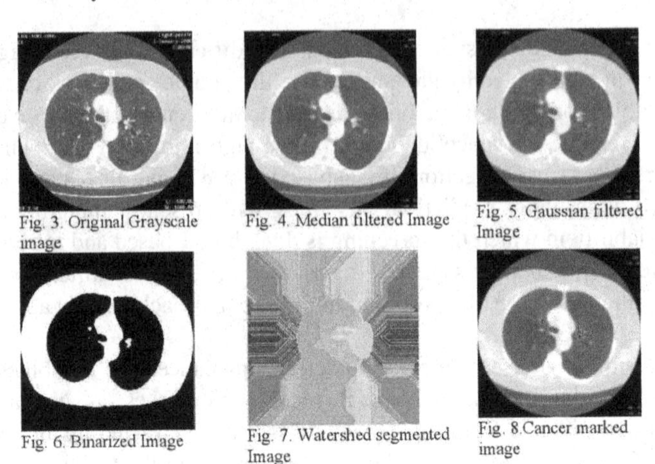

Fig. 3. Original Grayscale image

Fig. 4. Median filtered Image

Fig. 5. Gaussian filtered Image

Fig. 6. Binarized Image

Fig. 7. Watershed segmented Image

Fig. 8.Cancer marked image

3.2 Segmentation

In the second stage of the procedure, processing is performed to separate the pixels or voxels. The DICOM file is required to extract the pixel array and metadata, including the patient ID, patient DoB, instance number, slice number, and other associated information, which is delivered along with the CT scan image. The pixel array that has been extracted is used to create Hounsfield units. In order to generate internal, external, and watershed markers in the pixel array, and create markers that will be used by marker-controlled segmentation to perform picture segmentation, Hounsfield units must be calculated. Marker-controlled segmentation performs picture segmentation to extract lung segments using the markers created in the previous stage and the Sobel gradient.

3.3 VGG16 CNN Model

The convolutional neural network (CNN) is a popular and widely used algorithm for image classification problems or non-linear data. In our study, we applied CNN on CT scans to determine whether they are normal or if they indicate the presence of lung cancer. The CNN approach involves layers containing filters, also known as kernels, as well as a pooling layer used for max-pooling or average pooling to condense the feature presence. Finally, the fully connected layer provides us with classified images. We utilized the VGG-16 layer algorithm in deep learning, which involves initializing weights and specifying the number of filters and output dimensions for each layer. This approach consists of 16 layers. Ultimately, this method can assist radiologists in making diagnoses or in screening large populations for lung diseases.

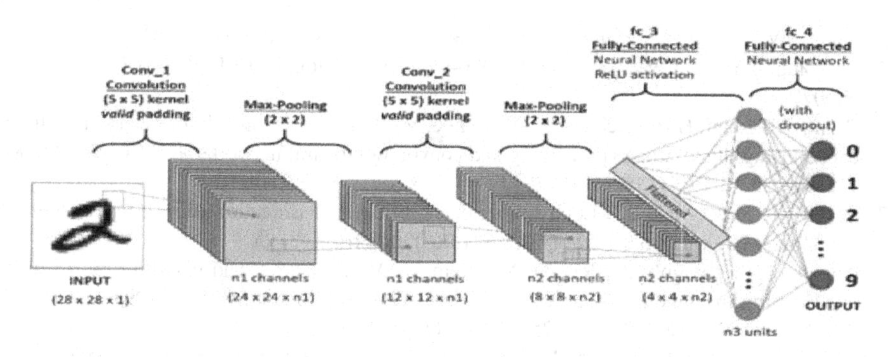

4 Conclusion

In recent decades, it has been observed how cancer impacts human life, especially lung cancer, being one of the deadliest diseases. To address this issue, we have implemented an interface designed for radiologists to evaluate patients' lung cancer detection using CT scans. The aim is to visualize the model's decision-making process to increase radiologists' trust and improve adoption. We use the Watershed method to construct the segmentation of the CT scan images and feed them into the VGG16 convolutional neural network for identifying lung cancer. The position of the nodules is determined after assessing the ratios in various regions, particularly after completing U-Net segmentation and boundary detection for nodule quantification. This procedure is faster and more accurate, and the staging of lung cancer will be enhanced by taking into account the symptoms of the patients described earlier. Finally, we summarize the results and create a medical profile of the patient. Clinicians can easily access this profile through various patient reports.

Acknowledgment. We extend our heartfelt appreciation to our mentor, Miss K. Haripriya, for her invaluable technical assistance and guidance, which culminated in the successful completion of the literature survey phase of the project.

References

1. Krishna, A., Srinivas Rao, P.C., Zeelan Basha, C.M.A.K.: Computerized classification of CT lung images using CNN with watershed segmentation -ICIRCA - 2020- IEEE Xplore Part Number: CFP20N67-ART
2. Primakov, S., Ibrahim, A., Janita E.: van timmeren automated detection and segmentation of non-small cell lung cancer computed tomography images
3. Marjolein, A., Heuvelmans a,b, Peter, M.A. Van Ooijen C , Sarim Ather, D.: Carlos francisco silva lung cancer prediction by deep learning to identify benign lung nodules
4. Sibille, L., et al.: 18F-FDG PET/CT uptake classification in lymphoma and lung cancer by using deep convolutional neural networks. Radiology **294**(2), 445–452 (2020). https://doi.org/10.1148/radiol.2019191114

5. Abdul, W.: An automatic lung cancer detection and classification (ALCDC) system using convolutional neural network -DeSE-2020-DOI: https://doi.org/10.1109/DeSE51703.2020.945077
6. Rohit, Y.B., Jani, H.P., Rachana, K., Gaitonde, V.R.: A novel approach for detection of lung cancer using digital image processing and convolution neural networks -ICACCS2019-978-1-5386-9533-3/19/$31.00 ©2019 IEEE
7. Tekade, R., Rajeswari, K.: Lung cancer detection and classification using deep learning -ICCUBEA-2018-978-5386-5257-2/18/$31.00©2018 IEEE
8. Chaturvedi, P., Jhamb, A., Vanani, M., Nemade, V.: Prediction and classification of lung cancer using machine learning techniques. IOP Conf Ser. Mater. Sci. Eng. **1099**(1), 012059 (2021). https://doi.org/10.1088/1757-899X/1099/1/012059
9. Sharma, K., Sonil, H., Agarwal, K.: Researched lung cancer detection in CT scans of patients using image processing and machine learning technique 1 System Level Solutions (I) Pvt. Ltd., V.U. Nagar 388121, Gujarat, Department of CSE, Sikkim Manipal Istitute of Technology, Sikkim, India
10. Nandyala, S.H., Duday, D.: Developed exhaled breath volatile organic compound analysis for the detection of lung cancer- a Syst. Rev. https://doi.org/10.4028/p-dab04
11. Alakwaa, W., Nassef, M.: AMR BADR has researched Lung Cancer Detection and Classification with 3D Convolutional Neural Network (3D-CNN). (IJACSA) Int. J. Adv. Comp. Sci. Appl. (2017). https://doi.org/10.14569/IJACSA.2017.080853
12. Ibrahim, M., et al.: University, gaza date written: March 2019 lung cancer detection using artificial neural network
13. Nasser, I.M., Abu-Naser SS 2019 Lung Cancer Detection Using Artificial Neural Network. Int. J. Eng. Inf. Syst. (IJEAIS) 17–23 (2019)
14. Xie, Y., et al.: Early lung cancer diagnostic biomarker discovery by machine learning methods. Transl. Oncol. **14**(1), 100907 (2021). https://doi.org/10.1016/j.tranon.2020.100907
15. Shanthi, S., Rajkumar, N.: Lung cancer prediction using stochastic diffusion search (SDS) based feature selection and machine learning methods. Neural Proc. Lett. **53**(4), 2617–2630 (2020). https://doi.org/10.1007/s11063-020-10192-0

Bi-directional Long Short-Term Memory with Gated Recurrent Unit Approach for Next Word Prediction in Bodo Language

Ajit Das[1(✉)], Abhijit Baruah[1], and Sudipta Roy[2]

[1] Department of Computer Science and Technology, Bodoland University Kokrajhar, Kokrajhar 783370, Assam, India
adas0078@rediffmail.com

[2] Department of Computer Science and Engineering, Assam University Silchar, Silchar 788011, Assam, India
sudipta.roy@aus.ac.in

Abstract. Next word prediction, an essential task in neural language processing, holds significant importance across various applications like text completion, suggestion system, and machine translation. Within this domain, Recurrent Neural Network (RNN), particularly the Bi-LSTM (Bidirectional Long Short-Term Memory) and GRU (Gated Recurrent Unit).In this study, we introduce a hybrid model that combines Bi-LSTM and GRU for next word prediction specifically in English. Our objective is to leverage the strengths of both architectures to enhance the model's capability to capture contextual dependencies, thereby improving the accuracy of predictions.

Word prediction is a commitment to prognosticate what word will come incontinently. It's one of the main tasks of NLP and has numerous operations. Our thing is to make this model prognosticate the coming word as snappily as possible in the minimal quantum of time. Since RNN is a long-term short-term memory, it'll understand the once textbook and prognosticate words, which can be helpful for druggies to make rulings, and this fashion uses letter-by-letter vaticination, which means it predicts letter-by-letter to form a word. Next-word vaticination is helpful for druggies and helps them type more directly and briskly.

Keywords: Bodo language · Bi-LSTM · GRU · NLP · RNN

1 Introduction

The Bodo language, native to northeastern India, particularly Assam, is primarily spoken by the Bodo people. It is also found in certain regions of West Bengal, Meghalaya, and neighboring states. Bodo is categorized within the Tibeto-Burman branch of the Sino-Tibetan language family. Historically, the Bodo script, also known as the "Deodhai" script, was used for writing. However, the Latin script has become the dominant writing system since the 1960s, although recent efforts have aimed to revive the use of the Bodo script.

P. Das et al. (Eds.): AMRIT 2023, CCIS 1953, pp. 221–231, 2024.
https://doi.org/10.1007/978-3-031-47224-4_20

Bodo has a relatively extensive range of consonant sounds and a smaller set of vowel sounds. It features a tonal system where the pitch or tone of a syllable can change the meaning of a word. The language follows an agglutinative structure, employing affixes to form new words from root words. The word order in Bodo is subject-object-verb (SOV). Nouns in Bodo are categorized based on factors like shape, size, and animacy. The Bodo vocabulary reflects the cultural and geographical context of the language, incorporating terms related to flora, fauna, agriculture, traditional practices, and kinship. This vocabulary helps preserve the unique cultural and linguistic heritage of the Bodo people in the region.

Next word prediction technology plays a significant role in this process by assisting users in creating documents more efficiently. It eliminates spelling errors and speeds up text entry, which is particularly beneficial for individuals with limited spelling skills. Furthermore, next word prediction can greatly benefit people with severe motor or oral disabilities, as well as those using handwriting recognition, mobile phones, or tablets for texting. Additionally, word sequence prediction, which is closely related to next word prediction, can serve as input for various natural language processing studies like speech recognition and handwritten recognition models.

While a considerable number of people use the Bodo language for communication, it's important to note that Bodo belongs to the northeastern Indian Language Group and is a Semitic language. It is spoken in certain regions of West Bengal, Meghalaya, and neighboring states. Bodo serves as the working language of the Bodo community. Given this, having an assistive Bodo text entry system can significantly enhance text entry speed and provide assistance to individuals requiring alternative communication methods. Next word prediction plays a vital role in this context as it predicts the following word based on the context of the sentence or phrase. This prediction process greatly helps to avoid typing errors or reduce the number of required keystrokes. The next word can be predicted based on the probability of the previous words and the overall context.

Colorful technologies have been developed in this field of deep literacy and machine literacy in moment's world. Every day we use colorful bias to write textbook information and shoot it to the other end, but it's useless to write the entire textbook over and over again. Thus, the vaticination of the coming word provides backing to the stoner in writing a judgment and reduces the time consumption of jotting. Utmost of the time, the speed of transmission of textual information in a certain time can be increased by prognosticating the coming word. We can also use it for other language. This helps in saving the stoner's keystrokes. This was extended to continue prognosticating several other words for a given sequence of words. Word vaticination is a abecedarian part of natural language generation. It's useful in correcting grammatical crimes and arranging the word in a particular process. Word vaticination can help beforehand learners, similar as scholars or neophyte experimenters, make smaller spelling miscalculations and increase codifying speed. Language models assign probabilities to sequences of words or phrases, indicating the likelihood or probability of the next word given a preceding sequence of words. It's useful in correcting grammatical crimes and arranging the word in a particular process. These models can be useful in colorful fields similar as spelling correction, speech recognition, machine literacy. This study includes natural language processing, N-gram modeling, intermittent neural networks, deep literacy, machine literacy, convolutional neural networks.

2 Related Work

The approach used to conduct the literature review is described in this section. This essay represents an examination of previous contributions to the academic community.

[1] Multi-window convolution (MRNN) algorithm is implemented in this paper. In addition, a residual connected minimal gate unit (MGU), a condensed version of long-term memory (LSTM), is created. By omitting several layers during training, cnn tries to reduce training time while maintaining high accuracy.

The authors of [2] proposed an LSTM model for the Assamese language that predicts the next word(s) based on a set of recently used terms. There is no longer a problem with words having multiple synonyms as the Assamese language has been transcribed using the International Phonetic Association (IPA). Their model uses LSTM to predict the next word with 88.20% accuracy when looking at a dataset of transliterated Assamese words. The result shows that LSTM is remarkably good at predicting the next word, even for a language with few sources.

A stochastic model vaticination of the next term for the Urdu language was described in Paper (3). The retired Markov model was used to predict the future state, and the present state and future sheltered state were both designed using the Unigram model. The N-gram model was then employed, with N set at 2. The programme was created to use this paradigm for Urdu Language (UL) and was put to the test using both conventional and new content pens to verify their increased writing speed.

A forecasting system for automated law completion was proposed in composition (4). The vocabulary term vaticination and the identification vaticination were both important criteria employed by the writers. An LSTM-based neural language model that considers word vaticination within the lexicon was presented. A concept built on a network of pointers is suggested for the vaticination of identifiers. The suggested system is estimated using the source legislation gathered in the judges' online system. The trial's findings demonstrate that the suggested approach is capable of accurately predicting both the upcoming word in the lexicon and the upcoming identifier. Zilly et al.

(5) proposed an intermittent Highway Network (RHN), which is an extension of LSTM, allowing for multiple updates of the sheltered state at each time step. In RHN, the mass matrix declination was eased so that the evaporating grade was resolved [6] The authors of [6] proposed an LSTM model for the Assamese language that predicts the subsequent word(s) given a collection of recently used terms. There is no longer a problem with words having several synonyms because the Assamese language has been transliterated using the International Phonetic Association (IPA). Their model use LSTM to predict the following word with an accuracy of 88.20% while looking over a dataset of transliterated Assamese words. The result shows that LSTM is remarkably good at predicting the next word, even for a language with few resources.

[7] An automatic code completion algorithm for the next word was presented in article [7]. Vocabulary word prediction and identifier prediction were the authors' two main strategies. A neural language model built on the LSTM network was proposed in

order to predict words from the vocabulary. A model built on a network of indicators is suggested for the prediction of identifiers. The proposed method is assessed using the source code gathered in the online judging system. The experiment's findings demonstrate that the suggested method is highly accurate at predicting both the following word in the vocabulary and the following identifier.

The utilization of long-term memory-recursive networks for generating real-time text sequences by predicting one data point at a time was discussed in a research paper [18]. In human comprehension, each word is understood based on the preceding words, allowing for continuous understanding and response. Traditional neural networks face the challenge of diminishing in performance under such circumstances. However, Recursive neural networks address this issue by providing a solution.

In a separate investigation [19], researchers introduced a neural language prediction model known as MCNN-ReMGU (Multi-Window Convolutional Neural Network and Redundant Connected Minimal Gate Unit). The unique connectivity within the MGU network was employed to address the challenges of gradient fading and network corruption. Through experiments conducted on the Penn Treebank and WikiText-2 datasets, the authors successfully showcased the superior performance of their proposed method in predictive applications.

In another paper [20], researchers explored various learning networks to create new communication between characters. This work provides a comparison between LSTMs, GRUs and bidirectional recurrent neural networks (RNNs). This model focuses on learning the repetitive sequence of symbols by accessing them from time to time.

3 Proposed Methodology/Algorithm

3.1 BLSTM

A mechanism called Bidirectional Long-Short-Term Memory (BLSTM) makes sure that each neural network has both a forward and a backward data ordering. By time stepping in both directions, the BLSTM utilises the majority of the data. Our input flows in two ways, which distinguishes BLSTM from conventional LSTM. We can make the input inflow for a typical LSTM go either backwards or forwards. To store both past and future information, we may still create an a bi-directional architecture that allows input to flow in both ways. In order to learn the aspects of the data that are applicable at each time step, BLSTM must learn the bidirectional relationship between sequenced data. The GRU cellular unit is the alternative subcaste of the suggested network. We may reduce the complexity of the network model by using a specific kind of RNN with fewer gates.

3.2 GRU

The LSTM intermittent neural network model is interpreted more simply by GRU. GRU only requires two gate vectors—a reset gate and an update gate—and one state vector. The intermittent unit carries out the LSTM-like functions of voice signal modelling, music modelling, and natural language processing. As compared to LSTM, the GRU model was designed to perform better with smaller datasets. A single gateway serves as

both the input and forgotten gates in GRU. GRU is less complicated than LSTM because of this gate configuration. The suggested model training algorithm may be explained in the manner below. Algorithm: Model proposed by BLSTM-GRU
Start

1. Enter the Bodo series of sentences
2. Produce the next word in the specified sequence. BLSTM layer
3. (hidden size, batch size) GRU layer
4. (hidden size, batch size)
5. The activation feature for Softmax

4 Methodology

The set of Bodo sentences with various lengths serves as the foundation for the suggested next word prediction model (number of tokens). We add a zero block at the end to equalise the length of the short phrases. The Bi-LSTM model's input layer is then given the coded text. Then it moves into the GRU input layer. Experiments are used to determine the proposed model's hyperparameters. The model learns the proper word order through practise sentences. The model may be used with programmes that produce Bodo words after it has been trained and stored. This eliminates typing and saves time.

5 Dataset Overview and Data-Preprocessing

To assess the new strategy we have proposed, we have gathered a significant amount of data in Bodo from TDIL (Technology Development for Indian Languages). The total amount of data collected is 63 thousand and has been gathered from various sources. After gathering the data from various sources, we cleaned it using the cleanup function, which was put in place to get rid of extraneous objects ("", (), /, !), and also helped to standardize the original dataset so that it could be used for other things in the future. Each time we want to predict a word or words, we typically input four words into our model. The output will always be one word no matter how many words are in the input word.

6 Tokenization

Tokenization is the process of breaking text into words. Tokenization can occur on any character, but the most common way to tokenize is to do it on a space character. Tokenization refers to the procedure of converting a sequence of letters or characters into a sequence of tokens. These tokens subsequently need to be converted into numerical vectors to facilitate comprehension by a neural network.

7 System Flowchart

See Fig. 1.

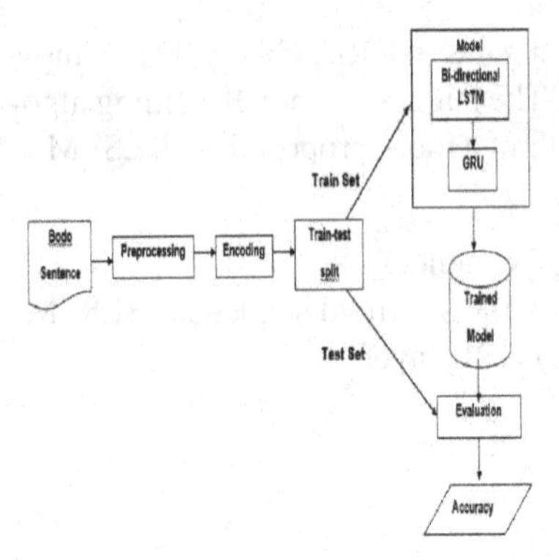

Fig. 1. System FlowChart

8 Model Architecture

Figure 2 and 3 shows the structure of our model which has two hidden layers named bilstm and gru with a total number of 22,667,753 (params). We used a hybrid model (Bi-LSTM and GRU) to train our previously created dataset and build one matching model.

After training our dataset, there is now one trained model for the three length inputs. The model takes four different word lengths as input and produces a single output, which is the next word most likely to follow the sequence of input words.

```
Model: "sequential"
_________________________________________________________________
Layer (type)                 Output Shape              Param #
=================================================================
embedding (Embedding)        (None, 4, 10)             45230

bidirectional (Bidirectiona  (None, 4, 2000)           8088000
l)

gru (GRU)                    (None, 1000)              9006000

dense (Dense)                (None, 1000)              1001000

dense_1 (Dense)              (None, 4523)              4527523

=================================================================
Total params: 22,667,753
Trainable params: 22,667,753
Non-trainable params: 0
```

Fig. 2. Model Architecture

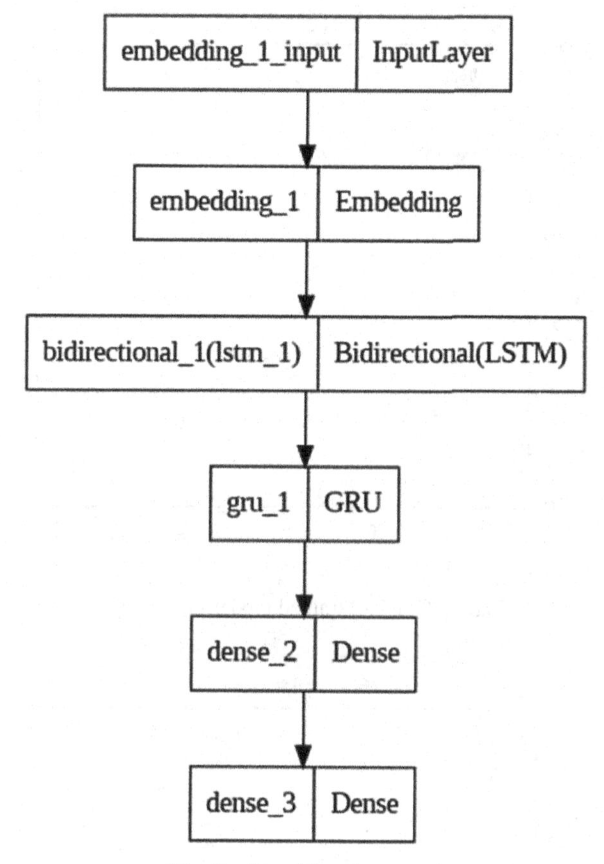

Fig. 3. Algorithm Structure

9 Result Analysis

Conducting experiments and carefully analyzing the results are required to validate the suggested approach. So, using a corpus dataset to train a model with identical structures for up to 70 epochs, we assessed our suggested method. The trained Hybrid model has an average accuracy of 97%, as shown in Figs. 4, 5 and 6.

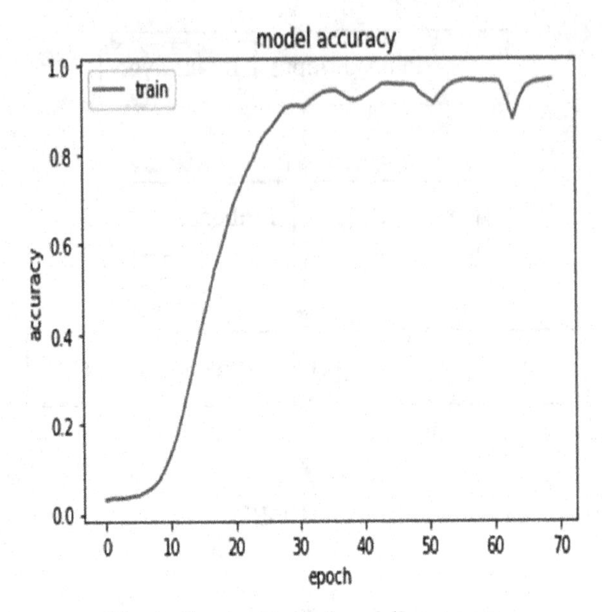

Fig. 4. Graph of trained model's accuracy

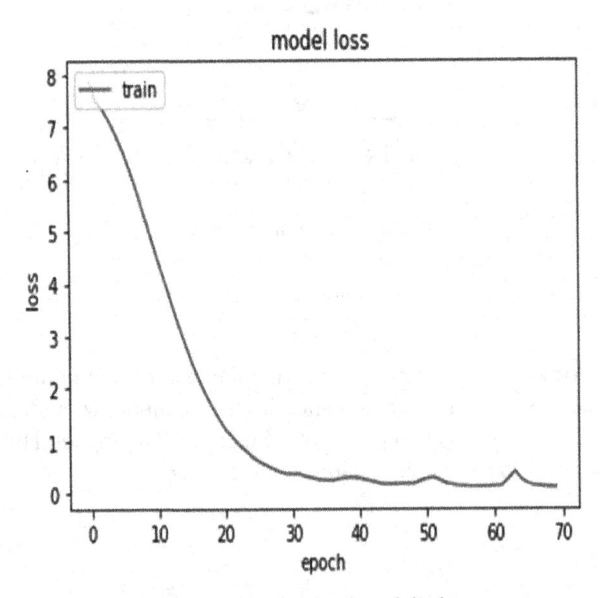

Fig. 5. Graph of trained model's loss

```
Epoch 63: loss improved from 0.01812 to 0.01740, saving model to /content/drive/MyDrive/BODO_WRD_PRED_HYBRIDE/next_words.h5
235/235 [==============================] - 6s 26ms/step - loss: 0.0174 - accuracy: 0.9914
Epoch 64/70
234/235 [============================>.] - ETA: 0s - loss: 0.0163 - accuracy: 0.9917
Epoch 64: loss improved from 0.01740 to 0.01650, saving model to /content/drive/MyDrive/BODO_WRD_PRED_HYBRIDE/next_words.h5
235/235 [==============================] - 6s 24ms/step - loss: 0.0165 - accuracy: 0.9915
Epoch 65/70
235/235 [==============================] - ETA: 0s - loss: 0.0164 - accuracy: 0.9913
Epoch 65: loss improved from 0.01650 to 0.01641, saving model to /content/drive/MyDrive/BODO_WRD_PRED_HYBRIDE/next_words.h5
235/235 [==============================] - 6s 26ms/step - loss: 0.0164 - accuracy: 0.9913
Epoch 66/70
233/235 [============================>.] - ETA: 0s - loss: 0.0160 - accuracy: 0.9918
Epoch 66: loss improved from 0.01641 to 0.01586, saving model to /content/drive/MyDrive/BODO_WRD_PRED_HYBRIDE/next_words.h5
235/235 [==============================] - 6s 25ms/step - loss: 0.0159 - accuracy: 0.9918
Epoch 67/70
235/235 [==============================] - ETA: 0s - loss: 0.0158 - accuracy: 0.9920
Epoch 67: loss improved from 0.01586 to 0.01580, saving model to /content/drive/MyDrive/BODO_WRD_PRED_HYBRIDE/next_words.h5
235/235 [==============================] - 6s 25ms/step - loss: 0.0158 - accuracy: 0.9920
Epoch 68/70
234/235 [============================>.] - ETA: 0s - loss: 0.0156 - accuracy: 0.9920
Epoch 68: loss improved from 0.01580 to 0.01560, saving model to /content/drive/MyDrive/BODO_WRD_PRED_HYBRIDE/next_words.h5
235/235 [==============================] - 6s 25ms/step - loss: 0.0156 - accuracy: 0.9920
Epoch 69/70
235/235 [==============================] - ETA: 0s - loss: 0.0158 - accuracy: 0.9917
Epoch 69: loss did not improve from 0.01560
235/235 [==============================] - 5s 21ms/step - loss: 0.0158 - accuracy: 0.9917
Epoch 70/70
```

Fig. 6. Snapshot of trained model

10 Conclusion

In this research, we present a method that combines the Bi-LSTM and GRU models to create a hybrid model for next word prediction in the Bodo language. Our architecture takes advantage of the bidirectional nature of Bi-LSTM, enabling it to capture contextual information from both preceding and succeeding words. Additionally, we incorporate the gating mechanism of GRU to enhance memory management and facilitate faster training convergence.

By conducting extensive experiments on benchmark datasets, we provide evidence of the superiority of our hybrid model when compared to standalone Bi-LSTM and GRU models in terms of prediction accuracy and contextual understanding. The results clearly demonstrate that our hybrid model effectively merges the strengths of Bi-LSTM and GRU, resulting in improved performance. To further enhance our model, we suggest the inclusion of attention mechanisms or fine-tuning on larger datasets. Moreover, exploring transfer learning techniques, domain adaptation, and multilingual scenarios could serve as intriguing avenues for future research. However, the overall delicacy of this approach would be indeed more emotional, If we could use the larger data set we formerly used in this work. It was delicate to use the Bodo corpus dataset because neither Bodo has a ready- made dataset, so we had to gather the data from colorful sources. To ameliorate the RNN's capability to prognosticate the coming Bodo word and judgment, we will work to amass a sizable data set in the coming months. This study will also be useful as a tool for sustainable technologies in assiduity because of how extensively it can be applied.

References

1. Yang, J., Wang, H, Guo, K.: Natural language word prediction model based on multi-window convolution and residual network. IEEE Access **8**, 188036–188043 (2020). https://doi.org/10.1109/ACCESS.2020.3031200
2. Barman, P.P., Boruah, A.: A RNN based approach for next word prediction in Assamese phonetic transcription. Procedia Comput. Sci. **43**, 117–123 (2018)
3. Muhammad, H., et al.: Effective word prediction in Urdu language using stochastic model. Sukkur IBA J. Comput. Math. Sci. **2**(2), 38–46 (2018)
4. Kenta, T., Yutaka, W.: Code completion for programming education based on deep learning. Int. J. Comput. Intell. Stud. **10**(2–3), 109–114 (2021).
5. Zilly, J.G., Srivastava, R.K., Koutník, J., Schmidhuber, J.: Recurrent highway networks. In: Proceedings of the International Conference on Machine Learning (ICML), pp. 4189–4198 (2016)
6. Li, S., Li, W., Cook, C., Zhu, C., Gao, Y.: Independently recurrent neural network (IndRNN): building a longer and deeper RNN. In: Proceedings of the IEEE/CVF Conference on Computer Vision and Pattern Recognition, pp. 5457–5466 (2018)
7. Prajapati, G.L., Saha, R.: REEDS: relevance and enhanced entropy based Dempster Shafer approach for next word prediction using language model. J. Comput. Sci. **35**, 1–11 (2019)
8. Stremmel, J., Singh, A.: Pretraining federated text models for next word prediction. In: Arai, K. (eds.) FICC 2021. AISC, vol. 1364, pp. 477–488. Springer, Cham (2021). https://doi.org/10.1007/978-3-030-73103-8_34
9. Qu, X., Kang, X., Zhang, C., Jiang, S., Ma, X.: Short-term prediction of wind power based on deep long short-term memory. In: 2016 IEEE PES Asia-Pacific Power and Energy Engineering Conference (APPEEC), pp. 1148–1152. IEEE (2016)
10. Sundermeyer, M., Schlüter, R., Ney, H.: LSTM neural networks for language modeling. In: Thirteenth Annual Conference of the International Speech Communication Association (2012)
11. Jozefowicz, R., Vinyals, O., Schuster, M., Shazeer, N., Wu, Y.: Exploring the limits of language modeling. arXiv preprint arXiv:1602.02410 (2016)
12. Bengio, Y., Schwenk, H., Senécal, J.S., Morin, F., Gauvain, J.L.: Neural probabilistic language models. Innov. Mach. Learn. **3**, 137–186 (2006)
13. Mikolov, T., Karafiát, M., Burget, L., Cernockỳ, J., Khudanpur, S.: Recurrent neural network based language model. In: Eleventh Annual Conference of the International Speech Communication Association (2010)
14. Chung, J., Gulcehre, C., Cho, K., Bengio, Y.: Empirical evaluation of gated recurrent neural networks on sequence modeling. arXiv preprint arXiv:1412.3555 (2014)
15. Gal, Y., Ghahramani, Z.: A theoretically grounded application of dropout in recurrent neural networks. In: Advances in Neural Information Processing Systems, pp. 1019–1027 (2016)
16. Graves, A., Mohamed, A.R., Hinton, G.: Speech recognition with deep recurrent neural networks. In: 2013 IEEE International Conference on Acoustics, Speech and Signal Processing, pp. 6645–6649 (2013)
17. Sundermeyer, M., Schlüter, R., Ney, H.: From feedforward to recurrent LSTM neural networks for language modeling. IEEE/ACM Trans. Audio Speech Lang.Process. **23**(3), 517–529 (2015)
18. Chen, X., et al.: Microsoft COCO captions: data collection and evaluation server. arXiv preprint arXiv:1504.00325 (2015)
19. Akash, K., Anjali, G., Anirudh, M., Amrita, J.: Text sequence prediction using recurrent neural network. Adv. Appl. Math. Sci. **20**(3), 377–382 (2021)

20. Yang, J., Wang, H., Guo, K.: Natural language word prediction model based on multi-window convolution and residual network. IEEE Access **8**, 188036–188043 (2020)
21. Mangal, S., Joshi, P., Modak, R.: LSTM vs. GRU vs. bidirectional RNN for script generation (2019)

Authorship Attribution for Assamese Language Documents: Initial Results

Smriti Priya Medhi[1(✉)] and Shikhar Kumar Sarma[2]

[1] Department of Computer Science and Engineering, Assam Don Bosco University, Guwahati, India
sp.medhi26@gmail.com

[2] Department of Information Technology, Gauhati University, Guwahati, India
sks@gauhati.ac.in

Abstract. Impact of Digital India in the creation of electronic content on the web is primarily acknowledged. However, critically observing, we also realize the problems, especially in the cases of identifying the creator of content. Also, there can be issues of false annotation or even plagiarism. Every person has a unique style of writing. This characteristic can be explored to further train a system to identify the accurate author of a content, also popularly known as the field of authorship attribution. This paper attempts to showcase the initial experimental results of author identification done on a manually collected and annotated assamese literary corpus. Assamese being a low-resourced language, the applications of NLP like authorship identification has not been explored till date. And this reported work can be marked as the first attempt of bringing to the world the research scopes of authorship attribution in assamese language.

Keywords: NLP · Authorship Attribution · Stylometry · Linguistics · Plagiarism

1 Introduction

The field of stylometry dwells in the concept that every written content has some differentiating factor that separates it from its contemporaries. Examples of some stylometric features may include average length of words and sentences, number of words, number of special characters used, misspellings, abbreviations etc. Research on stylometric features saw its birth in the mid of 19th century when the first work was reported by Mosteller et al. (1963) [1] on the statistical techniques behind language modelling of the Federalist papers. The field of stylometry thus has a major role to play in the world of Authorship Attribution. The essence of any write-up lies in its lexical, syntactic, and semantic populace. If we compare the literary work of any two personalities, we may end up discovering enumerable number of patterns hidden in their writing style. This is formally known as stylometry and was the distinguishing factor among the different literary legends. Hence, wouldn't it be interesting to learn about their writing styles and as such preserve it as a literary fossil to be valued by the coming generation?

Typically, AA can be considered as a text classification problem with respect to the authors/creators concerned. While some features may be easier to extract, some, on

P. Das et al. (Eds.): AMRIT 2023, CCIS 1953, pp. 232–242, 2024.
https://doi.org/10.1007/978-3-031-47224-4_21

the other hand like POS, context-specific features may require in-depth analysis. The general approach of the AA problem may be visualized as depicted in Fig. 1.

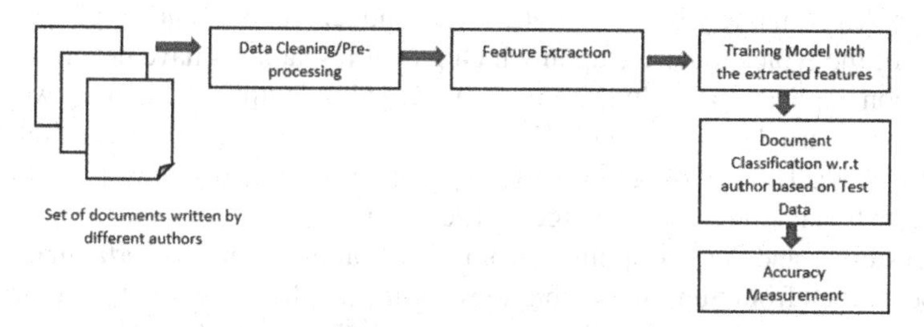

Fig. 1. Steps in an Author Attribution Process

Author Identification would not only help us in detecting the original author, but it will also aid in maintaining the Intellectual Property Rights and avoid theft of articles by unscrupulous individuals/organizations. Plagiarism or Content Theft is a global issue as stated in [5]. The trend of digitization has increased the impact of misattribution to literary contents popularly termed as "digital dilemma" [4]. As such there is a strong significance of the proposed study to understand and visualize the possible techniques of how the original author of an excerpt can be identified via the techniques of machine learning and artificial intelligence.

The legacy of assamese literature can be traced back to 9–10th century as stated by a famous author in [2]. Chronologically, it is divided into three periods namely, ancient, medieval, and modern era. The literary society of Assam named "Assam Sahitya Sabha" was founded in the year 1917, with the objectives of all-around development of the language, literature, and culture of the State.

With the advent of Digital India, the tech savvy generation were missing out reaching to some of the great literary masterpieces. Keeping this in mind, the society decided to launch a digital platform in the year 2022 containing around 500 old and precious books to preserve the works of great contributors of Assamese literature and take those legends across the globe in an electronic format [3]. Assamese being a low resourced language, there are a lot of challenges in exploring its field of AA. Limited dataset, robust parsers being some of the prime limitations in this regard. From this, we can understand the importance and need of developing an efficient and vigorous model that can correctly attribute an author given a random piece of text. A major requirement here will be a huge dataset in tunning the model from time to time. Assamese is free-word order and morphologically rich language and parsing it is still an open problem [23]. And as mentioned in [24], it is spoken by around 30 million speakers in Assam.

2 Literature Survey

The sole process of authorship identification depends upon learning certain author specific features and then using those features in identifying other literary works of the authors considered in the study [6]. Features are some specific characteristics that can help an application, set a distinguishing factor between different inputs. For any textual excerpt, the features may commonly include are lexical, syntactic, semantic, and even content-specific features. In this study, we have tried classifying the related work into four major classes namely based on dataset, technique, features, and language.

Some of the typical datasets upon which global researchers have developed author identification applications include books [10, 15, 16], scientific and literary works [12–14], newspaper articles [11], emails [7, 17], forum messages [7–9], buyer/seller feedbacks [7, 17] etc. The authors in [7] and [17] applied stylometric analysis on four types of online texts namely email messages retrieved from publicly available Enron email corpus, customer and vendor opinions from a popular e-commerce portal named eBay, code blocks taken from Sun Java Technology forum and the last set of data set was taken from IM chat forum called CyberWatch. Around 100 individual authors were selected from all the above datasets. The aim was to identify individuals based on their writing style. The extended feature set achieved an accuracy of 94%. The approaches they adopted included SVM to identify and classify the authors, PCA and KL Transforms to enhance the similarity detection. Their work focused mainly on lexical features such as special characters, letter frequency, content words, misspellings, and character n-grams. The datasets collected in their work were based on English language.

The task in [8] and [9] was to identify the authors of messages collected from web forums and discussion threads in newsgroups. The feature set considered here was a concoction of lexical, syntactic, structural, and content-specific features. The authors had typically considered the languages English and Arabic. They had adopted an SVM, and decision-tree based approach named C4.5. The accuracy scores were different for both the datasets. For the English dataset, they achieved an average score of 87.8% for the approach C4.5 and 93% for SVM and meanwhile for the Arabic dataset they achieved an average score of 67.5% for former and 91.9% for the later. The authors here had showcased in a comprehensive manner that how the different characteristics and number of features puts an effect on the overall accuracy of the classifiers.

Considering books as the test bed, the author in [10] consider a seven types of linguistic feature set. The features were extracted automatically using NLPWin system [18]. However, sentences more than 50 words are omitted by the system. Also, further the irrelevant features or features that occur only very few times or even only once were discarded using a simple frequency cut-off. The classification technique they adopted was linear SVM. Other research works which had successfully applied SVM in the problem of author identification include [19–21]. The problem in [10] were addressed using a Sequential Mining Optimizer [22], which has recorded results of functioning in an efficient manner in training a SVM with a large quadratic programming optimization problem. The accuracy score was obtained following two types of techniques. One by considering the features in a standalone manner and the other by combining the features altogether. The prior technique gave a score of 54% and the later gave 97.57% considering a frequency threshold of 75. They have also added some deep linguistic analysis features which has in turn helped in increasing their classification accuracy. Their work has mentioned on the fact that an intelligent thresholding technique will play a vital role in further improving the accuracy score of the classifier. In [15] considering the literary short texts of reputed novels of English language by authors Anne and Charlotte

Brontë, the author identification task was attempted using an SVM classifier. They had proposed bigrams of syntactic labels as a new classification feature which proved to be outperforming all other previous techniques using the same dataset. They have also put forward an observation that using short texts of little more than 200 words and considering diverse features, the accuracy of such classification tasks can be increased. The work in [16], involved designing a probabilistic context-free grammar to represent the linguistic style of authors. The grammar was used as a language model to train the classifiers in identification of a particular author.

In [11], the authors have focused on newspaper articles. Here, unlike the previous works, they have developed a non-lexical based author identification system wherein a set of 22 style markers are used to train the classification model. The markers consist of three levels, mainly token, phrase and analysis and was performed by an existing NLP tool named Sentence and Chunk Boundaries Detector (SCBD). They achieved an accuracy score of 81% which outperformed the lexical based approach by 7%. However, when they increased to set size of the style markers to 72, they did see an elevation in their accuracy score by 7% again. The results showed a strong inclination to the text lengths of the corpus. According to them, a text length of not less than 1000 words should be used if deep style markers are to be extracted. Among the marker levels chosen, the token level had the highest discriminating factor followed by the later levels. The said approach can be beneficial only for limited size dataset and cannot solve authorship identification tasks related to multiple authors contributing on a single document.

The authorship attribution tasks of [12, 13] and [14] were primarily on indic language texts of bengali, hindi and kannada respectively. The model in [12] applies some lexical n-grams as the discriminating factor. Their corpus consisted of 3,000 passages written by three Bengali authors, an end-to-end system for authorship classification based on character n-grams, feature selection for authorship attribution, feature ranking and analysis, and learning curve to assess the relationship between amount of training data and test accuracy.

The work in [13] explores supervised and un-supervised author classification approaches in hindi literature. The corpus was self-curated and consisted mainly of hindi novels written by famous authors namely Rabindranath Tagore, Vibhuti Narayan, Premchand, Sarat Chandra Chattopadhyay and Dhamarvir Bharati. In their approach, the feature vector was constructed based on lexical unigrams, bigrams, trigrams. The concept of MDA was also exploited by concatenation of the bigram and trigram feature vectors. The observations put forward in this work states that the granularity of a dataset highly puts an emphasis on the classification scores. More the number of diverse feature vectors, better the accuracy. Authors in [14] have dwelled upon a machine learning algorithm based on the author's profile to classify kannada text authors and have been able to achieve an accuracy score of 88%.

In study [26], an attempt was made to identify the author of a few written material by 20 different authors. The corpus was about economics, sports and literature subjects written in Brazilian language. For each of the author, around 30 pieces were considered. [27] on the other hand focuses in identifying 13 Nigerian authors who have contributed to their national daily newspaper. A total of 260 articles were taken into consideration with each author having 20 articles. The work of authorship attribution has had a global presence in the field of linguistic research. People had always the fascination of finding out the true owners of some relevantly early literature works. One such work is reported in [28] where excerpts of modern Hebrew were used to find out the respective authors. As mentioned in [29] Hebrew, like Phoenician and some other less well-known languages is a member of the Canaanite group and is one of the few Northwest Semitic languages to survive to the present day. A heterogeneous corpus was constructed from posts of web blogs and some early modern Hebrew literary works of nine different authors. They employed two different techniques including Markov chains and SVM respectively. The literary corpus with SVM showed a higher level of accuracy of 98% whereas the blog corpus showed only 75%. Speaking about the Persian language who is a member of the Indo-European family of languages, a notable attempt on authorship attribution was made by the authors in [30]. Here they have proposed a computational approach to identify anonymous authors of Persian language documents. The technique utilizes the TF-IDF scheme to chunk out the most differentiating terms of a particular document. The importance score further is used to find the similarity between a known and an unknown document. Their state-of-the-art technique has an accuracy score of 90.2% in english dataset and 93.1% in Persian dataset respectively. Another prominent member of the Indo-European language family is the Russian language. In [31], the authors have explored the attribution of authors for their online activity. They have tested their methodologies on Russian dataset which included literary excerpts and short comments extracted from their social media handles. Speaking about the methodology, they had undertaken both standalone and an ensemble of machine learning, neural networks in combination with their hybrid versions like Neural Networks, long short-term memory and even BERT and fastText. They had also conducted an experiment for selection of informative features using Genetic Algorithms. The ensemble technique gave them good accuracy scores and reduced time complexities. For literary texts, they got the highest accuracy score of 82.3% and for social media comments, the highest score was 73.2% using deep NNs. In the year 2007, another attempt to test the fitness of multivariate techniques in differentiating between Persian authors, the authors in [32] considered four Persian authors, two each from different periods of time. They had specifically exploited the contribution of function words in drawing out author-specific patterns from the dataset.

The Chinese language occupies a significant place among the two branches in the Sino-Tibetan family of languages. It is spoken by 95% of the Chinese inhabitants and majority of Chinese inhabitants across the globe [33]. For the Chinese language, the work in [34] describes how authorship attribution techniques were applied to identify the ownership of 20 poets in the Tang Dynasty. They have proposed a C-Transformer combination model. It is a collective of CNN, Transformer and a LDA model responsible to extract the global, content-specific and topic information from the corpus respectively. They have also compared their results with some of the SOTA techniques of AA. Out of which, their proposed model showed a greater accuracy as all the other techniques were only implemented on modern Chinese text rather than classical Chinese text as in this case.

3 Analysis and Discussion

The summarized information of the existing work in the field of Authorship Attribution and their future works to be addressed is included in Table 1.

Table 1. Summarized Review of Author Attribution in different languages

Author(s)	Dataset	Language	Features	Methods Adopted	Future Scopes/Limitations
Abbasi, A., & Chen, H. (2005), [8]	The corpus consisted of English and Arabic data sets extracted from Web forum messages. A total of 20 messages for each of 20 authors were extracted,	English and Arabic	For the english language a feature set of 301 features were taken including 87 lexical, 158 syntactic, 45 structural, and 11 content-specific feature. For the Arabic language, 418 features, including 79 lexical, 262 syntactic, 62 structural, and 15 content-specific features	C4.5 and SVM	Includes approaches to distinguish between a larger set of authors. Exploring the impact of geographical proximity and time play on group and individual author detection accuracies
Abbasi, A., & Chen, H. (2008), [7]	email messages, buyer/seller feedback from eBay, code snippets from sun java forum IM Chats. For each dataset, 100 authors were randomly extracted	English	Features: Lexical Features such as special characters, letter frequency, content words, misspellings, character n-grams	SVM for Author Identification and PCA, KL Transforms for Similarity Detection	Scale the proposed algorithm for a larger author set in a computationally efficient manner. Work on a dynamic individual author feature set that can adapt with the author's changing writing style
Argamon, S. et al. (2003), [9]	500 most recent postings from discussion threads in newsgroups	English	Features: Lexical, Orthographic, function words, net abbreviations	Expontiated Gradient Learning Algorithm	Expanding feature set to include other types of style markers, such as parts-of-speech and complexity metrics. Also, to investigate the use of automatic feature generation techniques to increase the accuracy of learning algorithms

(continued)

Table 1. (*continued*)

Author(s)	Dataset	Language	Features	Methods Adopted	Future Scopes/Limitations
Gamon, M. (2004)	Texts from Anne, Charlotte, and Emily Brontë were used	English	Features: avg length of sentences, noun phrases, function words, n-grams, POS trigrams, POS bigrams	SVM	Exploring the AA field with a wider set of writing style classification tasks and larger feature sets. Work on an ensemble of classifiers to assess their performance and accuracy scores with different feature sets
Hirst, G. et al. (2007)	Short texts from Anne, Charlotte, and Emily Brontë were used. Around 250000 words were used for each author	English	Avg word length, frequency of i-letter words, words frequency of i-syllable words, 40 function words, 20 punctuation marks, ratio of hapax legomena and dislegomena	SVM	Discriminating against multiple authors using multi-class SVMs. Also expanding the feature set with optimal features
Raghavan, S. et al. (2010)	A total of five different datasets were taken. The major categories were football, business, travel, cricket, and poetry. The number of authors in our datasets ranged from 3 to 6	English	Lexical and syntactic features	Trained an ensemble of PCFG model for each author for the classification task	To perform discriminative training of the PCFGs for each author and do a comparison with the state-of-the-art approaches
Jankowska, M. et al. (2013)	PAN 2013 Authorship Identification Competition Dataset	English, Spanish, Greek	n-gram character frequencies	Common N-Gram (CNG) dissimilarity measure	Investigating the performance of their proposed classifier with respect to the length and number of documents. Also studying the role of character n-grams in assessing the verification of authors of different genres and topical similarity of the documents

(*continued*)

Table 1. (*continued*)

Author(s)	Dataset	Language	Features	Methods Adopted	Future Scopes/Limitations
Reddy, P. et al. (2018)	Reviews Corpus. Author-specific document weighted approach is proposed	English	Document weight is taken as the document vector instead of features	Naïve Bayes, Logistic Regression, Random Forest	Consider the syntactic, domain characteristics, categorical features and semantic structure of the language while assigning weights to the documents
Santosh, K. et al. (2013)	PAN 2013	English, Spanish	Content, style and topic-based features	SVM, Max Ent, Decision Tree	Include sentiment analysis for more accurate author profiling
Stamatatos, E. et al. (2001)	Texts from Greek Weekly Newspaper was used	Greek	Features: Style Markers are automatically Updated	SCBD and Discriminant Analysis	Measuring the impact of vocabulary richness in author identification
Phani, S. et al. (2013)	3,000 passages written by three Bengali authors	Bengali	Lexical n-grams	SVM, Decision Trees, Naïve Bayes	Extend the approach to other forms of text, such as blogs, news articles, tweets, and online forum threads
Chandrika, C. P. et al. (2022)	Kannada documents	Kannada	Average Word Length, Average Sentence Length by Word, Average Sentence Length by Character, Special Character Count, Average character per Word, Punctuation Count	K-Means, SVM, NB, RF	Testing the proposed work on lengthy articles and exploring the impact of bigrams and trigrams on the accuracy scores

3.1 Experimental Results

The experiment was conducted on a manually collected and annotated assamese literary text corpus of 200 text snippets from two famous novels by the name "Shishu Sahitya Sambhar" by Assamese famous poet, novelist, and playwright Rasharaj Lakshminath Bezbarua and "Sasthya" by another noted Assamese poet and journalist Homen Borgohain. It was a labelled dataset which was pre-processed with all the punctuation removed. The stop words were also removed using a rule-based process. For the feature engineering step, bag of words model was used to vectorize the text snippets. For the classification process, we used the Multinomial Naïve Bayes Classifier as it has a past record of showing optimal results in cases of statistical analysis of any text/document classification tasks. With all this, we could achieve a training accuracy of 95% and validation accuracy of 82.5%. The confusion matrix with and without normalization for the same is depicted in Fig. 2.

3.1.1 Mathematical Model

In this work, we have made use of Multinomial Naïve Bayes Classifier for the said classification task. It considers its features to be independent of one another. It can do parameter estimation with a relatively small amount of training data. This classifier finds its mathematical basis in conditional probability. It combines the concept of conditional probability and a decision rule called maximum a posteriori (MAP) to perform the task of classification. The decision rule picks up the most probable class. This is done to minimize the rate of misclassifications. This model assigns a class label $\hat{y} = C_k$ for some k as follows:

$$\hat{y} = argmax_{k \in \{1,\ldots k\}} p(C_k) \prod_{i=1}^{n} p(x_i|C_k) \quad (1)$$

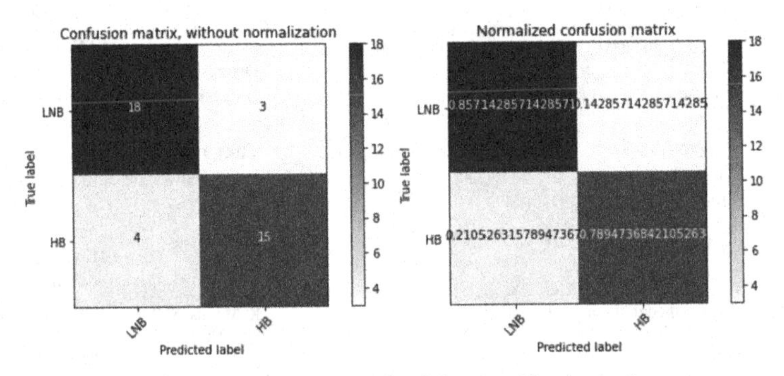

Fig. 2. Confusion matrix of the classification task

4 Conclusion and Future Work

The field of authorship attribution has a big significance in establishing the authority of a particular document. Also, this field has been very faintly explored in some regional languages like Hindi and Assamese. One of the sole reasons might be absence of well-established datasets to experiment on. Another reason about the challenges of processing such morphologically rich languages in the absence of standard tools and procedures. Hence there are valid research scopes in bringing out models that can establish author identification results for such low-resourced languages. Also from the above data, we can clearly conclude some findings such as the first author i.e., Rasharaj Lakshminath Bezbarua is most correctly classified. Hence, we can conclude that in comparison to Homen Borgohain, the former has a more unique style of writing.

Acknowledgements. The authors would like to extend their heartfelt gratitude to the NLP and Language Technology Lab, Department of IT, Gauhati University in extending their guidance in building the stand-alone corpus.

References

1. Mosteller, F., Wallace, D.L.: Inference in an authorship problem: a comparative study of discrimination methods applied to the authorship of the disputed Federalist Papers. J. Am. Stat. Assoc. **58**(302), 275–309 (1963)
2. Kakati, B.: Aspects of Early Assamese Literature-1953. Gauhati University (1953)
3. Times, S.: Assam Sahitya Sabha goes online with launch of digital archive, 14 February 2022. https://theshillongtimes.com/2022/02/14/assam-sahitya-sabha-goes-online-with-launch-of-digital-archive/. Accessed 24 Mar 2022
4. Comm. ON Intellectual Prop. Rights & THE Emerging Info. Infrastructure, Nat'L Research Council. The Digital Dilemma: Intellectual Property in the Information Age (2000)
5. Nwosu, L., Chukwuere, J.: The attitude of students towards plagiarism in online learning: a narrative literature review **18**, 93–106 (2020)
6. Mekala, S., Bulusu, V.V., Reddy, R.: A survey on authorship attribution approaches. Int. J. Comput. Eng. Res. (IJCER) **8**(8) (2018)
7. Abbasi, A., Chen, H.: Writeprints: a stylometric approach to identity-level identification and similarity detection in cyberspace. ACM Tran. Inf. Syst. (TOIS) **26**(2), 1–29 (2008)
8. Abbasi, A., Chen, H.: Applying authorship analysis to extremist-group web forum messages. IEEE Intell. Syst. **20**(5), 67–75 (2005)
9. Argamon, S., Šarić, M., Stein, S.S.: Style mining of electronic messages for multiple authorship discrimination: first results. In: Proceedings of the Ninth ACM SIGKDD International Conference on Knowledge Discovery and Data Mining, pp. 475–480, August 2003
10. Gamon, M.: Linguistic correlates of style: authorship classification with deep linguistic analysis features. In: COLING 2004: Proceedings of the 20th International Conference on Computational Linguistics, pp. 611–617 (2004)
11. Stamatatos, E., Fakotakis, N., Kokkinakis, G.: Computer-based authorship attribution without lexical measures. Comput. Humanit. **35**(2), 193–214 (2001)
12. Phani, S., Lahiri, S., Biswas, A.: Authorship attribution in Bengali language. In: Proceedings of the 12th International Conference on Natural Language Processing, pp. 100–105, December 2015
13. Kallimani, J.S., Chandrika, C.P., Singh, A., Khan, Z.: Authorship identification using supervised learning and n-grams for Hindi language. J. Comput. Theor. Nanosci. **17**(9–10), 4258–4261 (2020)
14. Chandrika, C.P., Kallimani, J.S.: Authorship attribution for Kannada text using profile based approach. In: Gunjan, V.K., Zurada, J.M. (eds.) Proceedings of the 2nd International Conference on Recent Trends in Machine Learning, IoT, Smart Cities and Applications. LNNS, vol. 237, pp. 679–688. Springer, Singapore (2022). https://doi.org/10.1007/978-981-16-6407-6_58
15. Hirst, G., Feiguina, O.G.: Bigrams of syntactic labels for authorship discrimination of short texts. Lit. Linguist. Comput. **22**(4), 405–417 (2007)
16. Raghavan, S., Kovashka, A., Mooney, R.: Authorship attribution using probabilistic context-free grammars. In: Proceedings of the ACL 2010 Conference Short Papers, pp. 38–42, July 2010
17. Burrows, J.: 'Delta': a measure of stylistic difference and a guide to likely authorship. Lit. Linguist. Comput. **17**(3), 267–287 (2002)
18. Heidorn, G.E.: Microsoft Research, Redmond, Washington. Handbook of Natural Language Processing, p. 181 (2000)
19. Joachims, T.: Text categorization with support vector machines: learning with many relevant features. In: Nédellec, C., Rouveirol, C. (eds.) ECML 1998. LNCS, vol. 1398, pp. 137–142. Springer, Heidelberg (1998). https://doi.org/10.1007/BFb0026683

20. Dumais, S.: Using SVMs for text categorization. IEEE Intell. Syst. **13**(4), 21–23 (1998)
21. Diederich, J., Kindermann, J., Leopold, E., Paass, G.: Authorship attribution with support vector machines. Appl. Intell. **19**(1), 109–123 (2003)
22. John, C.P.: Sequential minimal optimization: a fast algorithm for training support vector machines. MSRTR Microsoft Res. **3**(1), 88–95 (1998)
23. Saharia, N., Sharma, U., Kalita, J.: A first step towards parsing of assamese text. In: Special Volume: Problems of Parsing in Indian Languages (2011)
24. Deka, R.R., Kalita, S., Bhuyan, M.P., Sarma, S.K.: A study of various natural language processing works for assamese language. In: Dawn, S., Balas, V., Esposito, A., Gope, S. (eds.) ICIMSAT 2019. LAIS, vol. 12, pp. 128–136. Springer, Cham (2020). https://doi.org/10.1007/978-3-030-42363-6_15
25. Kakati, B.: Assamese, Its Formation and Development-Revised Ed. by Golockchandra Goswami. LBS Publication (1995)
26. Oliveira, W., Jr., Justino, E., Oliveira, L.S.: Comparing compression models for authorship attribution. Forensic Sci. Int. **228**(1–3), 100–104 (2013)
27. Ayogu, I.I., Olutayo, V.A.: Authorship attribution using rough sets-based feature selection techniques. Int. J. Comput. Appl. **152**(6), 38–46 (2016)
28. Gabay, D.: Authorship Attribution in Modern Hebrew (2008)
29. Sáenz-Badillos, A.: A History of the Hebrew Language. Cambridge University Press, Cambridge (1996)
30. Ramezani, R.: A language-independent authorship attribution approach for author identification of text documents. Expert Syst. Appl. **180**, 115139 (2021). ISSN 0957–4174. https://doi.org/10.1016/j.eswa.2021.115139
31. Reisi, E., Mahboob Farimani, H.: Authorship attribution in historical and literary texts by a deep learning classifier. J. Appl. Intell. Syst. Inf. Sci. **1**(2), 118–127 (2020)
32. Dabagh, R.M.: Authorship attribution and statistical text analysis. Adv. Methodol. Stat. **4**(2), 149–163 (2007)
33. Encyclopedia Britannica, Inc: Chinese languages summary. Encyclopedia Britannica (n.d.). https://www.britannica.com/summary/Chinese-languages. Accessed 5 Aug 2022

Load Balancing and Energy Efficient Routing in Software-Defined Networking

M. P. Ramkumar[1(✉)], J. Lece Elizabeth Rani[1], R. Jeyarohini[2], G. S. R. Emil Selvan[1], and Rahul Shiva Konar[1]

[1] Department of Computer Science and Engineering, Thiagarajar College of Engineering, Madurai, India
{ramkumar,emil}@tce.edu, rahulshivakonar@student.tce.edu

[2] Department of Electronics and Communication Engineering, PSNA College of Engineering and Technology, Dindigul, India

Abstract. Software-defined networking (SDN), a newly developed architecture that alters the route (paths) of traffic packets, is a flexible and logically centralized control plane that provides a reduction in the network's energy usage. SDN is designed in such a way that the switches and links can handle a large amount of traffic. Their energy use is not proportional to traffic. Due to the increase in network communication, SDN devices demand much more load and energy. The SDN controller has to be modeled efficiently. The system model of the controller is proposed which efficiently distributes the load. This model helps in traffic routing with minimal energy consumption for networking devices. In addition to the above, it balances the load based on the heterogeneous traffic distribution. The proposed method optimizes the energy consumed by the networking devices by means of using the few active switches, minimizing the number of device state transitions, and handling the volume of traffic. Load balancing is performed for multiple controllers. The switch migration technique is used for migrating the traffic from one controller to the other. Energy-efficient routing is done by the proposed algorithm with reduced device state transition.

Keywords: Load balancing · Power aware routing · Switch migration · Device state transition · Network monitoring

1 Introduction

There is an enormous amount of power consumption in the internet industry in the present world, including the production and purchase of devices and electricity consumption due to utilization. When the traffic in the network is low, the network devices are said to consume half of the total energy. This significant electricity consumption brings significant economic losses. Therefore, it is much needed to deploy power saving techniques in the network. While using traditional networks it will be difficult to manage the power-saving strategies as the control plane that controls the network and data plane at which flows in traffic are connected together make managing power-saving measures challenging.

P. Das et al. (Eds.): AMRIT 2023, CCIS 1953, pp. 243–261, 2024.
https://doi.org/10.1007/978-3-031-47224-4_22

SDN was introduced to overcome the above limitations by splitting Data and Control plane. The intelligence in SDN centralizes the network in its control plane, which serves as the network's "brain," while the data planes function as simple packet-forwarding devices. This is in contrast to the traditional approach of distributing network intelligence across multiple devices [1]. SDN can consistently drive the whole network with utmost flexibility and speed. As network applications are growing, there is a demand to manage heterogeneous traffic in the network [2]. Therefore, it becomes important that the load among all the different controllers is distributed properly. Effective resource management, throughput maximization, and delay reduction are the methods used in SDN controllers to address load balancing in controllers [2]. Generally, in a network, if a controller is highly loaded it is said to consume more energy which results in performance deterioration.

The significant contributions are provided below: (1) A network model with diverse controllers is an effective way to divide the controllers' overall traffic load during switch migration. (2) An energy-efficient routing algorithm that mainly aims in reducing power consumption using active and inactive paths. (3) Integrating the load balancing and energy-efficient for efficiently reducing the overall power in the network [3].

2 Review of Literature

2.1 Load Balancing

The primary controller is handled by a single controller that is monitored and balanced in the centralized load balancing method. Concerns regarding the decision to measure the load are stored on the primary node, and the data needed to do so is gathered from the other nodes upon demand or after a specified amount of time [4]. The controller with the large load is reminded to shift part of its burden to the controller with the relatively light load after gathering in the network, where the load is from all other remaining controllers. A command is transmitted by the controller to transfer the load to the one that is highly loaded, and a command to transfer the switch load to the controller that is least highly loaded. The apparent difference is that since data is not delivered freely, network traffic that is not necessary is decreased [5].

Each controller in the network measures and balances the load in Distributed Load Balancing Technique. Every controller in the network has a restriction called the threshold limit that specifies that load balancing is not necessary until the threshold value is met. In this case, load balancing is carried out not just on the master node but also on all slave nodes. When a controller's network load rises over the threshold level, that controller gathers the load of the other controllers in the network and is set up to balance the load. In this distributed load balancing technique, many controllers in the network will perform the load collection frequently because their load raises above the threshold limit. This causes a shortage of bandwidth as it consumes more bandwidth than should be required for the data plane because of the maximum exchange of messages to and from the controllers [5].

2.2 Energy Efficiency in SDN

Flow consolidation and sleep devices are inseparable. They are the most widely used ways to save energy. The authors presented a measure for energy efficiency. The measures of energy efficiency while being backed by utility link intervals are evaluated by Ratio for Energy Saving in SDN (RESDN). Also, they have provided a MaxRESDN method. It is a heuristic method. It is used to maximize RESDN [5].

Periodical Flow Monitoring, a classic network monitoring algorithm. It is used to achieve accurate network monitoring. It is done by the polling method. It polls the information about the flow from all switches. Modified Elastic Sketch, simply displays information about the flow from source switches and destination switches, and based on the traffic fluctuation, this Modified Elastic Sketch algorithm adjusts monitoring intervals [6]. Random Routing (RR), this algorithmic method selects a random routing path from all the available paths, and Power Greedy Routing (PGR) this method selects some path that has more energy consumption and reroutes the path to reduce power consumption.

3 Proposed Algorithm

A multiple-controller network model which includes the load balancing and power-aware routing framework is proposed. This framework is what each controller in the network makes use of. This model employs a self-adaptive load balancing scheme (SALBA) that self-adjusts when each controller's load increases to sustain load balance [4]. In this scheme, by moving some of the switches from one controller to the next, load balancing is accomplished. For migrating the switch using less energy consumption, a power optimization framework is proposed. After the switch migration, the controller becomes only lightly loaded, consuming less energy and performing at a high level. The SALBA's primary components are link reset, load migration, load balancing, and load broadcast [4]. However, the power optimization framework includes a power-aware routing approach. This methodology mainly focuses on the Sleep and Active switch and links which are the two most important operations. The multiple-controller network model provides an energy-optimized solution for managing the traffic flow in the network.

3.1 Network Monitoring

The information about the link is collected by the adaptive port monitoring algorithm and the information about the flow is collected by the period port monitoring algorithm. The topology data is acquired by the topology management module [7].

Port Monitoring

There are two port monitoring algorithms. Period port monitoring and an adaptive port monitoring method are the two options. As the period port monitoring, neither large intervals nor small intervals can be guaranteed to be accurate. The adaptive port monitoring algorithm adaptively sends the request messages to get information about the port. Hence the adaptive port monitoring algorithm is suggested. Equations (1)–(3), show the information about the link.

$$Bw = \frac{PC_{type}(t) - PC_{type}(t - interval)}{interval} \quad (1)$$

$$Ratio = \frac{Bw}{C} \quad (2)$$

$$Rw = C - Bw \quad (3)$$

$$\alpha = \frac{P_{type}(t) - P_{type}(t - interval)}{P_{type}(t - interval)} \quad (4)$$

where,

C - total capacity
Rw - Residual Bandwidth
Bw - Bandwidth
PC - Port Counter

In Eq. 1, PCtype(t) is the port counter's current bytes' value, and PCtype (t – Interval) denotes the stored byte's value at the last monitoring time. Interval denotes the initial interval.

In Eq. 4, The traffic fluctuation degree is for the traffic change [8]. The monitoring interval is acquired by Eq. 5.

The min_Interval represents the predefined minimum and the max_Interval represents the predefined maximum monitoring interval.

β1 which is greater than one and β2 which is greater than two represent the scaling factor of the interval.

$$\begin{aligned} \text{Interval} = {} & \min\{\text{Interval} * \beta 1,\ \text{max\_Interval}\} && \alpha \in (0, 0.05] \\ & \text{Interval} && \alpha \in (0.05,\ 0.2] \\ & \max\{\text{Interval}/\beta 2,\ \text{min\_Interval}\} && \alpha \in (0.2,\ +\infty) \end{aligned} \quad (5)$$

Algorithm 1 describes the controller that does parse the reply message which is received from the port and gets the information about the port in an extracted format. After acquiring the information about the port, the information about the link and the α value is calculated. The interval for monitoring is adaptively adjusted depending on the traffic fluctuation degree and after that, the time for monitoring is also updated.

```
Algorithm 1
Input: PortStatsReply
Output: PortStatsRequest,Bw,Rw,Ratio
1: Parse PortStatsReply and extract the port statistics
2: Calculate Bw, Rw, Ratio, and α by equations 1-4 3: if
α< 0.05 then
4: interval = min{interval*β1, max_interval} 5: else if
0.05 <= α <= 0.2 then
6: interval = interval 7: else
8: interval=max{interval/β1 ,min_Interval}
9: end if
10: p_time=t+interval
11: if current_time=p_time then
12: send PortStatsRequest to the corresponding switch 13:
end if
14: return PortStatsRequest,Bw,Rw,Ratio
```

Flow Monitoring

Upon implementing the adaptive flow approach, the monitoring interval becomes restricted which gets complicated and results in error-prone results. Therefore, Algorithm 2 is a kind of periodical flow monitoring algorithm. Each flow passing over the network is tracked by the Open Flow entry which is installed on every switch. The flow counter in each flow entry is set to store the information about the number of bytes and packets for every individual network flow. To obtain the flow statistics, it periodically sends a message to the flow counter, Flow Statistics Request.

Information on the flow level is obtained using such basic statistics and the flow rate is calculated by Eq. 6.

$$f_{rate} = \frac{f_{bytes}(t) - f_{bytes}(t - T)}{T} \tag{6}$$

$f_{bytes}(t - T)$ - the number of bytes stored at the final monitoring time
$f_{bytes}(T)$ - the current bytes carried by the flow

```
Algorithm 2
Input: FlowStatsReply, PacketIn, FlowRemoved Output:
FlowStatsRequest, f_bytes, f_duration, f_bytes 1: if msgi
= flowStatsReply then
2: Extract fbyte and interval
3: Calculate frate by Equation 6 4: ftime ← ftime +T
5: if currenttime = ftime then
6: Send FlowStatsRequest to the corresponding switch 7:
endif
8: Return FlowStatsRequest, 9: else if msgi = PacketIn
then
10: Extract the new flow information
11: Allocate the power-aware root using algorithm 4 12:
else if msgi = flow removed
13: Parse FlowRemoved and extract fduration,fdemand 14:
Return fduration, fdemand
```

The corresponding message fields are extracted from the message entering the controller. Based on the statistics obtained from FlowStatsReply, flow monitoring time and flow rate are obtained and calculated. In order to get the flow data derived from flow statistics for the next monitoring time, the controller transmits the FlowStatsRequest to the appropriate switches. Each new network flow is given a power-aware route after the controller has received the PacketIn. Suppose the flow removed message is got by the controller, it obtains the flow demand and flow duration.

An overall network picture is produced and topology, link, and flow-level information are acquired by putting the flow and port monitoring technique to use. The controller gets data from the whole network based on this global network view for further network power optimization.

3.2 Load Balancing

In the load balancing technique, the foundation threshold of a controller is primarily modeled using three parameters: clock, in PBW, and out PBW, where packet out BW is the bandwidth utilized to send packets out from the controller, the clock is the amount of time the controller needs to execute one packet request, and the bandwidth utilized to deliver packets into the controller is packet in BW. These three parameters are used to modify each controller's Base Threshold (BT). The ratio between the current threshold and base threshold is defined as $\alpha * \text{BT}$, known as Current Threshold (CT). The default α value is assigned as 1 and it fluctuates depending on the controller's load [7] materials and procedures. Provide enough details so that the task may be repeated. Only relevant modifications should be discussed when referencing previously published methods.

Load Measurement

In this module, the amount of load on each controller is measured at regular occurring intervals. The load is calculated using the rate at which each packet arrives at the controller from the linked switches. This is calculated as the total number of packets received from each switch attached to the controller $CL\mathrm{i} = \Sigma \mathrm{j} SL\mathrm{j}$.

Load Broadcast

The main idea behind this loading balancing technique is that each controller should know some information about every other controller in the network. Every controller broadcasts its current load (CL), only if there occurs some notable deviation from its previously stated load (PL) [9]. There comes a parameter δ called maximum acceptable load that can be deviated. Then calculate the CL's and PL's deviation if it is greater than δ, then the load broadcasting process is triggered by the scheme. The value of δ is now chosen to be 20% of BT, this helps to overcome the frequent transferring of broadcast messages between the controllers. This load broadcast is done when the switches are selected, which are to be drifted to the other lightly loaded targeted controller. Hence no other migration is triggered at the lightly loaded targeted controller. The highly loaded controller utilizes a time-out to assist gather broadcast data from the other controllers before proceeding on to the next phase, which is the load balancing approach. If a specific controller has problems broadcasting the information, such as packet drops, and the highly loaded controller does not get any messages from that specific controller, the controller is excluded from the load balancing process [10].

Load Balancing

Followed by the broadcast stage each controller in the network has base threshold (BT) and current load (CL) information in every controller through the network. With this information, the current threshold (CT) of every controller on the network determines another controller in CT. As mentioned in Algorithm 3, the load balancing operation is carried out by the controller that is under a heavy load. There are two instances [11]. When all of the network's controllers are heavily and equally loaded, there is no other choice the load must be distributed equally across all of the network's controllers. The high load is then to be distributed unevenly among all of the network's controllers, this occurs when some controllers are just lightly loaded. A parameter called $\beta \mathrm{i}$ is introduced which represents the high load in the ith controller and can be determined as CLi /BTi.

```
Algorithm 3: Load balancing Algorithm Input: ρ, C_Load,
B_threshold , C_threshold Output: α
1.  for p from 1 to n do
2.  if C_Load(i) > B_threshold then
3.  Send()
4.  if Load is Uniformly Distributed then
5.  if ((C_Load(i) - C_threshold(i))
6.  or (C_threshold(i) - C_Load(i)) > ρ) then
7.  αi = C_Load(i) /B_threshold
8.  BroadcastNewAlpha(αi)
9.  end if
10. else
11. Get_Controller_Switch_Pair();
12. power optimization();
13. LoadMigration(c,s,l);
14. if αi > 1 then
a.  BroadcastNewAlpha(1)
b.  αi = 1
15. end if
16. else
17. Receive()
18. receive (αi)
19. αmax = Findmax(αi)
20. C_threshold= αmax × B_threshold
21. end if and for
22. End
```

In the first approach, the α value in high load conditions is raised, which then increases the current threshold of the controller. Determine the difference between the current threshold (CT) and the CL assuming that the load is distributed uniformly [12]. To accept more load in the controller, a new α value is defined as the ratio between BT and CL, when the difference between CT and CL is greater than the new parameter value ρ. If the difference is the parameter ρ, no load balancing is needed [13]. In a lightly loaded condition, the value of α is reduced and defined as the ratio between CL and BT. In conditions like if CL is lower than BT, we set the value of α to 1. From this, the self-adaptive nature of this scheme is clearly understood, when the load in the controller is less it reduces the value of CT, on the other hand, if the load is more it raises the CT value to adjust the extra load in the network.

In the second approach, when heavily loaded controllers determine that a load is no longer distributed evenly across a network, then a controller-switch pair is selected and the load is drifted from the highly loaded controller to the lightly loaded one [14]. The main function of the controller is how it is paired, it constantly regulates the load of all highly loaded controllers and provides information on the controller-switch pair of all lightly loaded controllers, choosing the most direct route based on the following characteristics. (a) The drifted switch's load from a highly loaded controller should not be more than the CT of the targeted lightly loaded controller so that it avoids the back-to-back load migration. (b) The distance between the targeted lightly loaded control and the selected migration switch must be shorter [15].

Load Migration

The overloaded controller will then ask the selected controller to transmit a message to the selected controller switch pair that will acknowledge switching. The targeted controller, which is lightly loaded, sends a message to the highly loaded controller. After receiving verification from the switch that has been selected to drift, the switch migration process begins. Once the targeted overloaded controllers complete the switch's remaining tasks, the load of the targeted lightly loaded controller is transferred. The targeted lightly loaded controller then redoes its task and informs the drifting switch of its new task. At last, both the lightly loaded and the heavily loaded controller renew their own switch connection table [16]. As just the switch control is switched from one controller to the other while the load is constant, this unit has no impact on the data plane.

Link Reset

In this unit modified link is reset to the original state, it is done before the load migration process is performed as per Algorithm 4. Later in the load balancing process, switches are controlled by the controllers which are far away from them, though the load in the network is not more [17]. Because of this, the performance of the SDN networks is reduced. To acquire better performance, this Link Reset unit helps in resetting the drifted link of the switch to its original state. This unit works every day for a while. This unit works if it detects that the network is lightly loaded.

```
Algorithm4: Get_Controller_Switch_Pair()
Input: C_Load,S_Load, C_threshold,δ,
LL_Controller,β,HL_Controller
Output: Sk, Ci
1.  if LL_Controller.size==0 then return 1;
2.  end if
3.  for k ∈ SetOfSwitchesUnder_HL_controller do
4.  fori∈LL_Controllerandi∈HL_Controllerdo
5.  βi = (C_Load(i) +δ+SLk)/ C_threshold (i)
6.  if βi<βi&&βi< 1 then
7.  hcd=hc(Sk,Ci )-hc(Sk,Ci) mapping.push(hcd, Sk,Ci )
8.  End if and for
9.  End for
10. if mapping.size == 0 then return-1; end if
11. sort(mapping) pop(mapping)
12. return(Sk,Ci ) hc- hop_count
13. End
```

3.3 Power Optimization

Three algorithms are proposed to decrease the power used by the device while moving the load from one controller to another controller.

Power-Aware Routing

During the process of waking up, the device which is in idle mode uses a lot of energy and also gives a lot of problems [18]. The devices which are in active mode, transmit the packets using their remaining bandwidth. This process uses 8% of the energy. The energy consumed by this method is slower than the process of waking up the devices that are in idle mode. Here the Power-Aware Routing along with very few state transition algorithms to minimize the power is proposed. This event maintains the number of the device's mode transitions. This algorithm mainly focuses on the active devices that are it sends the packets through active devices instead of waking up the idle ones [5].

Two cases are used here. One is active paths and the other one is not active paths. Both cases are using the same network topology and configuration [19]. There are two modes. One is idle mode and the other one is active mode. Each device is in any one of the modes. The device in the active mode uses a lot of energy but the device in the idle mode uses less amount of power and it is negligible [5]. But switching the mode from idle to active consumes some amount of power [20]. The switch and link state are described below. For switch state (a), It is 1 if switch a is in the active mode or else 0. For link-state (a, b), It is 1 if the link (a, b) is in the active mode or else 0. The constraints in the power optimization.

Constraint of Non-negativity and Capacity

- The flow volume should be non-negative. Flow $(x, y) \geq 0, \forall(x, y) \in L$
- Link bandwidth should not exceed its value of capacity Bandwidth $(x, y) \leq$ Capacity $(x, y), \forall (x, y) \in L$

Switch State and Link State Constraints

If the links which are linked with the switch are in the idle mode, then the switch also will be in the idle mode and vice versa.

Constraints in the Power Consumption of the Link

The link energy usage is determined by the rate of transmission [21]. It is proportional to the link transmission rate. Algorithm 5 provides the steps of the power-reducing algorithm. This method selects the path that still has enough available bandwidth for each packet flow. The interrelationship among flows is considered, such that the probability of link congestion is greatly reduced [22]. By traversing the paths, the active paths are picked out. To transmit the packet flow, the algorithm chooses the active path, where switches are in active mode, and links with a few active switches. Alternatively, a path with a small number of switches that are in idle mode is selected to transfer the packet flow [23].

Table 1. State Transition

X (preState)	X (currentState)	X (preState) ⊛ X (currentState)
0	0	0
0	1	1
1	0	0
1	1	0

Table 1 shows the state transition, where ⊛-device state transition logical operator.

$$\begin{aligned} \text{cl (i, j)} &= \text{c1(Y preState (i, j)} \circledast \text{Y currentState (i, j))} \\ \text{cs (i)} &= \text{c2 (X preState (i)} \circledast \text{X currentState (i))} \end{aligned} \quad (7)$$

cl - It specifies how much power the link uses while switching from idle to active mode.

cs - It specifies how much power the switch uses while switching from idle mode to active mode

It requires no power while switching from active mode to idle mode. Further power consumption will not be obtained by the devices which do not perform state transition. It needs to minimize the energy consumed by the proposed algorithm [24] as mentioned in Eq. 7.

$$Min = \sum\nolimits_{x \in s} X_{currentState}(x)P_s(x) + C_s(x) + \sum\nolimits_{(x,y) \in L} Y_{currentState}(x, y)P_1(x) + C_1(x) \quad (8)$$

XcurrentState (x)PS(x) - It defines how much power the switch uses.

YcurrentState $(x, y)1(x, y)$ - It defines how much power the link used.

```
Algorithm 5
Input : Topo: network topology tf: Target flow in a
request
tf_bw: bandwidth demand for flow tf
Output : fpath: the path of flow tf
1: Swt_active <-- getActiveSwitches(Topo) 2: Link_active
<-- getActiveLinks(Topo) 3: Path_A, Path_B <-- NULL
4: Path_A <-- getAvailabePath(Topo,tf) 5: for path_a in
Path_A
6: Swt_a <-- getSwitches(path_a)
7: if Swt_a belongs to Swt_active then 8:
Add(path_a,Path_B)
9: end if and for
10: if Path_B!="NULL" then
11: for path_b in Path_B
12: i = 0
13: Swt_active[i]= getNumberOfActiveSwitches(path_a)
14: increment the the i value by 1
15: end for
16: index=getIndex(argmin(Swt_active))
17: return Path_B[index]
18: else
19: for path_a in Path_A 20: i=0
21: Swt_sleep[i]= getNumberOfSleepSwitches(path_b)
22: increment the the i value by 1
23: end if
24: index=getIndex(argmin(Swt_sleep))
25: return Path_A[index]
26: end if
27: adapt the link rates of Link_active using the
algorithm 7
```

Flow Rerouting

Algorithm 5 caused the link congestion, to avoid that Algorithm 5 is proposed to reroute the packet flow. The algorithm finds the link which is congested and selects the redirected flow which is the high demand for traffic [25]. The path to which the flow is to be forwarded is determined for each f_max. After f_max transmission, the link band-width is updated at the appropriate time. 90% of the link usage for most of the flow is much less than their maximum requirement, therefore, the link bandwidth using a 90% flow requirement is updated. If the usage of link congestion is less than the threshold value, then the flow redirecting will be terminated as narrated in Algorithm 6.

```
Algorithm 6
Input: Link_conj : congested link Output: forwarding path
step 1: while Bw <= δC
step 2: flows=getAllFlows(Link_conj)
step 3: f_max=getMaximumFlow(flows)
step 4: route f_mausingalgorithm1 step5:Bw=Bw- f_max
step 6: end while step7:Adapt the link rates
```

Link Rate Adaptation

The Algorithms 5 and 6, Link Rate Adaptation is proposed. It is mentioned in Sect. 3.3 that the rate of transmission determines how much energy is used by the link. It correlates with the link transmission rate. Therefore, this Algorithm 7 is used to reduce the power further by adapting the transfer rate of the link to the current bandwidth of the link.

```
Algorithm 7
Input: Link_active: active link set
Bw: bandwidth of the link link_act which belongs to
Link_active Output: rate: rate of the active link
link_act
1: for link_act in Link_active 2: if bw<=rate(10) then
3: rate(10) = rate
4: else if rate(10) < bw <= rate(100) then step 5:
rate(100) = rate 6: else
7: rate(1000)=rate
8: end if
```

4 Results and Discussion

The proposed model is implemented using mininet, a network emulator that generates virtual hosts, controllers, links, and switches to imitate any kind of network [26]. The algorithm used in this model is written in python and uses the Ryu controller. To simulate a real Time traffic scenario, scapy tool is used which is a packet manipulation tool. A sample Tree topology with three controllers and a common data center network Fat Tree topology are both utilized for simulation.

In this Self Adaptive Load Balancing Scheme(A), 20% of BT is the maximum load deviation, so that the controller can adapt more load before it could do load broadcast, which results in higher load balancing on every controller in the network, less band-width utilization and less number of switches to be migrated.

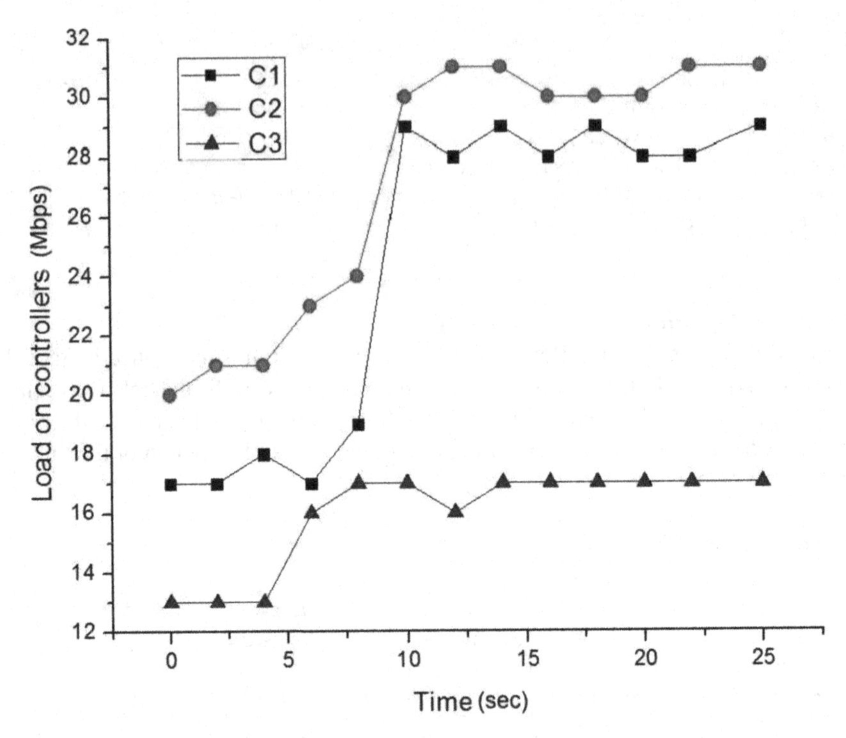

Fig. 1. Controller load without load balancing

In Fig. 1 and 2, each controller's load in the network without load balancing and after implementing the Self Adaptive Load Balancing (A) algorithm are compared. One can conclude that the load in each controller is greatly reduced after implementing our SALBA algorithm.

In Table 2, C2 represents the heavily loaded controller, C1 and C3 represent the lightly loaded controller. Hence, the controllers C1 and C3, the load from controller C2 must be transferred. The values of the switches connected to controller C2 are S1 = 300, S2 = 200, S3 = 150, and S4 = 250.

The switch S3 has been migrated from C2 to C3 by increasing the threshold value (20% of BT) and BT value. Thus, the Current Load (CL), which is the aggregate of all the packets received by each switch linked to the controller $CLi = \Sigma jSLj$, of all the controllers is less than the value of the Base Threshold (BT).

As the maximum load deviation in this SALBA algorithm is increased, the frequent load broadcast is avoided and hence the bandwidth utilization is reduced (Fig. 3) compared to the SALB scheme. One could see that the number of migrated switches in SALBA is very less (Fig. 4) compared to all other algorithms. This is because the controller could adapt more before it could be migrated. The overall discussion is highlighted in Table 3. M stands for the overall quantity of broadcasts and migration messages sent and received by controllers.

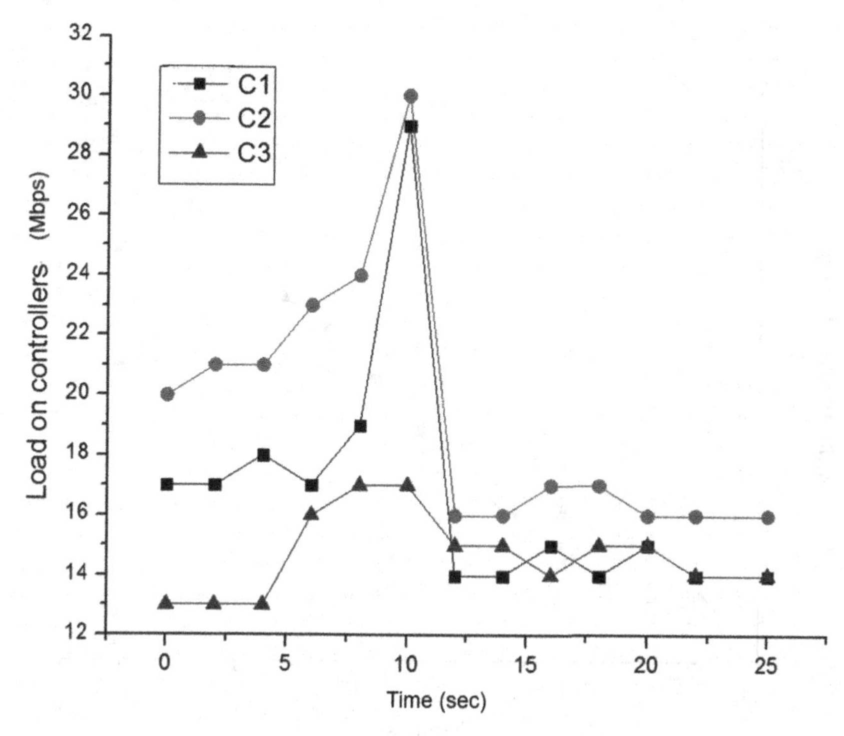

Fig. 2. Controller load with SALBA load balancing

Table 2. Metrics required for the migration of load between controllers

Parameters	Controller		
	C1	C2	C3
BT	800	850	850
CL	720	750	700

With this (Fig. 5) comparison, it is quite forward that POLST is using less energy for choosing the power-aware route in the network. It avoids changing the devices from idle mode to active mode which consumes more energy. Hence, in this load balancing method When the load on the controller is high, POLST is integrated such that it picks a power-aware route to perform switch migration.

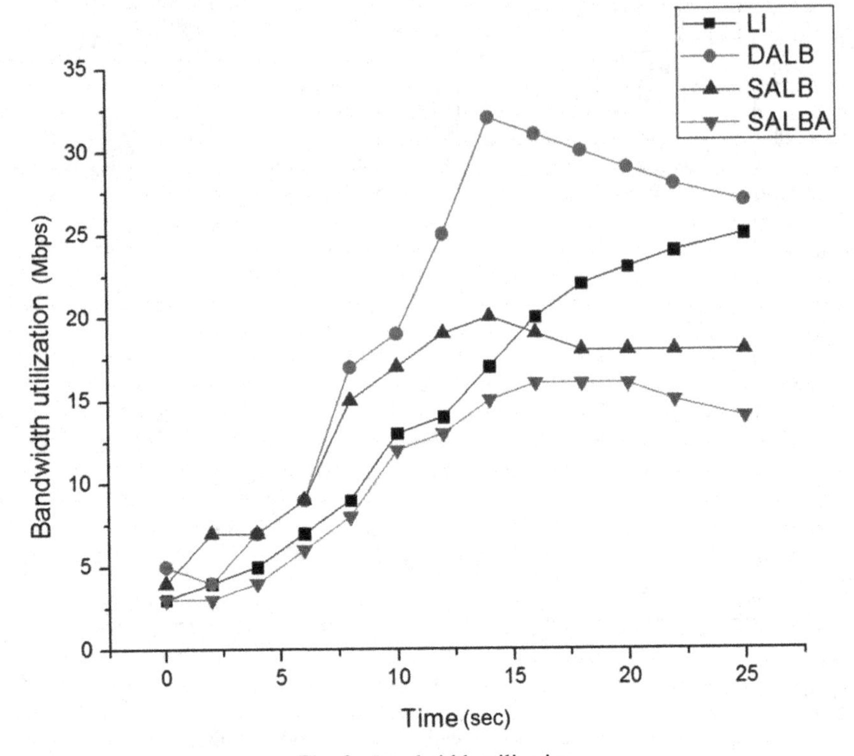

Fig. 3. Bandwidth utilization

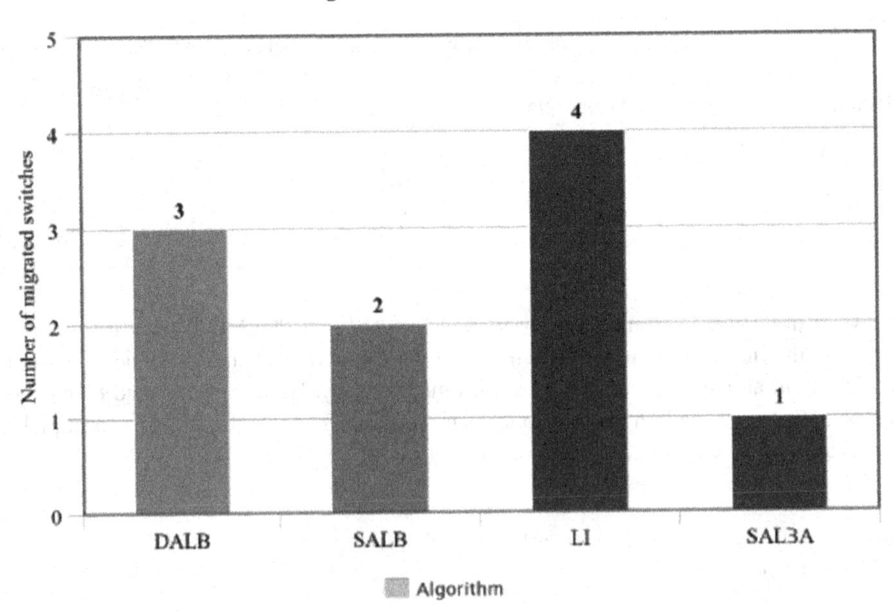

Fig. 4. Number of switches migrated

Table 3. Performance Comparison of Algorithms

Algorithm	Controller Overhead (% of Messages)	Time taken for Load Balancing (s)	Throughput (packets/s)	Loss Rate (% of Messages)
SALBA	1.5	0.37	13.7	0.0041
SALB	3	1.371	10.48	0.0132
LI	25	19.8	7.8	0.25
DALB	10	22.67	6.7	0.15

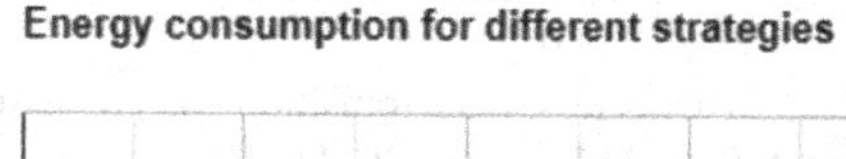

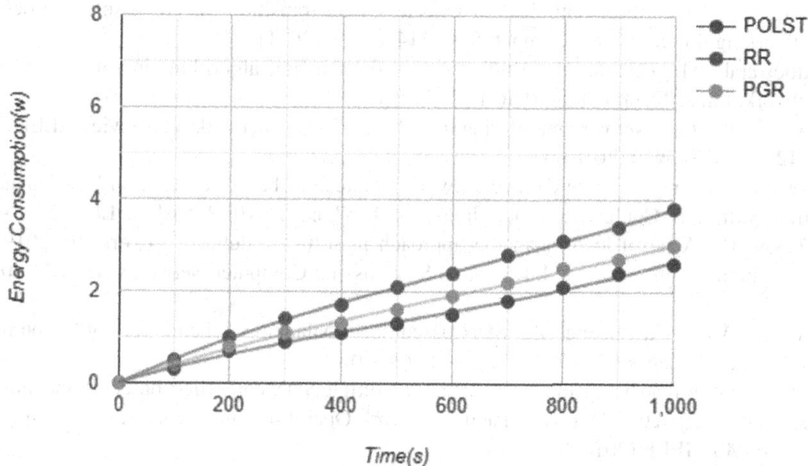

Fig. 5. Energy consumption for different strategies

5 Conclusion

An energy optimization framework and a self-adaptive load balancing technique are both built on a multi-controller SDN control plane. The integration of the above-mentioned techniques helps the controllers in balancing the load by using less power. In the first function, as a self-adaptive controller, each controller exposes its load to every other controller in the network. With that knowledge, switch migration is performed from a highly loaded controller to a controller which has comparatively less load. In the second function, Power optimization is performed to find a path for switch migration with less state transition. This proposed system improves the network's efficiency as measured by throughput, packet drop, and the amount of time needed to balance each network's load. As a part of our future work, load balancing on the controller could be improved further and can be implemented in real-time.

References

1. Bavani, K., Ramkumar, M. P., Emil Selvan, G.S.R.: Statistical approach based detection of distributed denial of service attack in a software defined network. In: 2020 6th International Conference on Advanced Computing and Communication Systems (ICACCS). IEEE (2020)
2. Priyadarsini, M., et al.: An energy-efficient load distribution framework for SDN controllers. Computing **102**, 2073–2098 (2020)
3. Wu, R., et al.: A MECS redistribution algorithm for SDN-enable MEC using response time and transmission overhead. Wirel. Netw. **27**, 1445–1457 (2021)
4. Etengu, R., et al.: AI-assisted framework for green-routing and load balancing in hybrid software-defined networking: proposal, challenges and future perspective. IEEE Access **8**, 166384–166441 (2020)
5. Priyadarsini, M., et al.: An adaptive load balancing scheme for software-defined network controllers. Comput. Netw. **164**, 106918 (2019)
6. Zhao, Y., et al.: Power optimization with less state transition for green software defined networking. Future Gener. Comput. Syst. **114**, 69–81 (2021)
7. Albowarab, M., Zakaria, N., Abidin, Z.: Load balancing algorithms in software defined network. Int. J. Technol. Eng. (IJRTE) **7** (2019)
8. Tsai, P.-W., et al.: Network monitoring in software-defined networking: a review. IEEE Syst. J. **12**(4), 3958–3969 (2018)
9. You, X., Wu, Y.: Software defined network architecture based research on load balancing strategy. In: AIP Conference Proceedings, vol. 1967. no. 1. AIP Publishing LLC (2018)
10. Thakur, D.: A novel load balancing approach in software defined network. In: 2019 4th International Conference on Information Systems and Computer Networks (ISCON). IEEE (2019)
11. Li, G., Wang, X., Zhang, Z.: SDN-based load balancing scheme for multi-controller deployment. IEEE Access **7**, 39612–39622 (2019)
12. Yu, J., et al.: A load balancing mechanism for multiple SDN controllers based on load informing strategy. In: 2016 18th Asia-Pacific Network Operations and Management Symposium (APNOMS). IEEE (2016)
13. Cui, J., et al.: A load-balancing mechanism for distributed SDN control plane using response time. IEEE Trans. Netw. Serv. Manag. **15**(4), 1197–1206 (2018)
14. Sams Aafiya Banu, S., et al.: SMOTE variants for data balancing in intrusion detection system using machine learning. In: Singh, P., Singh, D., Tiwari, V., Misra, S. (eds.) MISP 2022. LNEE, vol. 998, pp. 317–330. Springer, Singapore (2023). https://doi.org/10.1007/978-981-99-0047-3_28
15. Li, L., Xu, Q.: Load balancing researches in SDN: A survey. In: 2017 7th IEEE International Conference on Electronics Information and Emergency Communication (ICEIEC). IEEE (2017)
16. Deepa, S., et al.: Routing scalability in named data networking. In: 2022 4th International Conference on Advances in Computing, Communication Control and Networking (ICAC3N). IEEE (2022)
17. Zhang, S., et al.: Online load balancing for distributed control plane in software-defined data center network. IEEE Access **6**, 18184–18191 (2018)
18. Ramkumar, M.P., et al.: Single disk recovery and load balancing using parity declustering. J. Comput. Theor. Nanosci. **14**(1), 545–550 (2017)
19. Fernández-Fernández, A., Cervelló-Pastor, C., Ochoa-Aday, L.: Energy efficiency and network performance: a reality check in SDN-based 5G systems. Energies **10**(12), 2132 (2017)

20. Chai, R., et al.: Energy consumption optimization-based joint route selection and flow allocation algorithm for software-defined networking. Sci. China Inf. Sci. **60**, 1–14 (2017)
21. Maaloul, R., et al.: Energy-aware routing in carrier-grade Ethernet using SDN approach. IEEE Trans. Green Commun. Netw. **2**(3), 844–858 (2018)
22. Zheng, K., Wang, X., Liu, J.: DISCO: distributed traffic flow consolidation for power efficient data center network. In: 2017 IFIP Networking Conference (IFIP Networking) and Workshops. IEEE (2017)
23. Naeem, F., Tariq, M., Vincent Poor, H.: SDN-enabled energy-efficient routing optimization framework for industrial Internet of Things. IEEE Trans. Ind. Inform. **17**(8), 5660–5667 (2020)
24. Liu, C., Malboubi, A., Chuah, C.N.:. OpenMeasure: adaptive flow measurement & inference with online learning in SDN. In: 2016 IEEE Conference on Computer Communications Workshops (INFOCOM WKSHPS), pp. 47–52. IEEE, April 2016
25. Nam, T.M., et al.: Energy-aware routing based on power profile of devices in data center networks using SDN. In: 2015 12th International Conference on Electrical Engineering/Electronics, Computer, Telecommunications and Information Technology (ECTI-CON). IEEE (2015)
26. Kreutz, D., et al.: Software-defined networking: A comprehensive survey. Proc. IEEE **103**(1), 14–76 (2014)

Sentiment Analysis: Indian Languages Perspective

Abhishek A. Vichare[1,2](✉) and Satishkumar L. Varma[3]

[1] Department of Computer Engineering, Mukesh Patel School of Technology Management and Engineering, NMIMS University, Mumbai, India
vichare1@gmail.com

[2] Pillai College of Engineering, Navi Mumbai, India

[3] Department of Information Technology, Pillai College of Engineering, Navi Mumbai, India

Abstract. Sentiment is a feeling or opinion on the basis of perception of a particular person, product or situation. Process of sentiment analysis (SA) consists of identification and categorizing such sentiments. It also includes natural language processing as well as data analysis to get meaningful data from raw data. An abundance of research is done in Sentiment analysis area with respect to English language. But there have not been many experiments done in the Indian local/regional languages like Hindi, Marathi, Konkani, Gujarati. Since the last decade, there has been tremendous increase in digital data accessible in Indian dialects on world wide web from different sources like social media, blogs, reviews, opinions, news content and feedback systems. So, it has become crucial to investigate and process such data to get insights of it. This data can be very beneficial for government agencies, businesses, organizations or individuals.

This paper reviews research in the area of sentiment analysis in Indian regional languages especially Hindi and Marathi. It is also discussed in literature review about different sentiment analysis tools, challenges, approaches and techniques. The paper also provides an overview of research gap from existing research work.

Keywords: Sentiment analysis · NLP · Indian language · Marathi · Aspect based sentiment analysis

1 Introduction

Because of speedy evolution in internet advancement and social networking websites, sentiment analysis has become crucial method to extract opinions from given or available data [1]. As per the Cambridge advanced learner's dictionary definition 'sentiment' is a thought, opinion, or idea based on a feeling about a situation. It is also a way of thinking about something [2]. SA is field of research to analyze emotions, opinions, sentiments with respect to products, individual, company or incidents [3].

There is absolute need to detect individual's opinion as well sentiments about different parameters or subject like events, trends, elections, government policies; about participant/users convey sentiments about.

P. Das et al. (Eds.): AMRIT 2023, CCIS 1953, pp. 262–276, 2024.
https://doi.org/10.1007/978-3-031-47224-4_23

Data from market surveys, public relations, reviews as well feedback about goods and services can be very useful for commercial applications, business perspectives or government. There is a huge investigation required to recognize the aspects of Indian languages and to enhance the results of sentiment analysis [4].

The number of internet users in India will cross 829 million figure in the year 2021, as per the Cisco Visual Networking Index (VNI) report published in 2017. This report also mentioned that the number of internet-enabled devices will touch 2 billion benchmark in 2021. Reasons to pay heed to other languages apart from English are mushrooming because of a lack of resources, and many researchers are concerned about local, regional dialects in India [5]. Indian internet users are using vernacular languages for internet surfing.

According to Google-KPMG report 2017, the number of Indian regional language web users (234 million) has already surpassed the English language users (175 million) in 2016.

India is a multicultural nation of various religions and languages. As per the Constitution of India (8th Schedule), as of 28th December 2022; India has total twenty-two official languages, namely Assamese, Bengali, Bodo, Dogri, Gujarati, Hindi, Kannada, Kashmiri, Konkani, Maithili, Malayalam, Manipuri, Marathi, Nepali, Oriya, Punjabi, Sanskrit, Santhali, Sindhi, Tamil, Telugu and Urdu [6].

Previously large amount of research work is done on sentiment analysis considering English language but not much on the data available in Indian languages (dialects) [4, 16]. With Indian perspective, majority of the sentiment analysis research is done in Hindi and Bengali language.

As per the 2011 census, number of native speakers of Hindi and Marathi languages are five hundred twenty-eight million and eighty-three million, respectively. Marathi and Hindi language is subgroup of the Indo-Aryan language, which falls in Indo-European language family. Maharashtra state is second sub-national entity with respect to population across the globe and second state in India with respect to national population.

Marathi is the official language of Maharashtra state. It is jointly official language for the union territories of Diu, Daman, Dadra and Nagar Haveli. Marathi is third most spoken language in India as per census 2011 data. As per the Ethnologue (2019), encyclopedic work of worlds languages produced by SIL International (US based non-profit organization); Marathi is 10th ranked language. World widely Marathi language is having 83.1 million as first language speaker's. It is mentioned in Census 2011 that the population of Maharashtra is 11.24 Crores.

These numbers indicate that there is a lot of data available in Hindi and Marathi for the analysis, which could be used for business purposes or the welfare of local population. A comprehensive overview of current research and updates in the area of SA done for Marathi & English language is provided in this paper. Methods, algorithms and applications in SA field are investigated and presented in this paper.

2 Structure of the Paper

The Structure of this paper is as follows. In Sect. 1, Introduction of the topic is mentioned. Structure is discussed in Sect. 2. In Sect. 3, process of sentiment analysis, steps of SA, levels of SA with sub-tasks in SA and popular techniques used for SA are discussed. Section 4 focuses on milestones at global level in SA history. Section 5 focuses on milestones achieved in SA for Marathi and Hindi languages. In Sect. 6, recent research work done for SA in Marathi and Hindi language is discussed in details. Based on Sect. 5 and Sect. 6 the paper emphasizes the need for research in SA for the Marathi language.

3 Process of SA

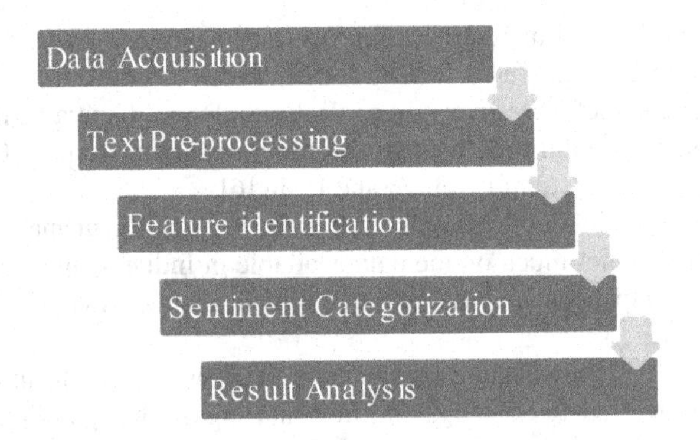

Fig. 1. Steps in Sentiment Analysis

SA method consists information acquisition, text processing, feature extraction, sentiment categorization and result analysis steps. General steps in sentiment analysis process are shown in Fig. 1. In general, sentiment analysis is considered as technique to get subjectivity and polarity from data. Semantic orientation provides polarity as well as strength of data particularly words [10].

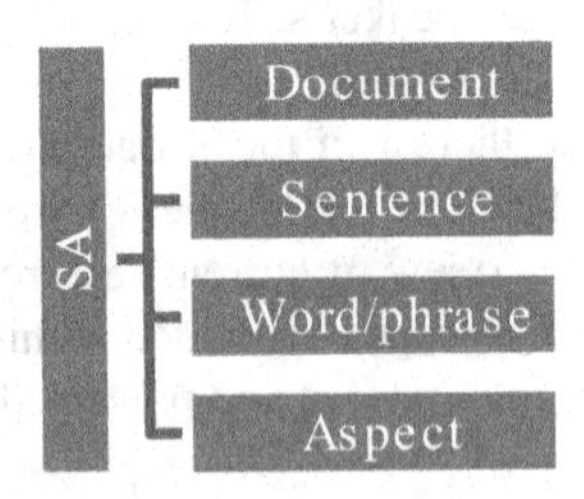

Fig. 2. SA levels

Figure 2 showcase general SA levels. SA can be conducted at four levels based on the granularity of input data.

Aspect Based

This type of SA detects the aspect of text; which discovers what exactly the user liked or did not liked. Sometimes, it is not enough to say whether a post has a "positive" or "negative" sentiment. The user may want to identify the aspects of the target present in the input and the sentiment mentioned for every aspect. 'I love Apple products but iPhone 7s is overhyped"; in this scenario there are two opinions, one with a positive note about Apple product and one with a negative note about iPhone 7s.

Subtasks in aspect based SA:

1. Aspect Term Extraction (ATE)
2. Aspect Term Sentiments (ATS)
3. Aspect Category Detection (ACD)
4. Aspect Category Sentiments (ACS)

Example: Ice-cream is nice as well.
ATE: ice-cream, ATS: Positive,
ACD: Food/Dessert, ACS: Positive

The objective of aspect based sentiment analysis is to find aspects and extract the sentiments associated.

Word/Phrase Level

SA is executed on phrases or words present in the input text. Sentiment of a phrase/entity is determined in this approach. Use of lexicon to find positive and negative weightage term is considered as general method [12].

Sentence Level

At this level differentiation is done between objective phrases which indicates facts and subjective phrases which indicates sentiments [11]. SA is performed for each sentence. Opinion about sentences can be gauged using sentence level SA.

Document Level

This level of SA classifies the entire document into positive, negative or neutral categories. In document level it is considered that the entire document is related to one entity.

From survey, it is observed that ML-based and lexicon-based methods are the most popular SA techniques. Figure 3 indicates Sentiment Analysis popular techniques.

Lexicon based method is referred as rule-based method. This method consists sentiment lexicon, which uses a repository of known and precompiled sentiments terms. Lexicon based method consists of two categories, dictionary based category as well corpus based category. Above mentioned categories use statistical or semantic techniques to identify the polarity of sentiments from given input [5].

In dictionary based method, set of sentimental words are created. This set is further expanded by exploring the WordNet to find synonyms and antonyms. Corpus

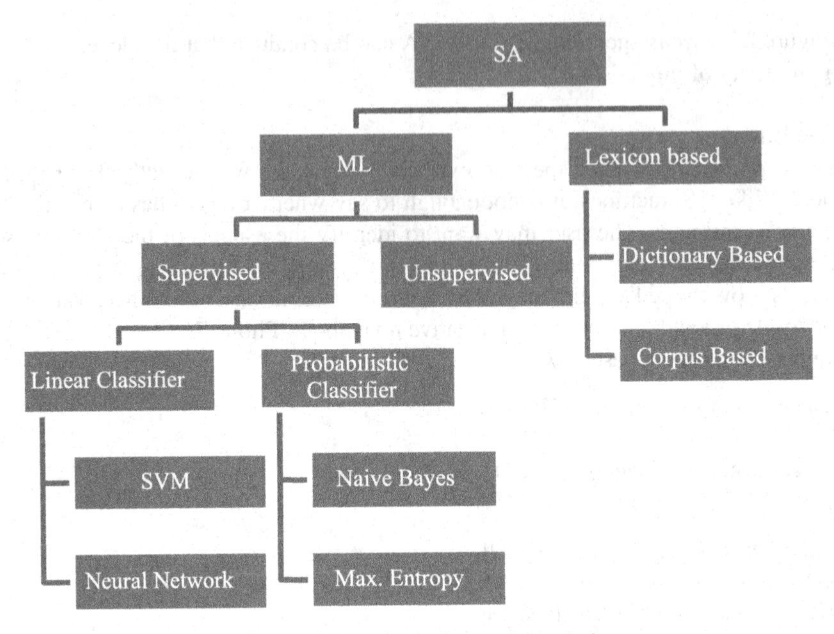

Fig. 3. Sentiment Analysis popular techniques

based method is used to extract domain relevant sentiment words with respect to their orientation [4, 5].

Machine learning approach has two methods, supervised and unsupervised learning. Supervised method consists huge number of labelled trained dataset. To train the model, Naïve Bayes, Support Vector Machine are extensively used classifiers instead of other classifiers like Maximum entropy, Neural network, K-Nearest Neighbor. This trained model can classify data as per sentiments. Unsupervised approach is applied when trained datasets are not available.

Naive Bayes (NB) classifier belongs to probabilistic classifiers, as shown in Fig. 3. This classifier is considered as simple and effective method which combine efficiency with considerable accuracy [13]. NB discover the probability of an event using probability of another already happened event.

$$\mathrm{P(c|x)} = \frac{P(x|c)P(c)}{P(x)} \quad (1)$$

P(c|x): Posterior Probability
P(x): Predictor Prior Probability
P(c): Class Prior Probability
P(x|c): Likelihood

In simple terms it can be expressed as

$$Posterior = \frac{Prior \times likelihood}{Evidence} \quad (2)$$

SVM algorithm defines N-dimensional hyper-plane that categorizes information in two separate groups. SVM accept the input data, then it provides prediction of class. Aim of the training procedure is to find hyperplane

$$(\underset{w}{\rightarrow}.\underset{x}{\rightarrow}) = \sum_i y_i\, \alpha_i\, (\underset{x_i}{\rightarrow}.\underset{x}{\rightarrow}) + b \tag{3}$$

Here $\vec{x}_i$ is input vector, yi is output class, $\vec{w}_i$ is weight vector, α_i is lagrange multiplier.

4 Milestones in History of Sentiment Analysis Research

Peter D. Turney (2002)
Turney (2002) showed that it is possible to classify reviews using average semantic orientated words to label them as positive or negative [7]. This paper demonstrates result for rating a review by using simple unsupervised learning algorithm. Major issues for this work were required time and the level of accuracy for dataset.

Wilson T. et al. (2005)
For this research work, authors have presented phrase level SA method. First, this analysis finds whether phrase is neutral or has polarity value. Then disambiguates the polarity of the such expressions [8]. Previous research in SA classify document on the basis of sentiments, and this paper classify individual words and phrases instead of entire document.

Bo Pang and Lillian Lee (2008)
This paper published by Bo Pang and Lillian Lee (2008) is considered as milestone in SA research field. This paper highlighted on importance of understanding opinions of others. Paper also covers techniques to handle sentiment analysis challenges with comparison to traditional systems i.e. fact based analysis [9]. This work creates interest in researchers about intellectual richness and breadth of the SA area.

M. Taboada, J. Brooke, M. Tofiloski, K. Voll, and M. Stede (2011)
This paper presents a lexicon based method to identify sentiment from input [10]. Earlier research used adjectives from sentences, but here authors used the Semantic Orientation CALculator (SO-CAL) for the remaining speech data. They refined the approach to negation by using an intensifier.

F. Iqbal et al. (2019)
In this research paper, authors proposed an integrated framework to combine lexicon based and ML based techniques for greater accuracy and scalability [1]. To overcome the problem of scalability because of increase in feature-set, GA (Genetic algorithm) based feature reduction method is suggested in this paper. Three approaches (SentiWordnet, Machine Learning and Machine learning with Genetic algorithm) are experimented with SA.

Akhtar et al. (2020)
Akhtar et al. proposed an ensemble framework. In this framework they developed DL

based models on the basis of CNN, LSTM and GRU. Proposed method was used for emotion recognition and sentiment analysis. Results achieved are comparatively better than other ensemble framework like Gradient boosting, Adaboost and Bagging.

Singh et al. (2020)
It is observed that cross domain sentiment analysis can provide better behavioural aspects of the customer or user. Singh et al. highlighted need of research work in cross domain opinion classification due to lack of proper solutions in this area.

5 Milestones in SA Research with Marathi and Hindi Languages

In the field of Indian dialects SA, majority of the R&D work is done for Hindi and Bengali language [4].

Das and Bandyopadhyay (2009)
In 2009, Das and Bandyopadhyay introduced first work of sentiment analysis in an Indian language, which was for Bengali language. Das and Bandyopadhyay used SentiWordNet and subjectivity lexicon to generate a lexical resource with sentiments with POS tags mapped with English words [14]. Lexical transfer technique is applied to SentiWordNet and Subjectivity word list. For each English word, set of Bengali words were produced using dictionary search. Sentiment score (positive or negative) for the Bengali words is referred from equivalent English words.

Aditya Joshi et al. (2010)
Aditya Joshi et al. (2010) reports first research on SA for Hindi language. In this work they proposed three approaches [15]. First approach involves Hindi training corpus for categorizing sentiments in Hindi document. In second approach after translating document in English, trained classifier applied to categorize document. In third approach, author build a lexical resource named Hindi-SentiWordNet (HSWN) and executed a majority score based technique to categorize documents in Hindi.

Balamurali A R et al. (2012)
In this paper authors have presented method to cross-lingual SA that uses WordNet senses as features of a supervised sentiment categorization. This approach was applied to two popular languages in India Marathi and Hindi [17].

Akhtar et al. (2016)
Akhtar et al. (2016) presented hybrid deep learning architecture for SA. This was first attempt where DL model such as CNN was used for SA in Hindi language [16]. This work includes only aspect term sentiment classification, which is sub type of aspect based sentiment analysis.

R. Bhargava et al. (2019)
R. Bhargava et al. proposed technique to forecast the sentiments of tweets published in Hindi, Bengali as well Tamil language [18]. Use of neural network to predict sentiments

of tweets make this work distinguishable. RNN, LSTN and CNN (neural network layers) are used to create sequential models.

Removal of emoticons, exclamations makes this work less accurate. More precise sentiment to a sentence (extremely negative, positive or negative) is not considered for computing.

Shelke et al. (2022)
Shelke et al. proposed Marathi SentiWordNet, which consists polarity score on the basis of Hindi SentiWordNet. Authors have performed sentiment analysis on movie review data using Marathi SentiWordNet with Bi-LSTM.

6 Recent Research for SA in Hindi and Marathi Language

This section consists the illustrations of recent research and new trends in SA area. Table provides a comprehensive overview of research done for SA in Hindi and Marathi language (Tables 1 and Table 2).

Table 1. Recent research for SA in Hindi

Citation, Author	Dataset used	Parameters used for evaluation	Description of research
[15] Aditya Joshi et al.	Movie Review	Accuracy achieved 78.14%	Sentence level SA performed. Fallback strategy was used, developed own lexical resource HSWN (Hindi SentiWordNet)
[17] Balamurali A R et al.	Travel Dataset	Accuracy achieved 79%	Sentence level SA performed
[19] Patra et al.	Tweets	Maximum accuracy achieved for Hindi language is 55.67%	Sentence level SA performed. This is the first attempt in the context of Indian language tweets
[16] Akhtar et al.	Twits Hindi, Movie review Hindi	Accuracy for Twitter 62.52%, Movie review 44.88 %	Sentence level SA performed. First time CNN used for SA for Hindi language
[20] Bakliwal et al.	Product review	Accuracy 79.03% on product review	Author proposed a graph based technique to produce the Hindi subjectivity lexicon

(*continued*)

Table 1. (*continued*)

Citation, Author	Dataset used	Parameters used for evaluation	Description of research
[21] Sharma P et al.	Election tweets	Accuracy for NB 62.1%	Unsupervised lexicon method used for classification of Hindi tweets
[22] Sharma R et al.	Movie Review	Accuracy 65%	Hindi dictionary was developed by authors to explore the polarity of movie reviews in Hindi
[23] Mittal N et al.	Movie Review	Accuracy 80.21%	Author proposed coverage of HSWN, also effect of negation and discourse rules are discussed considering Hindi language
[24] M. Yadav et al.	HSWN database, Product review	Accuracy [approach 1–52%, approach 2–71.5%, approach 3–70.27%]	Authors proposed three approaches. First categorization is achieved by applying NN Prediction with the help of pre classified word. In second method, categorization is done with the help of IIT- Bombay Hindi SentiWordNet (HSWN). Last method proposed use of pre-classified sentences as labeled data for classification
[25] Bhargava R. et al.	Dataset of Mixed language (English language and Indian languages)	Overall Precision for Hindi 0.553	This work involves identifying the four Indian languages in a code mix sentences and categorizing sentiments after translating document in Indic language

(*continued*)

Table 1. (*continued*)

Citation, Author	Dataset used	Parameters used for evaluation	Description of research
[26] Seshadri et al.	Data provided by SAIL 2015	72.01% for Hindi language	Recurrent neural network is used for classifying sentiments in the tweets
[35] K. Yadav et al.	IIIT Code-mixed dataset	73% accuracy	Authors proposed a bidirectional LSTM based method
[36] Mariya A Ali, S.B. Kul-karni	Movie review Dataset by IIT Bombay	Not available	Discussed about pre-processing of data, feature extraction and deep belief network classification

Table 2. Recent research for SA in Marathi

Citation, Author	Dataset used	Corpus Size	Parameters used for evaluation	Description of research
[17] Balamurali A.R	Travel Review	Not available	84% for Marathi	Authors presented method to execute cross-lingual SA with the help of WordNet synset identifiers as features of a supervised classifier
[27] Mhaske N	Not available	Not available	Not available	In this paper problems in the opinion mining of Marathi content are discussed. General issues in language like sentence structure, negation, ironic statements are discussed in this paper

(*continued*)

Table 2. (*continued*)

Citation, Author	Dataset used	Corpus Size	Parameters used for evaluation	Description of research
[28] Deshmukh S. et al.	General (online text file)	Not mentioned	Not available	Sentence level SA performed. Polarity of each word is calculated using English sentiwordnet Discrepancies in the results due to: a. Limited corpus words b. Limited scope of English SentiWordNet c. Non acceptance of special characters
[30] Bolaj P et.al	Marathi documents	Not mentioned	Not mentioned	Author mentioned that ontology based classification can be used to perform SA
[31] Pawar S. et al.	General	200 words	Not mentioned	lexicon based approach is applied to calculate polarity of words
[33] Shrivastava et al.	Marathi Sentence	Not mentioned	Accuracy for UNL method 82%	This paper proposed paraphrase detection for Marathi text by applying mixture of statistical as well semantic resemblance
[34] Chaudhari C. et al.	User Reviews	Not mentioned	Not mentioned	This paper proposed approach for Sentiment Analysis of Marathi text uses Gate Processor (Natural language processor) to find the overall polarity of the document

(*continued*)

Table 2. (*continued*)

Citation, Author	Dataset used	Corpus Size	Parameters used for evaluation	Description of research
[37] Atharva Kulkarni et al.	L3CubeM ahaSent	16,000 distinct tweets	Accuracy for IndicBERT (INLP) 84.13	Authors present the Marathi Sentiment Analysis Dataset L3CubeMahaSent CNN, LSTM, ULMFiT, and BERT based DL models are used
[38] Bafna, Prafulla & Saini, Jatinder	Own dataset	1206 Marathi text documents	Accuracy for DSMM 80%	Author proposed approach of DSMM corpus. Analysis of DSMM and DTMM is provided in paper

7 Conclusion

There is absolute need to perform more experiments associated with natural language processing like SA in regional languages of India. Majority of sentiment analysis work is done in Hindi language in Indian scenario. Need to perform SA for Marathi language is highlighted in this paper due to large set of Marathi language speaking population. Traditional machine learning algorithms need to be applied along with deep learning techniques to improvise the results. There are many challenges like lack of annotated corpora, unbalanced datasets, unavailability of lexicon tools to conduct SA in Indian regional languages.

Sentiment analysis with codemixed languages, phrase level SA and aspect based sentiment analysis are some unexplored research area for Marathi language. Cross domain sentiment analysis is emerging as hot topic for research in English as well regional languages.

References

1. Iqbal, F., et al.: A hybrid framework for sentiment analysis using genetic algorithm based feature reduction. IEEE Access **7**, 14637–14652 (2019). https://doi.org/10.1109/ACCESS.2019.2892852
2. Cambridge dictionary. https://dictionary.cambridge.org/dictionary/english/sentiment. Accessed 16 May 2023
3. Mahtab, S., Nazmul, R., Mahfuzur, Md.: Sentiment analysis on Bangladesh cricket with support vector machine. In: 2018 International Conference on Bangla Speech and Language Processing, ICBSLP, Bangladesh, pp. 1–4. IEEE (2018). https://doi.org/10.1109/ICBSLP.2018.8554585

4. Rani, S., Kumar, P.: A journey of Indian languages over sentiment analysis: a systematic review. Artif. Intell. Rev. **52**, 1–48 (2019). https://doi.org/10.1007/s10462-018-9670-y
5. Medhat, W., Hassan, A., Korashy, H.: Sentiment analysis algorithms and applications: a survey. Ain Shams Eng. J. **5**(4), 1093–1113 (2014). https://doi.org/10.1016/j.asej.2014.04.011
6. Eighth Schedule, The Constitution of India. https://www.mha.gov.in/sites/default/files/Eighth Schedule_19052017.pdf. Accessed 16 May 2023
7. Turney, P.: Thumbs up or thumbs down? Semantic orientation applied to unsupervised classification of reviews. In: Proceedings of the 40th Annual Meeting on Association for Computational Linguistics (ACL 2002), pp. 417–424. Association for Computational Linguistics, USA (2002). https://doi.org/10.3115/1073083.1073153
8. Wilson, T., Wiebe, J., Hoffmann, P.: Recognizing contextual polarity in phrase-level sentiment analysis. In: Proceedings of HLT/EMNLP, Canada, pp. 347–354. Association for Computational Linguistics (2005). https://doi.org/10.3115/1220575.1220619
9. Pang, B., Lee, L.: Opinion mining and sentiment analysis. Found. Trends Inf. Retr. **2**, 1–135 (2008). https://doi.org/10.1561/1500000011
10. Taboada, M., Brooke, J., Tofiloski, M., Voll, K., Stede, M.: Lexicon-based methods for sentiment analysis. Comput. Linguist. **37**, 267–307 (2011). https://doi.org/10.1162/COLI_a_00049
11. Behdenna, S., Barigou, F., Belalem, G.: Document level sentiment analysis: a survey. EAI Endorsed Trans. Context Aware Syst. Appl. **4**, e2 (2018). https://doi.org/10.4108/eai.14-3-2018.154339
12. Gamallo, P., Garcia, M.: Citius: a Naive-Bayes strategy for sentiment analysis on English tweets. In: Proceedings of the 8th International Workshop on Semantic Evaluation (SemEval 2014), Ireland, pp. 171–175 (2015). https://doi.org/10.3115/v1/S14-2026
13. Das, A., Bandyopadhyay, S.: Subjectivity detection in English and Bengali: a CRF-based approach. In: Proceedings of ICON-2009, India (2009)
14. Joshi, A., Bhattacharyya, P., Balamurali, R.: A fall-back strategy for sentiment analysis in Hindi: a case study. In: Proceedings of the 8th International Conference on Natural Language Processing, India (2010)
15. Akhtar, M.S., Kumar, A., Ekbal A, Bhattacharyya, P.: A hybrid deep learning architecture for sentiment analysis. In: Proceedings of the 8th International Conference on Natural Language Processing, Osaka, Japan, pp. 482–493. COLING (2016)
16. Balamurali, A.R., Joshi, A., Bhattacharyya, P.: Cross-lingual sentiment analysis for indian languages using linked WordNets. In: Proceedings of the International Conference on Computational Linguistics, Mumbai, pp. 73–82. COLING 2012 (2012)
17. Bhargava, R., Arora, S., Sharma, Y.: Neural network-based architecture for sentiment analysis in Indian languages. J. Intell. Syst. **28**, 361–375 (2019). https://doi.org/10.1515/jisys-2017-0398
18. Patra, B.G., Das, D., Das, A., Prasath, R.: Shared task on sentiment analysis in Indian languages (SAIL) tweets - an overview. In: Prasath, R., Vuppala, A., Kathirvalavakumar, T. (eds.) MIKE 2015. LNCS, vol. 9468. Springer, Cham (2015). https://doi.org/10.1007/978-3-319-26832-3_61
19. Bakliwal, A., Arora, P., Varma, V.: Hindi subjective lexicon: a lexical resource for Hindi polarity classification. In: Proceedings of the 8th International Conference on Language Resources and Evaluation, Istanbul, Turkey, pp. 1189–1196. European Language Resources Association (2012)
20. Sharma, P., Moh, T.S.: Prediction of Indian election using sentiment analysis on Hindi Twitter. In: Proceedings IEEE International Conference on Big Data, USA, 2016, pp. 1966–1971 (2016). https://doi.org/10.1109/BigData.2016.7840818
21. Sharma, R., Nigam, S., Jain, R.: Polarity detection of movie reviews in Hindi language. Int. J. Comput. Sci. Appl. **4**, 49–57 (2014). https://doi.org/10.5121/ijcsa.2014.4405

22. Mittal, N., Agarwal, B., Chouhan, G., Bania, N., Pareek, P.: Sentiment analysis of Hindi review based on negation and discourse relation. In: Sixth International Joint Conference on Natural Language Processing, Japan, pp. 45–50 (2013)
23. Yadav, M., Bhojane, V.: Semi-supervised mix-Hindi sentiment analysis using neural network. In: 9th International Conference on Cloud Computing, Data Science & Engineering (Confluence), India, pp. 309–314 (2019). https://doi.org/10.1109/CONFLUENCE.2019.8776943
24. Bhargava, R., Sharma, Y., Sharma, S.: Sentiment analysis for mixed script Indic sentences. In: International Conference on Advances in Computing, Communications and Informatics, ICACCI, India, pp. 524–529 (2016). https://doi.org/10.1109/ICACCI.2016.7732099
25. Seshadri, S., Kumar, M., Soman, Kp.: Analyzing sentiment in Indian languages micro text using recurrent neural network. IIOAB J. **7**, 313–318 (2016)
26. Mhaske, N., Patil, A.: Issues and challenges in analyzing opinions in Marathi text. Int. J. Comput. Sci. Issues IJCSI, 19–25 (2016). https://doi.org/10.20943/01201602.1925
27. Deshmukh, S., Patil, N., Nunes, J.: Sentiment analysis of Marathi language. Int. J. Res. Publ. Eng. Technol. IJRPET, 93–97 (2017)
28. Govilkar, S., Bakal, W., Rathod, S.: Part of speech tagger for Marathi language. Int. J. Comput. Appl. **119**, 29–32 (2015). https://doi.org/10.5120/21169-4245
29. Bolaj, P., Govilkar, S.: Text classification for Marathi documents using supervised learning methods. Int. J. Comput. Appl. **155**, 6 (2016). https://doi.org/10.5120/ijca2016912374
30. Pawar, S., Mali, S.: Sentiment analysis in Marathi language. Int. J. Recent Innov. Trends Comput. Commun. **5**(8), 21–25 (2017). https://doi.org/10.17762/ijritcc.v5i8.1160
31. Ansari, M.A., Govilkar, S.: Sentiment analysis of mixed code for the transliterated Hindi and Marathi texts. Int. J. Nat. Lang. Comput. 15–28 (2018). https://doi.org/10.2139/ssrn.3429694
32. Srivastava, S., Govilkar, S.: Paraphrase identification of Marathi sentences. In: International Conference proceeding on Intelligent Data Communication Technologies and Internet of Things (ICICI), India, pp. 534–544 (2018). https://doi.org/10.1007/978-3-030-03146-6_59
33. Chaudhari, C.V., Khaire, A.V., Murtadak, R.R., Sirsulla, K.S.: Sentiment analysis in Marathi using Marathi WordNet. Imp. J. Interdiscip. Res. **3** (2017)
34. Yadav, K., Lamba, A., Gupta, D., Gupta, A., Karmakar, P., Saini, S.: Bi-LSTM and ensemble based bilingual sentiment analysis for a code-mixed Hindi-English social media text. In: 2020 IEEE 17th India Council International Conference (INDICON), India, pp. 1–6 (2020). https://doi.org/10.1109/INDICON49873.2020.9342241
35. Ali, M.A., Kulkarni, S.: Preprocessing of text for emotion detection and sentiment analysis of Hindi movie reviews. In: International Conference on IoT Based Control Networks and Intelligent Systems (ICICNIS 2020), pp. 848–856 (2021). https://doi.org/10.2139/ssrn.3769237
36. Kulkarni, A., Mandhane, M., Likhitkar, M., Kshirsagar, G., Joshi, R.: L3CubeMahaSent: a Marathi tweet-based sentiment analysis dataset. In: Proceedings of the Eleventh Workshop on Computational Approaches to Subjectivity, Sentiment and Social Media Analysis, pp. 213–220. Association for Computational Linguistics (2021)
37. Bafna, P., Saini, J.: Marathi document: similarity measurement using semantics-based dimension reduction technique. Int. J. Adv. Comput. Sci. Appl., 138–143 (2020).https://doi.org/10.14569/IJACSA.2020.0110419
38. Shelke, M.B., Alsubari, S.N., Panchal, D.S., Deshmukh, S.N.: Lexical resource creation and evaluation: sentiment analysis in Marathi. In: Zhang, Y.D., Senjyu, T., So-In, C., Joshi, A. (eds.) Smart Trends in Computing and Communications. LNNS, vol. 396, pp. 187–195. Springer, Singapore (2023). https://doi.org/10.1007/978-981-16-9967-2_19
39. Akhtar, M.S., Ekbal, A., Cambria, E.: How intense are you? Predicting intensities of emotions and sentiments using stacked ensemble. IEEE Comput. Intell. Mag. **15**(1), 64–75 (2020). https://doi.org/10.1109/MCI.2019.2954667

40. Shelke, M., Sawant, D., Kadam, C., Ambhure, K., Deshmukh, S.: Marathi SentiWordNet: a lexical resource for sentiment analysis of Marathi. Concurr. Comput. Pract. Exp. **35**(2) (2023). https://doi.org/10.1002/cpe.7497
41. Singh, R.K., Sachan, M.K., Patel, R.B.: 360-degree view of cross-domain opinion classification: a survey. Artif. Intell. Rev. **54**(2), 1385–1506 (2021). https://doi.org/10.1007/s10462-020-09884-9

Author Index

A
Abbas, Ali I-1, I-169
Acharjee, Tapodhir II-1
Adhikary, Nabanita I-57
Agarwal, Shreya I-1, I-169
Aich, Utathya II-209
Alladi, Suryatheja I-213
Amiripalli, Shanmuk Srinivas II-147
Arun Karthick, S. I-157
Aruna Devi, K. I-96
Ashwitha, U. I-11

B
Barbhuiya, Nurulla Mansur II-126
Baruah, Abhijit I-221
Baruah, Mandwip I-46
Bekal, Anush II-70, II-83
Bhajanka, Vaibhav I-105
Bharathi, D. S. I-11
Binodini, Pebam II-187
Bohre, Aashish Kumar I-57
Bonthu, Sridevi II-236

C
Chakraborty, Ishita I-88
Chakravorty, Hitesh II-273
Chakravorty, Rahul II-247
Chaudhary, Aman II-1
Chawngsangpuii, R. II-229

D
Das, Ajit I-221
Das, Prodipto II-273
Das, Purnendu II-126
Dasika, Surayanarayana II-256
Dayal, Abhinav II-236
Debnath, Aritri I-23
Debnath, Somen I-181
Devi, Laishram Munglemkhombi I-88
Dey, Barnali I-23
Dey, Lopamudra I-116
Divya, O. M. I-96
Dutta, Sandip II-139
Dwivedi, Shivam II-157

E
Emil Selvan, G. S. R. I-157, II-37

G
Gogoi, Munmi I-181
Goje, Saidesh I-213
Goswami, Paromita I-181
Gupta, Sumit II-236

H
Hari Priya, K. I-213
Harish, H. I-11
Hulke, Nainesh I-194
Hussain, S. Mahaboob II-256

I
Inja, Shiva Sai Pavan II-147

J
James, Nikhil II-61
Jeevan Samrudh, L. H. II-198
Jeyarohini, R. I-157, I-243
Jha, Prajna I-1, I-169

K
Kakkar, Rajan II-170
Kalita, Bikash I-137
Kandula, Narasimharao II-256
Karkera, Thejas II-83
Kasera, Rohit Kumar II-1
Kaur, Gurmeet II-100
Kaur, Ravpreet II-112
Kaushik, Eva II-25
Kayser, Tanvir I-31
Kayyali, Mustafa II-47
Khan, Ajoy Kumar I-181
Khanra, Anurag II-198
Konar, Rahul Shiva I-243

P. Das et al. (Eds.): AMRIT 2023, CCIS 1953, pp. 277–278, 2024.
https://doi.org/10.1007/978-3-031-47224-4

Krishna, Arvind II-198
Kumar, Ankur II-157
Kumar, Kunal I-57
Kumar, Prince I-57
Kumar, Vineet II-1
Kushwah, Anand II-25
Kushwaha, Ravindra Singh II-170

L
Lalruatfeli, Angelina II-229
Lece Elizabeth Rani, J. I-157

M
M, Srinivas P. II-70
Makhija, Rahul D. II-198
Malhotra, Varun II-157
Mamatha, M. I-11
Manasa, A. I-11
Mayanglambam, Sushilata D. I-194
Medhi, Shilpita II-61
Medhi, Smriti Priya I-232, II-61
Mehta, Aditya II-157
Mohapatra, Sudhir Kumar I-127
Mukherjee, Debdeep I-116
Mushahary, Jaya Rani I-46

N
Nama, Himanshu I-213
Nandy, Taniya I-88
Nongmeikapam, Kishorjit II-187

P
Pamula, Rajendra I-194
Pandey, Anubhav II-14
Pandey, Satwik II-25
Pandey, Saurabh II-14
Parui, Rupak I-116
Pathania, Nahita I-204
Paul, Hrituparna II-14
Paul, P. K. II-47
Pradhan, Nitisha I-105
Prajwal, P. II-83
Prasad, V. R. Badri II-198
Prasanna, S. S. II-37
Prasanthi, B. V. II-256

R
Raj, Anu II-247
Rajora, Tushar II-25
Ramkumar, M. P. I-157, I-243, II-37
Rani, J. Lece Elizabeth I-243
Ranjitha, M. I-96
Rao, Arla Lakshmana II-236
Rao, Supriya B. II-70
Reddy, M. Nithin I-213
Reddy, Viswa Ajay II-147
Rongala, Surya II-147
Roy, Anandarup II-209
Roy, Bipul I-46
Roy, Bishwa Ranjan II-126
Roy, Debashis II-209
Roy, Soumen II-139
Roy, Sudipta I-221
Roy, Utpal II-139, II-209

S
Saavedra, Ricardo II-47
Sahil, Koppala Somendra II-147
Sahu, Suren Kumar I-127
Sarkar, Sunita II-187
Sarma, Pankaj Kumar Deva I-73, I-137
Sarma, Rajkamal I-73
Sarma, Shikhar Kumar I-232
Satapathy, Santosh Kumar I-127
Selvan, G. S. R. Emil I-243
Shetty, Shailesh S. II-70
Shetty, Shailesh II-83
Shill, Pintu Chandra I-31
Shriwastav, Kunal Kumar II-61
Shruthi, M. G. I-11
Siddiqui, Tanveer J. I-1, I-169
Singh, Chandra II-70, II-83
Singh, Divyansh II-25
Singh, Neha II-247
Singh, Sarbjeet II-100, II-112
Srivastava, Jyoti II-247

U
Uppalapati, Padma Jyothi II-256

V
Vaishali I-204
Varma, Satishkumar L. I-262
Vichare, Abhishek A. I-262

Y
Yadav, Deepak II-1

Printed in the United States
by Baker & Taylor Publisher Services

Printed in the United States
by Baker & Taylor Publisher Services